John Mairhudon

GEORGE B. THOMAS, JR.
Massachusetts Institute of Technology

JOHN K. MOULTON
Brookline High School, Brookline, Massachusetts

MARTHA ZELINKA
Weston High School, Weston, Massachusetts

ELEMENTARY CALCULUS FROM AN ADVANCED VIEWPOINT

ADDISON-WESLEY PUBLISHING COMPANY

Reading, Massachusetts
Palo Alto
London
Don Mills, Ontario

This book is in the
ADDISON-WESLEY SERIES IN
SCIENCE AND MATHEMATICS EDUCATION
Consulting Editors:
Richard S. Pieters
George B. Thomas, Jr.
John Wagner
Paul Rosenbloom

Copyright © 1967
by Addison-Wesley Publishing Company, Inc.
All rights reserved.
This book, or parts thereof,
may not be reproduced in any form
without written permission of the publisher.
Printed in the United States of America.
Published simultaneously in Canada.
Library of Congress Catalog Card No. 67–13404.

PREFACE

PREFACE

For many years certain high schools and independent schools have offered an introduction to calculus to students in their last pre-college year. In the decade from 1956 to 1966 the number of students taking the College Entrance Examination Board Advanced Placement examination in calculus has grown from 386 to 9643. There is still being carried on a healthy debate about the merits of calculus, probability and statistics, vectors, modern algebra, or analytic geometry for the advanced high school course. It is not the purpose of this book to contribute to that debate; rather it is addressed, in part, to those teachers who find themselves teaching (or about to begin teaching) calculus—and are seeking to strengthen their understanding of the logical structure of the subject. To mathematicians looking for the strictest observance of logical development we offer no apology—our book is not written for them. Their needs are better met by books on the theory of functions of a real variable, or the more theoretical books on advanced calculus. Nor do we go through the well-known construction of the real numbers from Peano's postulates with Dedekind cuts or regular sequences to carry us from the rational number system to the reals. That, too, has been admirably done in many other books. Our attempt is to begin with things that we believe will be familiar to the majority of our readers (how calculus is used for finding the slope of a curve or the area under a curve) and then to show what the most immediate supporting theory is. This naturally entails at least some of the theory of limits with theorems and proofs. Chapter 2 treats limits of sequences, and Chapter 3 deals with limits of functions that are not sequences.

In Chapter 4, the limit theory is applied to deduce the standard formulas for derivatives of the rational functions and the circular functions. Chapter 5 shows how the derivative is used to search for maximum or minimum values of a function. Here is a payoff that is understood and appreciated by teachers and students alike. The subtler question of whether or not a maximum or minimum exists is part of the content of Chapter 6. Now the theoretical aspects begin to dominate the scene: technical definitions, theorems, and proofs of various properties of continuous functions on a closed bounded interval occupy most of our attention. Theorem 6–8 is the Heine-Borel covering theorem, which is first stated and used to prove boundedness of the range of a function continuous on an interval $[a, b]$. The reader who wishes to follow the logical order (as opposed to what we feel is a more strongly motivated psychological order) can read Section 6–4 before Section 6–3, or immediately after the formal statement of Theorem 6–7 in Section 6–3. More than one reading is recommended in any event.

Additional properties of continuous functions are treated in Chapters 7 and 8 (uniform continuity, intermediate values, composite functions, inverse functions). Most of the material in Chapters 6, 7, and 8 is also available in a separate pamphlet, *Continuity*, by one of the authors (G.B.T., Jr.) Addison-Wesley Publishing Co., Inc., 1965.

Rolle's theorem, the mean-value theorem, and Taylor's theorem with remainder are the main features of Chapter 9. Chapter 10 deals with derivatives of composite functions and inverses of functions, and Chapter 11 with inverse trigonometric functions in particular.

Chapter 12 is devoted primarily to the theory of Riemann integration of continuous functions. The existence of the integral of a function that is continuous on a closed bounded interval is proved, and is used in Chapter 13 to study the natural logarithm function and its inverse. We have included some computations (ln 2, ln 10, e), and have proved that e is irrational. Chapter 13 also includes a heuristic approach to the natural logarithm to motivate the definition in terms of area under the hyperbola $y = 1/t$ over $[1, x]$.

Essentially all of the material in Chapters 1 through 11, and parts of the material in the last two chapters, formed the subject matter for an in-service course sponsored by the Association of Advanced Placement Mathematics Teachers, meeting one night per week for about 25 weeks during the 1964–1965 academic year. In total hours, this was roughly the same as a three-semester-hour college course. However, this same material is not easily found in a regular college course because it comes between the standard first course in calculus and the more highbrow advanced calculus or real variables courses. It is our hope that this book will help those who desire to renew or to strengthen their knowledge of elementary calculus, as well as those who wish to probe at greater depth. It could be used in a variety of ways: (1) as a supplement to a first course in calculus, (2) as a text or reference for in-service programs for teachers, (3) as a text for teachers preparing for the teaching of advanced placement mathematics or otherwise renewing their study of mathematics, (4) as a reference for students in courses in advanced calculus or for beginning students of real variables who want more illustrative examples or more elementary explanations than they find in their regular texts.

The authors express their thanks to Phyllis Ruby and Fay Thomas for typing the manuscript and for assembling the earlier version of notes on which this book is based. We acknowledge with pleasure the help of the staff and consulting editors of the Addison-Wesley Publishing Co., Inc. Readers who find errors or obscurities in the book are cordially invited to send their comments to the authors.

November 1966

G. B. T., Jr.
J. K. M.
M. Z.

CONTENTS

CONTENTS

1 BEGINNINGS OF THE CALCULUS 1

    1–1 Tangent 2
    1–2 Velocity 4
    1–3 Derivatives 6
    1–4 Fundamental problems of differential calculus 10
    1–5 The area problem 11
    1–6 Absolute value 13
    1–7 Functions 17

2 LIMITS OF SEQUENCES 25

    2–1 Introduction 26
    2–2 Definition of limit of a sequence 28
    2–3 Some properties of limits 31
    2–4 Graphical representation of sequences and limits; monotone sequences 45

3 LIMITS OF FUNCTIONS THAT ARE NOT SEQUENCES . . 55

    3–1 Definitions and theorems 56
    3–2 Problems of the differential and integral calculus 67

4 THEORY OF LIMITS APPLIED TO FORMAL DIFFERENTIATION 75

    4–1 Definition of derivative and example 76
    4–2 Theorems 77
    4–3 Derivatives of the trigonometric functions 84

5 MAXIMA AND MINIMA 88

    5–1 Examples and terminology 89
    5–2 Definitions: maxima and minima 90
    5–3 Critical values 92
    5–4 Tests for extrema 97

6 CONTINUOUS FUNCTIONS. EXISTENCE OF MAXIMUM AND MINIMUM VALUES 102

    6–1 Introduction, definitions, and examples 103
    6–2 Existence of a maximum or minimum of a function. Examples . 110
    6–3 Boundedness and existence of maximum and minimum. Upper and lower bounds 112
    6–4 The Heine-Borel covering theorem 118

7 UNIFORM CONTINUITY AND INTERMEDIATE VALUES . 123

    7–1 Uniform continuity 124
    7–2 Intermediate values 129

8	COMPOSITE FUNCTIONS AND INVERSE FUNCTIONS	137
	8–1 Composite functions	138
	8–2 Continuity of composite functions	142
	8–3 Inverse functions	145
	8–4 Continuity of inverse functions	153
9	DIFFERENTIABLE FUNCTIONS	158
	9–1 Differentiability and continuity	159
	9–2 Rolle's theorem	163
	9–3 Extensions of the mean-value theorem	167
10	ADDITIONAL DIFFERENTIATION FORMULAS	176
	10–1 Increment of a function	177
	10–2 Derivatives of composite functions: the chain rule	180
	10–3 Derivatives of inverse functions	184
	10–4 Algebraic functions and their derivatives	193
11	DIFFERENTIATION OF TRIGONOMETRIC FUNCTIONS AND THEIR INVERSES	197
	11–1 Differentiation of the trigonometric functions	198
	11–2 Inverse trigonometric functions	201
12	AREA AND INTEGRATION	214
	12–1 Introduction	215
	12–2 Integration of continuous functions	220
	12–3 Integrals evaluated directly from the definition	235
	12–4 Trapezoidal rule for approximating integrals	247
	12–5 Properties of the Riemann integral	251
13	LOGARITHMIC AND EXPONENTIAL FUNCTIONS	263
	13–1 Introduction	264
	13–2 Heuristic approach: in search of the logarithm	265
	13–3 The natural logarithmic function	270
	13–4 The exponential function	278
	ANSWERS TO EXERCISES	291
	TABLE 1 Natural trigonometric functions	329
	TABLE 2 Exponential functions	330
	TABLE 3 Natural logarithms of numbers	331
	INDEX	333

1
BEGINNINGS OF THE CALCULUS

BEGINNINGS OF THE CALCULUS

The part of mathematics called the differential and integral calculus, or, more briefly, calculus, was developed over many centuries. Two geometrical problems, the tangent problem and the area problem, illustrate the two branches of calculus. The fact that these two problems have an intimate relationship is at the heart of what is called the fundamental theorem of (integral) calculus (Theorem 12–2). The working out of that relationship was one of the highlights in the history of mathematics.

1–1 TANGENT

Let us begin with an example. Consider the graph of the equation $y = 1/x$, $x \neq 0$. We know that when x is positive, so also is its reciprocal. Hence, to the right of the y-axis the graph lies above the x-axis. Moreover, when x is small in absolute value, its reciprocal is large, and when x is large its reciprocal is small. Suppose we plot the points $(\frac{1}{10}, 10)$, $(1, 1)$, $(2, \frac{1}{2})$, $(4, \frac{1}{4})$, $(10, \frac{1}{10})$, and sketch a smooth continuous curve through them. (We shall have more to say about both smoothness and continuity at a later time.) We obtain a portion of the familiar *rectangular hyperbola* shown in Fig. 1–1. We can also get the portion of the curve to the left of the y-axis and below the x-axis by plotting the points $(-x, -1/x)$ for the same values of x used above. The curve is symmetric with respect to the origin, because $y = 1/x$ is equivalent to $-y = 1/(-x)$ (see Section 1–7).

Suppose, to be explicit, that we are interested in constructing a line tangent to the curve at the point $(2, \frac{1}{2})$. Let x be a positive real number that is different from 2. Then the point $P(2, \frac{1}{2})$ and the point $Q(x, 1/x)$ together determine a line. The slope of this line is

$$m_{PQ} = \frac{y_Q - y_P}{x_Q - x_P} = \frac{1/x - \frac{1}{2}}{x - 2}$$

$$= \frac{(2 - x)/2x}{x - 2} = -\frac{1}{2x}. \quad (1)$$

So long as $x \neq 2$, Eq. (1) is meaningful and gives the slope of the secant line through P and Q. If we take $x = 3$, $Q = (3, \frac{1}{3})$, the slope is $-\frac{1}{6}$. If $x = 2.01$, $Q = (2.01, 1/2.01)$, then the slope is $-1/4.02$. Or, if we let $x = 1.999$, then $Q = (1.999, 1/1.999)$, and the slope is $-1/3.998$.

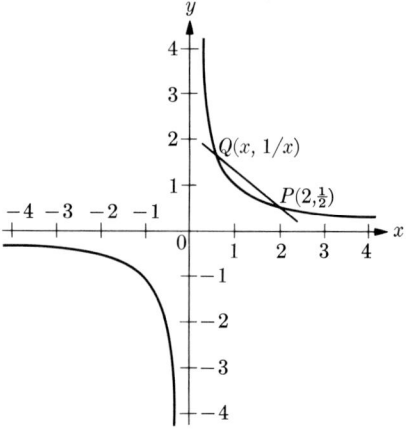

FIG. 1–1. Portion of the graph of $y = 1/x$, $x \neq 0$.

Clearly, when x is very near 2, the slope is very near $-\frac{1}{4}$. It seems reasonable to *define* the slope of the tangent at $P(2, \frac{1}{2})$ to be the limit, as x approaches 2, of the slope of the secant line PQ:

$$m_{\tan P} = \text{limit, as } Q \to P, \text{ of } m_{PQ} = \lim_{x \to 2} (-1/2x) = -\tfrac{1}{4}.$$

In subsequent sections we shall have more to say about the whole subject of limits. The present example is intended mainly to introduce the fundamental idea of the slope of a tangent to a curve and show that it leads us directly to the consideration of limits. If we accept $-\frac{1}{4}$ as the slope of the tangent to the curve $y = 1/x$ at $P(2, \frac{1}{2})$, then we can easily construct the tangent by going to the right 4 units and down 1 unit to the point $T(6, -\frac{1}{2})$. We then draw the line through P and T. Because it has slope $-\frac{1}{4}$ and it passes through P, it is defined as the tangent to the curve at P.

As a slight variation on this example, suppose x_1 is any real number different from zero, and we wish to construct the tangent to the hyperbola at $P_1(x_1, y_1)$, where $y_1 = 1/x_1$. We proceed as before: let $Q(x_2, y_2)$ be any other point on the hyperbola. Then $y_2 = 1/x_2$, $x_2 \neq x_1$, and the slope of the secant line P_1Q is

$$m_{\sec} = \frac{y_2 - y_1}{x_2 - x_1} = \frac{(1/x_2) - (1/x_1)}{x_2 - x_1} = -\frac{1}{x_2 x_1}. \qquad (2)$$

To get the slope of the tangent line at P_1 we hold x_1 fixed and let x_2 approach x_1 as limit. As we do so, $-1/x_2 x_1$ approaches $-1/(x_1)^2$ as its limit, and we define this limit to be the slope of the tangent line at P_1:

$$m_1 = \text{limit, as } x_2 \to x_1, \text{ of } -1/x_2 x_1 = -1/(x_1)^2. \qquad (3)$$

Of course, if we take $x_1 = 2$, then Eq. (3) gives us $-\frac{1}{4}$ as the slope of the tangent at $P(2, \frac{1}{2})$, in agreement with our earlier work. And, beyond that, it gives the slope of the tangent at *any* point on the curve because we started the discussion by assuming that x_1 is any nonzero real number. Given the slope $-1/(x_1)^2$ and the point $P_1(x_1, 1/x_1)$ on the curve, it is easy to construct the line of proper slope through P_1, and thus we get the line tangent to the curve at any point on it.

EXERCISES 1–1

1. Draw the graph of $y = x^2$. Take point $A(1, 1)$ and find the slopes of secant lines joining point A with the points having abscissas 2, $1\frac{1}{2}$, $1\frac{1}{4}$, $1\frac{1}{10}$.

2. Repeat Exercise 1 with points having abscissas 0.5, 0.7, 0.8, 0.9.

3. Do the same for the secant joining the points $P_1(x_1, y_1)$ and $P_2(x_2, y_2)$. What is the limit of the slope as P_2 approaches P_1? Now let $x_1 = 1$ and evaluate the limit.

1-2 VELOCITY

If an object moves along a straight line, its position s at time t may be given by an equation of the form

$$s = f(t), \qquad a \leq t \leq b, \tag{1}$$

where f represents a function* whose domain includes the closed interval $[a, b]$. From such an equation of motion, we can find the *displacement*

$$s_2 - s_1 = f(t_2) - f(t_1) \tag{2}$$

over an interval of time from t_1 to t_2 if $a \leq t_1 < t_2 \leq b$. The *average velocity* from time t_1 to time $t_2 \neq t_1$ is defined to be

$$v_{av} = \frac{s_2 - s_1}{t_2 - t_1} = \frac{f(t_2) - f(t_1)}{t_2 - t_1}. \tag{3}$$

For example, if the law of motion is

$$s = 64t - 16t^2, \qquad 0 \leq t \leq 4, \tag{4a}$$

then

$$v_{av} = \frac{s_2 - s_1}{t_2 - t_1} = \frac{(64t_2 - 16t_2^2) - (64t_1 - 16t_1^2)}{t_2 - t_1}$$

$$= \frac{64(t_2 - t_1) - 16(t_2^2 - t_1^2)}{t_2 - t_1}$$

$$= 64 - 16(t_2 + t_1). \tag{4b}$$

Thus, if $t_1 = 1$ and $t_2 = 2$, the average velocity from t_1 to t_2 is

$$64 - 16(2 + 1) = 16.$$

But, if $t_1 = 1$ and $t_2 = 1 + h$, and $|h|$ is very small but not zero, then the average velocity from t_1 to t_2 is

$$64 - 16(2 + h) = 32 - 16h \approx 32.$$

Although an average velocity can be computed only over an *interval* of time, we can, if we wish, imagine that interval to be of extremely short duration. If the motion is continuous and smooth for intervals of time around t_1, it may happen that the average velocity from t_1 to t_2 is practically the same for all values of t_2 near t_1. In other words, the average

---

* The reader probably recalls that a function is a mapping from a domain to a range. See Section 1-7 for an extended discussion and formal treatment of functions.

velocity may approach a definite limit, as t_2 approaches t_1. If so, we define that limit to be the *velocity at* t_1:

$$v_1 = \lim_{t_2 \to t_1} \frac{s_2 - s_1}{t_2 - t_1} = \lim_{t_2 \to t_1} \frac{f(t_2) - f(t_1)}{t_2 - t_1}. \tag{5}$$

In Eq. (5), t_1 is held fixed during the calculation of the average velocity $(s_2 - s_1)/(t_2 - t_1)$ and its limit as $t_2 \to t_1$. Although we have talked about an interval of time *from* t_1 *to* t_2, thus implying that $t_2 > t_1$, we could equally well compute an average velocity from t_2 to t_1, with $t_2 < t_1$. In fact, the limit in Eq. (5) does not exist unless the same answer is obtained when t_2 approaches t_1 from the left and from the right.

For example, Eq. (4b) gives the average velocity from t_1 to t_2 (or, from t_2 to t_1), for the motion described by Eq. (4a). If we let $t_2 = t_1 + h$, and allow h to be either positive or negative, Eq. (4b) gives

$$v_{av} = (64 - 32t_1) - 16h. \tag{6}$$

Since $|16h|$ can be made as small as we wish by making $|h|$ one-sixteenth that large, the average velocity in Eq. (6) has a limit, as $t_2 \to t_1$ and $h \to 0$, given by

$$v_1 = \lim_{h \to 0} [(64 - 32t_1) - 16h] = 64 - 32t_1. \tag{7}$$

From Eq. (7) we learn that the velocity is 64 at time $t_1 = 0$, is positive for $0 \leq t_1 < 2$, is zero at $t_1 = 2$, and is negative for $t_1 > 2$. Equations (4a) and (7) describe the motion of a particle projected vertically upward with an initial velocity of 64 ft/sec (if s is measured in feet and t in seconds). Gravity produces a negative acceleration. The particle comes to rest at the end of two seconds, then falls back to the ground during the next two seconds. Its greatest distance above the ground is $s_{max} = 64(2) - 16(2)^2 = 64$, attained at the instant the velocity is zero.

EXERCISES 1–2

According to the laws in physics, the distance through which a body travels in t seconds, neglecting air resistance, is the following:

i) if it drops from rest,

$$s = \tfrac{1}{2}gt^2 \quad \text{or} \quad s = 16t^2;$$

ii) if dropped with initial velocity v_0,

$$s = v_0 t + 16t^2;$$

iii) if projected upward with initial velocity v_0,

$$s = v_0 t - 16t^2.$$

1. If a body falls from rest, what is the average velocity for each of the following time intervals?
   a) $t = 1$ and $t = 2$
   b) $t = 1$ and $t = 1.5$
   c) $t = 1$ and $t = 1.2$
   d) $t = 1$ and $t = 1.1$
   e) $t = 1$ and $t = 1.01$
   f) $t = 1$ and $t = 1.001$
   g) $t = t_1$ and $t = t_2$
   h) Let $t_2 \to t_1$ in (g). Interpret the result.

2. If a body is dropped with the initial velocity $v_0 = 80$ ft/sec, what are the average velocities for the same time intervals as in Exercise 1?

3. If a body is projected upward with the initial velocity $v_0 = 160$ ft/sec, what are the average velocities for the same time intervals as in Exercise 1?

4. Use the result of Exercise 3(h) and find the velocity for $s = 160t - 16t^2$ and $t = 2, 3, 4, 5, 6$; find the corresponding values of s. For what values of t is $s = 0$? From these answers describe the motion of the body.

1–3 DERIVATIVES

The slope of the tangent to a curve, and the instantaneous velocity of a moving object, are examples of *derivatives*. For the slope of the tangent we have

$$m = \lim_{x_2 \to x_1} \frac{y_2 - y_1}{x_2 - x_1} = \lim_{x_2 \to x_1} \frac{f(x_2) - f(x_1)}{x_2 - x_1} \tag{1a}$$

and for velocity,

$$v_1 = \lim_{t_2 \to t_1} \frac{s_2 - s_1}{t_2 - t_1} = \lim_{t_2 \to t_1} \frac{f(t_2) - f(t_1)}{t_2 - t_1}, \tag{1b}$$

provided the equation of the graph is $y = f(x)$ for Eq. (1a), and the equation of motion is $s = f(t)$ for Eq. (1b). In the first case, we use the variables x and y; in the second, t and s. Otherwise, the calculations are identical. So we could also say that the velocity at time t_1 gives the slope of the graph of the equation of motion $s = f(t)$ at the point (t_1, s_1). And,

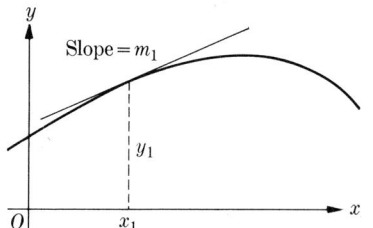 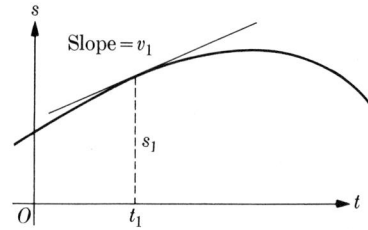

FIG. 1–2. Slope and velocity.

if we think of velocity as "the instantaneous rate of change of displacement (s) with respect to time (t)," then, by analogy, we can also say that the slope of the curve $y = f(x)$ is "the instantaneous rate of change of y with respect to x." (See Fig. 1–2.)

The entire set of calculations indicated by the notation

$$\lim_{x_2 \to x_1} \frac{f(x_2) - f(x_1)}{x_2 - x_1}$$

constitute a single operation that we call *finding the derivative of the function f at x_1*. We denote the result by $f'(x_1)$, read "f-prime at x_1," and call it *the derivative of f at x_1*. It should be noted that taking a derivative is a local operation; the derivative is computed at a single point.

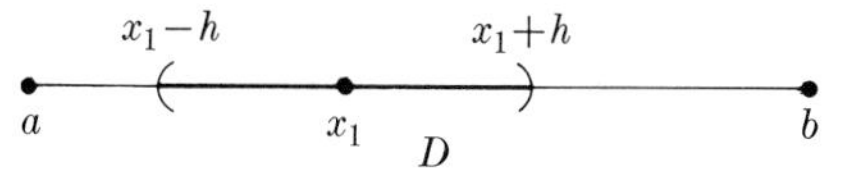

FIG. 1–3. Inner point x_1 and neighborhood.

Definition of Derivative. Let f be a function,

$$f : x \to y = f(x), \qquad x \in D,$$

whose domain D includes x_1 as an interior point. If, as $x_2 \to x_1$, the following limit exists, then f is said to be *differentiable* at x_1 and that limit is the *derivative* of f at x_1:

$$f'(x_1) = \lim_{x_2 \to x_1} \frac{f(x_2) - f(x_1)}{x_2 - x_1}. \qquad (2)$$

Remark 1. When we say "x_1 is an interior point of D" we mean that there is some open interval, centered at x_1, that is contained in D. In other words (Fig. 1–3), D contains all those values of x that satisfy

$$x_1 - h < x < x_1 + h \qquad \text{for some } h > 0.$$

Thus, if D is an interval $a \leq x \leq b$, then the endpoints a and b are not interior points of D, but all other points of D are. The restriction to interior points x_1 allows x_2, in Eq. (2), to approach x_1 from the left or from the right. Thus, differentiability at x_1 involves a two-sided kind of limit. At a boundary point of D, for instance at $x_1 = a$ when D is the interval $a \leq x \leq b$, only a one-sided limit might exist:

$$\lim_{x_2 \downarrow a} \frac{f(x_2) - f(a)}{x_2 - a}, \qquad (3a)$$

BEGINNINGS OF THE CALCULUS

where $x_2 \downarrow a$ means "x_2 decreases toward a," or "x_2 approaches a from the right," or "$x_2 \to a$, and $x_2 > a$." The one-sided limit in (3a), if it exists, is called the "right-hand derivative of f at a." Similarly, one might have

$$\lim_{x_2 \uparrow b} \frac{f(x_2) - f(b)}{x_2 - b} = \lim_{x_2 \uparrow b} \frac{f(b) - f(x_2)}{b - x_2}, \qquad (3b)$$

and call this the "left-hand derivative of f at b."

Remark 2. A function f may have a right-hand derivative and a left-hand derivative at a point x_1 that is interior to the domain of f. But f is not differentiable at x_1 unless these one-sided derivatives are equal. For example, consider the question of differentiability of the absolute-value function

$$f(x) = |x| = \begin{cases} -x & \text{if } x < 0, \\ x & \text{if } x \geq 0, \end{cases}$$

at $x_1 = 0$. Then $f(x_1) = 0$ and $f(x_2) = |x_2|$. Hence

$$\frac{f(x_2) - f(x_1)}{x_2 - x_1} = \frac{|x_2| - 0}{x_2} = \begin{cases} -1 & \text{if } x_2 < 0, \\ 1 & \text{if } x_2 > 0. \end{cases}$$

Therefore,

$$\lim_{x_2 \downarrow 0} \frac{f(x_2) - f(x_1)}{x_2 - x_1} = 1 \quad \text{and} \quad \lim_{x_2 \uparrow 0} \frac{f(x_2) - f(x_1)}{x_2 - x_1} = -1.$$

The right-hand derivative at zero is 1, the left-hand derivative is -1. Therefore the absolute-value function is not differentiable at $x = 0$. This means that the graph of $y = |x|$ (see Fig. 1–4) fails to have a tangent at $(0, 0)$.

Remark 3. The Derived Function. The collection of ordered pairs $(x_1, f'(x_1))$ for all those values of x_1 in the domain of f at which the limit called for in the definition of the derivative (Eq. 2) exists is called the *derived function* for f. We denote the derived function, or derivative, of f by f'. The domain of f' is always a subset of the domain of f.

In Eq. (2) we used x_1 to represent an arbitrary number in the domain of f'. The purpose of the subscript was to emphasize that once a value of x_1 is chosen, that value is held fixed as x_2 approaches x_1. It is not convenient to prolong this use of subscripts, especially when we develop general formulas for derivatives. Thus, for the hyperbola $f(x) = 1/x$, $x \neq 0$, we found that $f'(x_1) = -1/(x_1)^2$, $x_1 \neq 0$, which we can just as well write

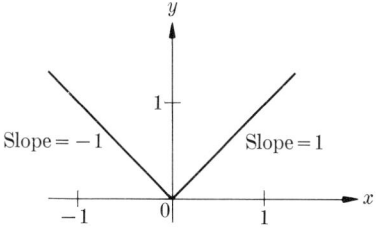
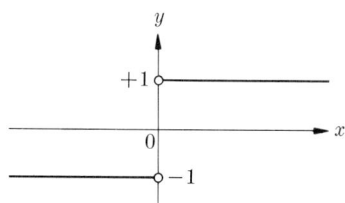

FIG. 1-4. Portion of the graph of $y = |x|$.

FIG. 1-5. Graph of the derived function for $f(x) = |x|$.

as $f'(x) = -1/x^2$, $x \neq 0$, omitting the subscripts. Here the domain of f' is all of the domain of f.

For the absolute-value function $f(x) = |x|$, the domain of f consists of all real numbers, but the domain of f' does not include zero. The graph of the derived function

$$f'(x) = \begin{cases} -1 & \text{if } x < 0, \\ 1 & \text{if } x > 0, \end{cases}$$

is shown in Fig. 1-5. There is no point on the graph for $x = 0$.

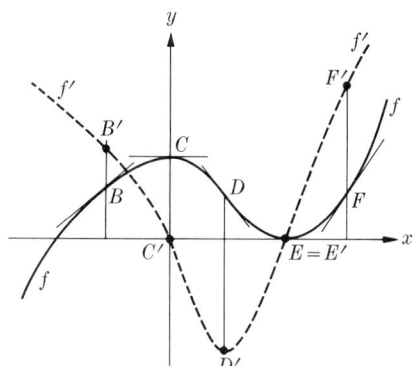

FIG. 1-6. Portion of the graph of a function f and an approximation to its derived function f'.

Example 1. Figure 1-6 shows a portion of the graph of a function and a graphical approximation to its derived function f'. At B the slope of f is positive. The slope decreases to zero at C, is negative between C and E, and has a minimum at D. To the right of E the slope is again positive and increasing. The dashed f' curve shows these features. The values of the slope of the f curve are estimated (roughly) at points B, C, D, E, and F. These slopes are the ordinates at points B', C', D', E', F'.

EXERCISES 1-3

For each of the following, sketch the graph of the derived function f' as suggested in Example 1 of this section.

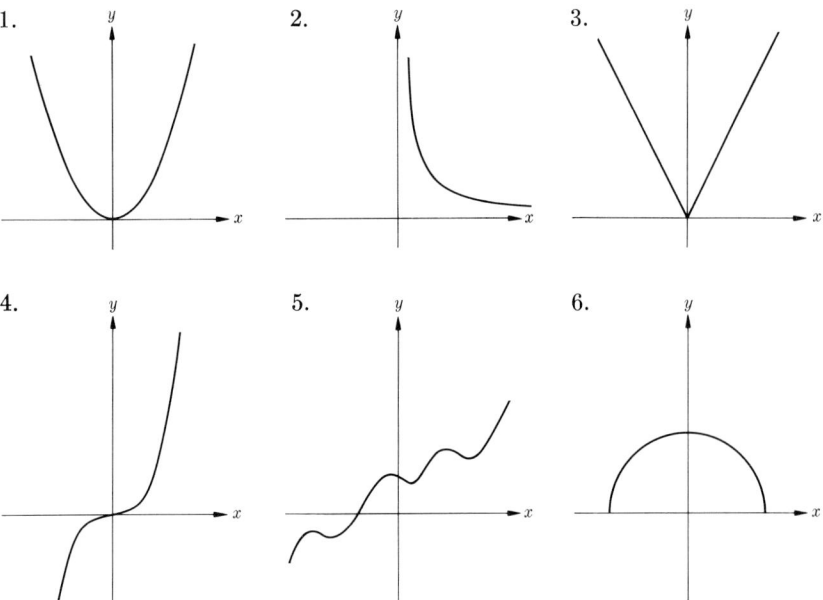

1-4 FUNDAMENTAL PROBLEMS OF DIFFERENTIAL CALCULUS

Problem 1. Given a function f and a number x_1 in the domain of f, determine whether f is differentiable at x_1.

Problem 2. If f is differentiable at x_1, what is $f'(x_1)$?

As students, we are also interested in meaningful applications and interpretations. We have already noted slope and rate of change as two interpretations of the derivative. Applications include curve-tracing, maximum and minimum problems, and analysis of equations of motion (velocity and acceleration).

Problem 3. Study broad categories of functions (for example, polynomials, rational functions, trigonometric functions) to determine where they are differentiable, and develop simple rules for calculating their derivatives.

Problems 1 and 2 relate to finding the derivative of a particular function f at a particular point x_1 in the domain of f. Equation (2) of Section 1-3 is the tool. It is also the tool for attacking the broader Problem 3. In order to apply that tool with some facility and with confidence, we take up the subject of limits in the next two chapters. We have already computed some limits informally. For example, in order to find the slope of the hyperbola $y = 1/x$ at (x_1, y_1) we found the following limit in Eq. (3) of Section 1–1,

$$\lim_{x_2 \to x_1} \frac{-1}{x_2 x_1} = -\frac{1}{(x_1)^2}.$$

In Eq. (7) of Section 1–2 we computed the instantaneous velocity

$$v_1 = \lim_{h \to 0} [(64 - 32t_1) - 16h] = 64 - 32t_1.$$

In the same informal way, the reader is encouraged to apply Eq. (2) of Section 1–3, the definition of the derivative, to find $f'(x_1)$ for each of the functions in the following set of exercises.

EXERCISES 1–4

Using the definition of the derivative [Eq. (2) of Section 1–3], find $f'(x_1)$ for the following functions for arbitrary real values of x_1 unless otherwise indicated.

1. $f(x) = 2x + 3$
2. $f(x) = ax + b$
3. $f(x) = \dfrac{1}{ax + b}, \quad x_1 \neq -b/a$
4. $f(x) = x^2$
5. $f(x) = ax^2 + bx + c$
6. $f(x) = x^3$
7. $f(x) = ax^2 + bx^3$
8. $f(x) = \sqrt{x}, \quad x_1 > 0$
9. $f(x) = \sqrt{x + 1}, \quad x_1 > -1$
10. $f(x) = 1/x^2, \quad x_1 \neq 0$

1–5 THE AREA PROBLEM

In this section, we consider the problem of finding the area of a region like $abQP$ shown in Fig. 1–7. (A more thorough treatment of the area problem will be given in Chapter 12.) The region is bounded above by

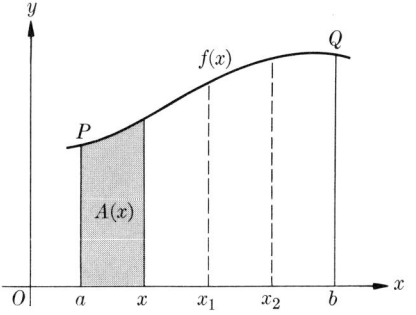

FIG. 1–7. Region $abQP$ is bounded above by the curve PQ and below by the interval $[a, b]$.

the graph of $y = f(x)$, below by the interval $[a, b]$ of the x-axis, and on the sides by the lines $x = a$ and $x = b$.

Let $A(x)$ be the function whose value for any x between a and b is the area of the region bounded above by a portion of the graph PQ, below by the interval $[a, x]$, and on the sides by straight lines perpendicular to the x-axis. This region is shaded in Fig. 1–7. If the area function has a derivative at x_1, then

$$A'(x_1) = \lim_{x_2 \to x_1} \frac{A(x_2) - A(x_1)}{x_2 - x_1}. \tag{1}$$

We can visualize $A(x_2) - A(x_1)$ as the area of the strip under the curve PQ that lies between the lines $x = x_1$ and $x = x_2$. In the particular situation shown in Fig. 1–7, that strip has more area than

$$f(x_1)(x_2 - x_1) \tag{2a}$$

and less area than

$$f(x_2)(x_2 - x_1), \tag{2b}$$

because (2a) is the area of an inscribed rectangle with base $[x_1, x_2]$, and (2b) the area of a circumscribed rectangle. Therefore, for this illustration,

$$f(x_1) < \frac{A(x_2) - A(x_1)}{x_2 - x_1} < f(x_2). \tag{3}$$

Now hold x_1 fixed and let $x_2 \to x_1$. For a continuous curve PQ, when x_2 approaches x_1, $f(x_2)$ also approaches $f(x_1)$, and the inequalities (3) imply that

$$A'(x_1) = \lim_{x_2 \to x_1} \frac{A(x_2) - A(x_1)}{x_2 - x_1} = f(x_1). \tag{4}$$

Although the inequalities (3) would be reversed if the graph of PQ sloped downward instead of upward over the interval $[x_1, x_2]$, the conclusion (4) would still hold. If the curve PQ oscillated up and down, Eq. (4) would also still be valid (as we shall see in more detail in Chapter 12.) Thus *the area function $A(x)$ and the function $f(x)$ whose graph is the upper boundary of the region are related by*

$$A'(x) = f(x). \tag{5}$$

We are therefore led to the following problem:

Given the function f, find a function A whose derivative is f, for $a < x < b$.

Such a function is called an *antiderivative* of f, or *primitive* of f, or *integral* of f. This problem represents just the opposite of finding the derived function. Once an appropriate primitive $A(x)$ has been found, such that $A(a) = 0$, we solve the original area problem by computing $A(b)$.

EXERCISES 1–5

1. In the preceding paragraph, why should it be true that $A(a) = 0$?

2. If $f(x) = mx$, where m is a positive constant, the graph of f is a line through the origin. If $0 < a < b$, the region that corresponds to $abQP$ in Fig. 1–7 is the interior of a trapezoid. Find its area by using geometrical formulas.

3. In Exercise 2, find the area by observing that $A(x) = \frac{1}{2}mx^2 + C$ satisfies the equation $A'(x) = mx$ for any choice of the constant C. What value of C makes $A(a) = 0$? Using this value of C, compute the corresponding value of $A(b)$. Compare with the result obtained geometrically.

1–6 ABSOLUTE VALUE

In this section we review some important properties of absolute values.

Definition 1
$$|a| = \begin{cases} a & \text{if } a \geq 0, \\ -a & \text{if } a < 0. \end{cases}$$

We can now prove some useful theorems.

Theorem 1–1. $|a| = \sqrt{a^2}$.

Proof. Remembering that $\sqrt{N}$ means the nonnegative number whose square is N, we have the following possibilities:

if a is any positive number, $\sqrt{a^2} = a$;

if a is any negative number, then $-a$ is positive, and
$$\sqrt{a^2} = \sqrt{(-a)^2} = -a;$$
if $a = 0$,
$$\sqrt{a^2} = \sqrt{0} = 0 = a.$$

Therefore, by Definition 1, for any real number a, $|a| = \sqrt{a^2}$. Q.E.D.

Theorem 1–2. $|a| = \max(a, -a)$.

Proof. If $a > 0$, then $-a < 0$ and $-a < a$; hence $\max(a, -a)$ is a, and $|a| = a$. If $a < 0$, then $0 < -a$ and $a < -a$; hence $\max(-a, a)$ is $-a$, and $|a| = -a$. Finally, if $a = 0$, $-a = 0$ and $|a| = |-a| = 0$.
Q.E.D.

Theorem 1–3. If a and b are any two real numbers, then

$$|ab| = |a|\,|b| \tag{1a}$$

and

$$\left|\frac{a}{b}\right| = \frac{|a|}{|b|}, \qquad b \neq 0. \tag{1b}$$

Proof. Applying Theorem 1–1 to the left-hand side of Eq. (1a), we have

$$|ab| = \sqrt{(ab)^2} = \sqrt{a^2 b^2} = \sqrt{a^2}\sqrt{b^2} = |a|\,|b|.$$

Equation (1b) can also be proved directly from Theorem 1–1:

$$\left|\frac{a}{b}\right| = \sqrt{\left(\frac{a}{b}\right)^2} = \sqrt{\frac{a^2}{b^2}} = \frac{\sqrt{a^2}}{\sqrt{b^2}} = \frac{|a|}{|b|}, \qquad b \neq 0. \qquad \text{Q.E.D.}$$

Theorem 1–4. If a and b are any two real numbers, then

$$|a + b| \leq |a| + |b|. \tag{2}$$

Proof. For any real numbers a and b

$$a \leq |a| \quad \text{and} \quad b \leq |b|,$$

by Theorem 1–2. Adding, we obtain

$$a + b \leq |a| + |b|.$$

Similarly, $-a \leq |a|$ and $-b \leq |b|$, and

$$-a - b \leq |a| + |b|.$$

Since both $a + b$ and $-(a + b)$ are less than or equal to $|a| + |b|$, their maximum, which is $|a + b|$, must be less than or equal to $|a| + |b|$. Therefore

$$|a + b| \leq |a| + |b|. \qquad \text{Q.E.D.}$$

This property of absolute values, known as the "triangle inequality," will be especially useful. Another method of proving it is suggested in Exercise 5 at the end of this section.

Examples

$$
\begin{aligned}
a = 7, \quad & b = 5 \quad & |a+b| = 12 \quad & = |a| + |b| \\
a = -7, \quad & b = 5 \quad & |a+b| = 2 < 12 & = |a| + |b| \\
a = 7, \quad & b = -5 \quad & |a+b| = 2 < 12 & = |a| + |b| \\
a = -7, \quad & b = -5 \quad & |a+b| = 12 \quad & = |a| + |b|
\end{aligned}
$$

Geometric Interpretation of Absolute Values. If a is any real number, $|a|$ is the distance between 0 and a on the number line. Thus both 3 and -3 are three units from 0, and $|3| = |-3| = 3$.

If a and b are any two real numbers, we can represent them by points on the number line, and the distance between them is given by $|b - a| = |a - b|$. Thus the distance between -3 and -5 is $|-3 - (-5)| = |-5 - (-3)| = 2$, and the distance between -3 and 5 is $|-3 - 5| = |5 - (-3)| = 8$. (See Fig. 1-8.)

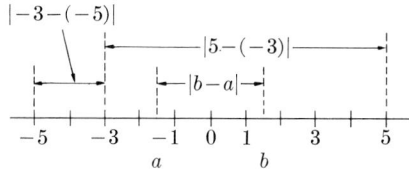

FIG. 1-8. Distance between a and b is $|b - a| = |a - b|$.

Absolute values and inequalities are used extensively. For example, if a is any real number and h is positive, consider the set of all numbers x for which

$$|x - a| < h. \tag{3a}$$

The inequality (3a) has the form

$$|x'| < h,$$

where $x' = x - a$. Those numbers x' whose absolute values are less than h are just the numbers between $-h$ and h:

$$-h < x' < h. \tag{3b}$$

The set of real numbers x for which

$$-h < x - a < h$$

is the open interval

$$a - h < x < a + h. \tag{3c}$$

Both formulas (3a) and (3c) represent the interval from $a - h$ to $a + h$, excluding the endpoints (Fig. 1-9).

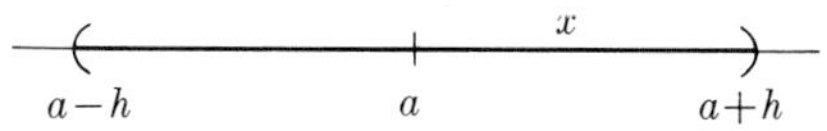

FIG. 1-9. $|x - a| < h$ is the same as $a - h < x < a + h$.

It is useful to think of (3a) as saying "the distance between a and x is less than h," or "x is within h units of a." Thus (3a) represents an interval with center at a and radius h, and (3c) is the same interval. Such an interval is called a neighborhood of a. We sometimes denote it by $N_h(a)$:

$$N_h(a) = \{x : |x - a| < h\}. \quad (4)$$

EXERCISES 1–6

1. For what values of x, if any, is it true that
   a) $\sqrt{x^2} = x$?
   b) $\sqrt{x^2} = -x$?
   c) $|x - 5| = x - 5$?
   d) $|x - 1| = -3$?

2. For what values of x, if any, is it true that
   a) $|x - 3| < 5$?
   b) $|3x - 6| \geq 3$?
   c) $|2x - 5| \leq 0.02$?
   d) $|x - 5| < -1.1$?

3. Sketch graphs of the following functions:
   a) $f(x) = \max(x^2, 4 - x^2)$
   b) $f(x) = x + |x|$
   c) $f(x) = 2 + |x^2 - 2|$
   d) $f(x) = |x|^2 - |x| - 6$

4. Let a and b be real numbers, and let $c = \frac{1}{2}(a + b)$, $d = \frac{1}{2}|a - b|$.
   a) If a, b, c are plotted on the x-axis, how are c and d related geometrically to a and b?
   b) For which values of x is $|x - c| \leq d$?
   c) Give a geometric reason why it is true that $c + d$ equals $\max(a, b)$ and $c - d$ equals $\min(a, b)$.
   d) Prove algebraically (or by appealing directly to the definition of $|x|$) that

   $$\max(a, b) = \frac{a + b}{2} + \frac{|a - b|}{2}$$

   and

   $$\min(a, b) = \frac{a + b}{2} - \frac{|a - b|}{2}.$$

5. Prove Theorem 1–4 ($|a + b| \leq |a| + |b|$) by noting that $2ab \leq 2|a||b|$, and then using Theorem 1–1 and the identity

   $$a^2 + b^2 = |a|^2 + |b|^2.$$

6. Prove that, if a and b are any two real numbers, then
   a) $|a - b| \leq |a| + |b|$,
   b) $|a - b| \geq |a| - |b|$.

7. If $a, b,$ and c are any real numbers, how does $|a + b + c|$ compare with $|a| + |b| + |c|$? Prove your answer.

1-7 FUNCTIONS

We have stated the problems of differential and integral calculus and sketched their solutions briefly. Whether we examine the slope of a secant of a curve and the limiting position of the secant, the velocity of a particle moving along a given path, or the area of a region bounded by a curve, we are concerned with a function of a real variable and its behavior over a given interval. For the sake of completeness we will therefore discuss in this section real-valued functions of a real variable and develop those concepts necessary to the understanding of the topics in the subsequent chapters.

Let us recall some important relations from algebra and analytic geometry. A one-to-one correspondence can be established between the ordered pairs of real numbers, (x, y), and the points in the xy-plane. Compare the following equations and the sets of points defined by them:

$y = mx + b$ straight line,
$y = x^2$ parabola,
$y = \sqrt{x}$ half a parabola,
$x^2 + y^2 = r^2$ circle,
$b^2 x^2 + a^2 y^2 = a^2 b^2$ ellipse,
$x^2 - y^2 = a^2$ or $xy = k$ hyperbola.

In each of the above examples there is a set, X, of all real numbers or a subset of the real numbers, whose elements are paired with the elements of a second set, Y, according to the defining equation. The set of all such ordered pairs (x, y) is called a *relation* from X to Y.

In the following examples, we denote the set of all real numbers by R.

Example 1. Given $x^2 + y^2 = 16$,

$$X = \{x : |x| \leq 4 \text{ and } x \in R\}, \quad Y = \{y : |y| \leq 4 \text{ and } y \in R\}.$$

The points with coordinates $(0, 4)$, $(0, -4)$, $(2, 2\sqrt{3})$, $(2, -2\sqrt{3})$, $(-2, 2\sqrt{3})$, $(-2, -2\sqrt{3})$, etc., are on the graph of the equation (Fig. 1–10), since the sum of the squares of the coordinates is 16 in each case.

Among the equations listed above, there are some which assign to each element of X a *unique* element of Y. Such equations describe *functions*.

Definition 2. Let X and Y be nonempty sets. Let f be a collection of ordered pairs (x, y) with $x \in X$ and $y \in Y$. Then f is a function from X to Y if to every $x \in X$ there is assigned a *unique* $y \in Y$.

BEGINNINGS OF THE CALCULUS   1–7

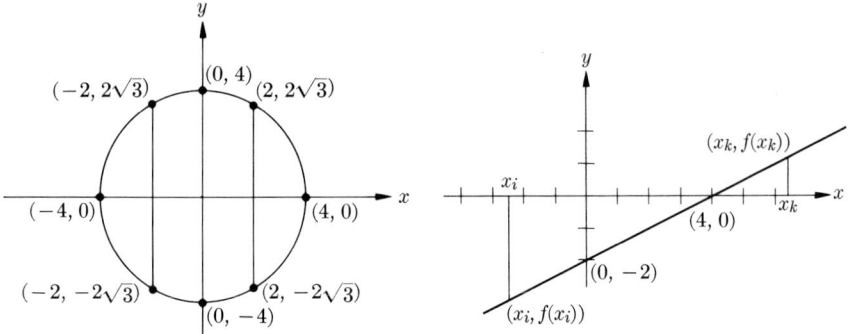

FIG. 1–10. Graph of $x^2 + y^2 = 16$.   FIG. 1–11. Graph of $y = \frac{1}{2}x - 2$.

Example 2. If $y = \frac{1}{2}x - 2$, $X = \{x : x \in R\}$, $Y = \{y : y \in R\}$, then to each x there corresponds exactly one y (Fig. 1–11).

The collection of all first elements x of the pairs (x, y) in f is called the *domain* of f, and will be denoted by D_f. The *range* of f, denoted by R_f, is the set of all second elements y of the pairs (x, y) in f. The definition states that every x in X is the first entry of exactly one pair (x, y) in f.

Example 3. Let X be the set of all triangles and let Y be the set of all real numbers. Let f be the set of ordered pairs (x, y), where

x is any triangle, and

y is the area enclosed by that triangle, measured in appropriate units.

To each triangle x there corresponds a unique real number y that represents its area; so f is a function. The domain of f is all of X; the range consists of all positive real numbers.

Perhaps less familiar but very important in the study of mathematics are the functions defined by

$$y = a^x, \quad a > 0, \quad a \neq 1,$$

and

$$y = \log_a x, \quad x > 0, \quad a > 0, \quad a \neq 1,$$

and whose graphs are shown in Fig. 1–12. These functions will be discussed in detail in Chapter 13.

Example 4. If $y = \ln x$, $X = \{x : x > 0\}$, $Y = \{y : y \in R\}$, then to each positive number x there is assigned a unique y. Note that $\ln x = \log_e x$ is called the *natural logarithm* of x.

18

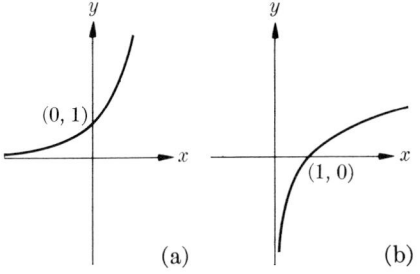

FIG. 1-12. (a) Portion of the graph of $y = a^x$, $a > 1$. (b) Portion of the graph of $y = \log_a x$.

Remark 1. We also speak of a function f from X to Y as a *mapping* that assigns to any element x in its domain a *unique* element y in its range such that the pair (x, y) belongs to f. This unique value of y that is thus associated with the given value of x is also expressed as $y = f(x)$, read "y equals f of x," or "y is the value of f at x." We say that f maps x onto $f(x)$, and it maps the domain D_f onto the range R_f (see Fig. 1-13).

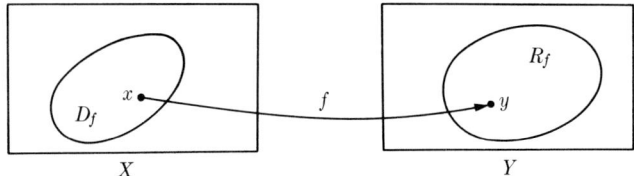

FIG. 1-13. A function f maps the domain D_f onto the range R_f. The image of x is $y = f(x)$.

Definition 3. A function f whose domain and range are sets of real numbers is said to be a real-valued function of a real variable.

For the remainder of this book, we shall always mean a real-valued function of a real variable when we say merely *function*.

Example 5. Let X and Y be the set of all real numbers. For each x in X let $f(x) = x^2$. Then

$$f = \{(x, y) : y = x^2, \ -\infty < x < \infty\}.$$

The domain of f is all of X, the range is the set of nonnegative real numbers. There are several ways of representing f:

1. By a table of corresponding values of x and x^2. This would be only a portion of f because such a table could not list all real values of x and their squares.
2. By corresponding number scales, as on a slide rule (Fig. 1-14(a)). This, too, is incomplete.

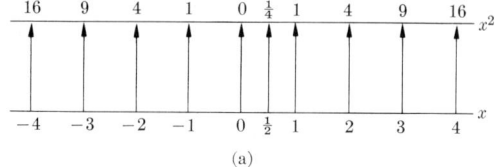

FIG. 1-14. (a) To each real number x on the lower scale corresponds the number x^2 on the upper scale. (b) The path from P to Q to R gives the mapping $f:x \to x^2$.

3. By the simple formula $f(x) = x^2$, which says, in effect, "choose any number x and square it." Thus

$$f(3) = 3^2 = 9, \qquad f(-2) = (-2)^2 = 4,$$
$$f(a + h) = (a + h)^2 = a^2 + 2ah + h^2.$$

When the domain is understood to be the largest set of real numbers for which the formula makes sense, we shall often resort to such a shortened version as "the function $f(x) = x^2$" in place of a more exact statement in terms of ordered pairs.

4. By a graph of the equation $y = x^2$. This is a curve C (Fig. 1-14(b)) consisting of points with coordinates (x, x^2). The domain of f is represented by the x-axis, the range by the nonnegative part of the y-axis. For any point P with coordinate x in the domain, we find the image by following the arrows, first from P up to $Q(x, x^2)$ on the curve, and then from Q over to the point R with coordinate x^2 on the y-axis.

Another notation is used frequently:

$$f:x \to x^2,$$

read "f sends x into x^2," to represent the function f which maps x onto x^2. This notation emphasizes the action of mapping, and, unless something is said to indicate otherwise, it is understood that the domain is the set of all real numbers to which the mapping can be applied. Thus, if

$$f:x \to x^2 + 2x - 3, \qquad \text{the domain is } -\infty < x < \infty,$$

but if

$$f:x \to 1/x, \qquad \text{the domain excludes 0.}$$

It may happen that there is no formula to express the value of $f(x)$ in terms of algebraic operations on x. Sometimes special notation is invented as in the next example.

20

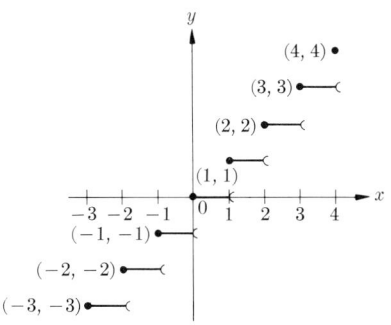
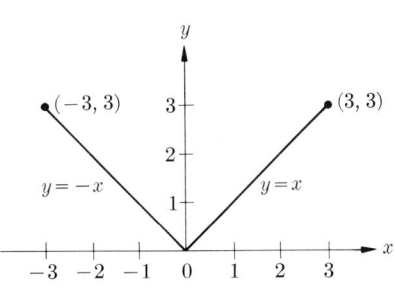

FIG. 1–15. Portion of the graph of $y = [x]$, for $-3 \leq x \leq 4$.

FIG. 1–16. Graph of $y = |x|$, for $-3 \leq x \leq 3$.

Example 6. The *greatest-integer function* maps any real number x onto that unique integer which is the largest among all integers that are less than or equal to x. This, the *greatest integer in x*, is represented by putting x in square brackets:

$$[x] = \text{greatest integer that is } \leq x.$$

With this convention, we can represent the greatest-integer *function* as the set of ordered pairs

$$f = \{(x, [x]) : -\infty < x < \infty\}$$

or as the mapping

$$f : x \to [x].$$

A part of the graph of $y = [x]$ is shown in Fig. 1–15. It resembles a set of stairsteps without any risers, and where the tread of each step starts at a point (n, n) with n an integer. For $n \leq x < n + 1$, we have $[x] = n$. Thus,

$$[-2] = -2, \quad [-1.8] = -2, \quad [0.2] = 0, \quad [3.99] = 3.$$

Each tread is a line segment one unit long, closed at its left end and open at its right end.

Example 7. Another function for which a special symbol has been introduced is the absolute-value function, $|x|$. Figure 1–16 shows that part of the graph of $y = |x|$ for $-3 \leq x \leq 3$. Note, in particular, that for all values of x from -3 to 3 we have $|x| \leq 3$. That is, if the domain of the absolute-value function is restricted to the interval from -3 to 3, then the range is the interval from 0 to 3. Conversely, if we restrict the range to $|x| \leq 3$, then the domain is restricted to $-3 \leq x \leq 3$.

At the beginning of this section we listed a number of equations describing relations. Some of these equations describe functions; others do not. The graph of a function has at most one point on each vertical line. Sometimes by taking a subset of a relation that is not a function we can get a function, as in the following example.

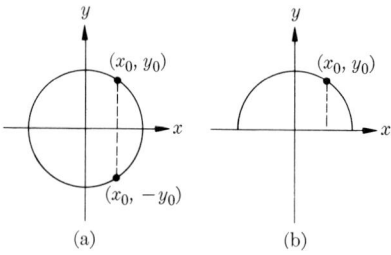

FIG. 1–17. (a) Graph of the relation $x^2 + y^2 = r^2$. (b) Graph of $f: x^2 + y^2 = r^2$ and $y \geq 0$.

Example 8. $f = \{(x, y) : x^2 + y^2 = r^2 \text{ and } y \geq 0\}$. This is the equation of a semicircle, and, since to every x in the domain there corresponds a unique y in the range, we have a function. Compare the graphs in Fig. 1–17.

Next, let us examine the following four functions and their graphs (Fig. 1–18):

$$f_1 = \{(x, y) : x^2 + y^2 = r^2 \text{ and } -r \leq x \leq 0 \text{ and } y \geq 0\},$$
$$f_2 = \{(x, y) : y = x^2 \text{ and } x \geq 0\},$$
$$f_3 = \{(x, y) : y = x^3 \text{ and } x \in R\},$$
$$f_4 = \{(x, y) : y = e^x \text{ and } x \in R\}.$$

In each of these four functions, to every abscissa in the domain there is assigned exactly one element in the range; and to every element in the range there is assigned exactly one element in the domain. Such functions are called one-to-one mappings. Note also that the ordinate increases as the abscissa increases as we move from left to right. If, instead, we had

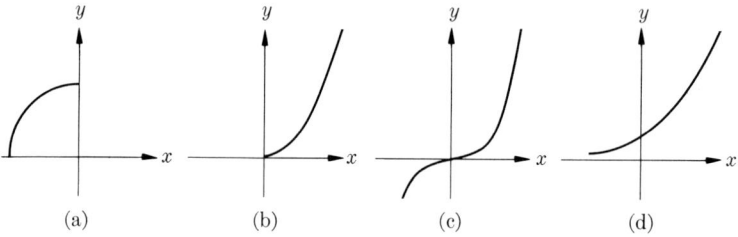

FIG. 1–18. (a) Graph of f_1. (b) Graph of f_2. (c) Graph of f_3. (d) Graph of f_4.

drawn the graphs of the functions defined by

$$g_1 : x \to x^2 \text{ and } x \leq 0, \qquad g_2 : x \to -x^3, \qquad g_3 : x \to e^{-x},$$

we would observe that, as x increases, $g(x)$ decreases.

A function is "one-to-one" if $f(x_1) \neq f(x_2)$ whenever $x_1 \neq x_2$. Also, we say that a function is strictly increasing if $f(x_2) > f(x_1)$ whenever $x_2 > x_1$. Similarly, a function g is strictly decreasing if $g(x_2) < g(x_1)$ whenever $x_2 > x_1$. In the study of functions and their inverses (Chapter 8) we will establish that a function has an inverse that is also a function if and only if it is one-to-one.

Finally, curve-tracing plays an important role in the analysis of a function or relation. The following properties of symmetry are often useful in this task:

1. Two points, A and B, are symmetric with respect to a line if the line is the perpendicular bisector of the segment AB.

2. Two points, A and B, are symmetric with respect to a point P if P is the midpoint of segment AB.

Therefore:

i) A curve is symmetric with respect to the y-axis if x can be replaced by $-x$; for example $y = x^2 \Leftrightarrow y = (-x)^2$ ($\Leftrightarrow$ means "is equivalent to");

ii) A curve is symmetric with respect to the x-axis if y can be replaced by $-y$; for example $y^2 = x \Leftrightarrow (-y)^2 = x$;

iii) A curve is symmetric with respect to the line $y = x$ if for every point $A(a, b)$ on the curve there exists a point $B(b, a)$ on the curve;

iv) A curve is symmetric with respect to the origin if x can be replaced by $-x$ and y by $-y$; for example, $y = x^3 \Leftrightarrow -y = (-x)^3$.

The converses of these statements are also true, and the reader will be asked to verify them in the exercises of this section.

EXERCISES 1-7

1. Suppose $X = \{1, 2, 3\}$ and $Y = \{0, 1\}$.
   a) Is the set $f = \{(1, 1), (2, 0), (3, 1)\}$ a function? What is $f(1)$?
   b) Is the set $\{(1, 1), (1, 0), (2, 1)\}$ a function? Why?
   c) How many different functions are there with domain X and range Y? List some of them.

2. Which of the following define functions? For each f that is a function, describe the range.
   a) $f(x) = 2x, \ 0 \leq x \leq \frac{1}{2}$ \qquad b) $f(x) = 1 - 2x, \ 0 \leq x \leq \frac{1}{2}$

c) $f(x) = 1 - 2x, \quad 0 \le x \le 1$
d) $f(x) = x^2, \quad 0 \le x \le 1$

e) $f(x) = \begin{cases} 2x & \text{if } 0 \le x \le \frac{1}{2} \\ 2x - 1 & \text{if } \frac{1}{2} < x \le 1 \end{cases}$
f) $f(x) = x - [x], \quad x \in R$

g) $f(x) = [2x - 3], \quad x \in R$
h) $f(x) = 2[x] - 3, \quad x \in R$

3. Draw the graph of each of the following equations:
   a) $9x^2 + 16y^2 = 144$
   b) $x^2 + y^2 = 25$
   c) $y^2 - x = 0$

4. For each equation in Exercise 3 make the necessary restrictions so that a function is defined. Give the domain and range of the function and draw its graph. Is there more than one choice?

5. Make the necessary restrictions for the equations in Exercise 3 so that one-to-one functions are defined. How many possibilities are there for each?

6. Draw the graph of $y = 2^x$ for $-3 \le x \le 3$.

7. Draw the graph of $y = (\frac{1}{2})^x$ for $-3 \le x \le 3$.

8. What is the relation between the graphs of 2^x and 2^{-x}? (Note: $(\frac{1}{2})^x = 2^{-x}$.)

9. For what values of m and k will $y = mx + k$ be symmetric about the line $y = x$?

10. Given the point $P(-3, 7)$, give the coordinates of points symmetric to P with respect to the
    a) y-axis
    b) x-axis
    c) origin
    d) line $y = x$

11. State what symmetries, if any, the graphs of each of the following have:
    a) $x^2 y = 36$
    b) $y = 9x - x^3$
    c) $y^2 = x^3$
    d) $x^2 y + xy^2 = 0$

12. Draw the graph of each equation in Exercise 11. Make use of the results of Exercise 11.

13. Prove: If point $A(a, b)$ is on the graph of $xy = k$, point $B(b, a)$ is also on the graph. Show that the line $y = x$ is the perpendicular bisector of segment AB.

14. Segment AB joins two points of the graph $y = x^3$ and is bisected by $O(0, 0)$. Given point $A(a, b)$, find the coordinates of point B. Prove your answer.

15. On the graph for the equation in Exercise 11(a), find the coordinates of point B, symmetric to point $A(-2, 9)$.

16. Repeat Exercise 15 for the equation in Exercise 11(d), if the coordinates of A are (r, s).

17. Repeat Exercise 15 for 2^x and 2^{-x} and $A(3, 8)$.

2
LIMITS
OF
SEQUENCES

2-1 INTRODUCTION

The notion of a limit arose early in mathematics in the calculation of the area inside a circle. Suppose the region enclosed by a circle is cut into a large number n of congruent pie-shaped segments, as in Fig. 2-1(a), and these are spread out to form a sawtoothed pattern as in Fig. 2-1(b). The area enclosed by the circle is equal to the sum of the areas of the segments.

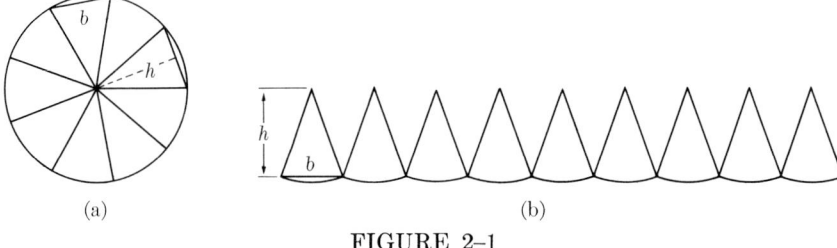

FIGURE 2-1

Each segment is approximately a triangle, of altitude h and of base b. If the number of segments is n, the total area is approximately $n \times \frac{1}{2}bh$. Let us write this as A_n to show its dependence on n; thus

$$A_n = n \times \frac{1}{2} bh = \frac{h}{2}(nb). \tag{1}$$

In the last member of Eq. (1), we recognize nb as the perimeter of the regular polygon of n sides inscribed in the circle and we feel intuitively that, if we take n to be very large, the perimeter of this inscribed polygon should be very nearly equal to the circumference of the circle:

$$nb \approx C.$$

And the altitude h should be very nearly equal to the radius of the circle r. Hence, $A_n = (h/2)(nb)$ should be very nearly equal to $(r/2)C$. One purpose for introducing the notion of a *limit* is to make more precise the meaning of the phrase "very nearly equal to," and to replace such statements as

$$nb \approx C, \quad h \approx r$$

by

$$\lim_{n \to \infty} h = r, \quad \lim_{n \to \infty} (nb) = C$$

and

$$\lim_{n \to \infty} A_n = \lim_{n \to \infty} \frac{h}{2}(nb) = \frac{r}{2} C.$$

The laws of algebra tell us that if $a = b$ and $c = d$, then $ac = bd$ and $a + c = b + d$. We would hope that similar laws would apply to limits if, in some sense, a "limit" is to replace "very nearly equal to." In other words, suppose we temporarily agree to use the approximation symbol, $\approx$, as a synonym for "is very nearly equal to." Then, if $a \approx b$ and $c \approx d$, we hope that it is true that $ac \approx bd$ and $a + c \approx b + d$. So one goal in studying limits will be to see under what conditions it will be true that the limit of a product (or sum) is equal to the product (or sum) of the separate limits.

In the circle example, if $\lim_{n \to \infty} h = r$ and

$$\lim_{n \to \infty} (nb) = C,$$

is it also true that

$$\lim_{n \to \infty} \frac{h}{2} = \frac{r}{2}$$

and

$$\lim_{n \to \infty} \frac{h}{2} (nb) = \left[\lim_{n \to \infty} \frac{h}{2}\right] \cdot \left[\lim_{n \to \infty} (nb)\right]?$$

To anticipate some later work, we answer "yes," and, assuming that $C = 2\pi r$ is a known formula for the circumference of a circle, we deduce

$$A = \frac{r}{2} \cdot 2\pi r = \pi r^2$$

for the enclosed area.

As a second example of a limit, consider the problem of summing a geometric series:

$$a + ar + ar^2 + \cdots + ar^n + \cdots \qquad (2)$$

Here a is the first term and r is the common multiplier that takes us from any term to the next. Let us denote the sum of the first $n + 1$ terms by S_n:

$$S_n = a + ar + ar^2 + \cdots + ar^n. \qquad (3)$$

Then

$$rS_n = ar + ar^2 + ar^3 + \cdots + ar^{n+1},$$

so that

$$S_n - rS_n = a - ar^{n+1} = a(1 - r^{n+1}),$$

or

$$(1 - r)S_n = a(1 - r^{n+1}). \qquad (4)$$

If $r = 1$, Eq. (4) becomes $0 = 0$, which is true but not helpful. However, in this case, Eq. (3) becomes

$$S_n = (n + 1)a \quad \text{if} \quad r = 1. \qquad (5a)$$

LIMITS OF SEQUENCES

If $r \neq 1$, we divide both sides of Eq. (4) by $1 - r$ and get

$$S_n = a \frac{1 - r^{n+1}}{1 - r}. \tag{5b}$$

Example. Find the sum of the first $n + 1$ terms of the series

$$\tfrac{3}{10} + \tfrac{3}{100} + \tfrac{3}{1000} + \cdots \tag{6}$$

Solution. Take $a = \tfrac{3}{10}$, $r = \tfrac{1}{10}$ and apply Eq. (5b) to get

$$S_n = \frac{3}{10} \frac{1 - (0.1)^{n+1}}{1 - 0.1} = \frac{3}{9}[1 - (0.1)^{n+1}].$$

Since $(0.1)^{n+1}$ is very nearly equal to zero when n is large, we would naturally conclude that

$$\lim_{n \to \infty} S_n = \tfrac{3}{9}[1 - 0] = \tfrac{1}{3}.$$

We are especially prepared to accept this result because it agrees with the repeating decimal expansion

$$\tfrac{1}{3} = 0.3333 \cdots = \tfrac{3}{10} + \tfrac{3}{100} + \tfrac{3}{1000} + \cdots$$

EXERCISES 2-1

Make reasonable guesses of the following limits.

1. $\lim\limits_{n \to \infty} \left(2 + \dfrac{1}{n}\right)$

2. $\lim\limits_{n \to \infty} \dfrac{2n + 1}{n}$

3. $\lim\limits_{n \to \infty} \dfrac{2 + n}{n^2}$

4. $\lim\limits_{n \to \infty} \left(1 + \dfrac{1}{2} + \dfrac{1}{4} + \cdots + \dfrac{1}{2^n}\right)$

2-2 DEFINITION OF LIMIT OF A SEQUENCE

In the previous section, we introduced the idea of "limit" as a kind of synonym for "very nearly equal to," in the sense that we write

$$\lim_{n \to \infty} S_n = L$$

if S_n is *very nearly equal to* L *when* n *is large*. We now want to make a more precise definition. To do so, we need a measure of the difference between S_n and the number L that is supposed to be very nearly equal to it. Since S_n might be either larger or smaller than L, we measure their nearness by the absolute value of the difference,

$$|S_n - L|. \tag{1}$$

In the geometric series example, we had
$$S_n = \tfrac{3}{9}[1 - (\tfrac{1}{10})^{n+1}]$$
and we guessed the limit to be
$$L = \tfrac{1}{3}.$$
The absolute value of the difference is
$$|S_n - L| = |\tfrac{3}{9} - \tfrac{3}{9}(\tfrac{1}{10})^{n+1} - \tfrac{3}{9}| = \frac{1}{3 \times 10^{n+1}}.$$
When $n = 1$,
$$|S_n - L| = \tfrac{1}{300};$$
when $n = 2$,
$$|S_n - L| = \tfrac{1}{3000},$$
and so on; the larger n is, the smaller the difference. Imagine that we are playing a game against an opponent, and he challenges us to make the difference less than 0.0001. We observe that this amounts to choosing n so large that
$$\frac{1}{3 \times 10^{n+1}} < 0.0001 = \frac{1}{10^4},$$
and clearly any $n \geq 3$ will do the job.

Consider a new example in which
$$S_n = \frac{n+2}{n+3} = \frac{(n+3)-1}{n+3} = 1 - \frac{1}{n+3}.$$
We see that when n is large, $1/(n+3)$ is small and S_n is close to 1. Hence we guess that the limit is $L = 1$. To back up our guess, we consider the absolute value of the difference:
$$|S_n - L| = \left|\frac{n+2}{n+3} - 1\right| = \left|-\frac{1}{n+3}\right| = \frac{1}{n+3}.$$
Imagine that our opponent challenges us to make this difference less than $0.0007 = 7/10^4$. Can we do it? That is, can we satisfy
$$\frac{1}{n+3} < \frac{7}{10^4}?$$
Or, what amounts to the same thing, can we satisfy
$$n + 3 > \frac{10^4}{7} = \frac{10{,}000}{7} = 1428\tfrac{4}{7}?$$

Yes; all we need do is take $n \geq 1426$, and for any such value of n we can be sure that
$$|S_n - L| < 0.0007.$$

These examples should help us see why the following definitions are widely accepted.

Definition 1. A *sequence* $\{S_n\}$ is a function that associates with each positive integer n a number S_n.

The formulas
$$S_n = \tfrac{1}{3}[1 - (\tfrac{1}{10})^{n+1}] \quad \text{and} \quad S_n = \frac{n+2}{n+3}$$
give examples of sequences.

Definition 2. The sequence $\{S_n\}$ is said to *approach the number L as a limit* when n tends to infinity provided that to each positive number ϵ (pronounced *epsilon*) there corresponds an integer N such that
$$|S_n - L| < \epsilon \quad \text{whenever} \quad n \geq N. \tag{2}$$

When (2) is satisfied, we also write
$$\lim_{n \to \infty} S_n = L.$$

For the example
$$S_n = \frac{n+2}{n+3} \quad \text{with} \quad L = 1,$$
we have
$$|S_n - L| = \frac{1}{n+3}$$
and we can make this less than any preassigned positive number ϵ by making
$$1/(n+3) < \epsilon \quad \text{or} \quad n+3 > (1/\epsilon).$$

Hence, if we take N to be any positive integer that is greater than $(1/\epsilon) - 3$, then we have $|S_n - L| < \epsilon$ whenever $n \geq N$.

For the geometric series example,
$$S_n = \tfrac{1}{3}\left[1 - \frac{1}{10^{n+1}}\right], \quad L = \tfrac{1}{3},$$
we found
$$|S_n - L| = \frac{1}{3 \times 10^{n+1}},$$

and we can make this less than any given positive number ϵ by making

$$\frac{1}{3 \times 10^{n+1}} < \epsilon,$$

or

$$3 \times 10^{n+1} > \frac{1}{\epsilon},$$

or

$$10^{n+1} > \frac{1}{3\epsilon},$$

or

$$(n + 1) > \log_{10}\left(\frac{1}{3\epsilon}\right).$$

So we can take N to be any positive integer greater than

$$\log_{10}\left(\frac{1}{3\epsilon}\right) - 1,$$

and be sure that

$$|S_n - L| < \epsilon \quad \text{whenever} \quad n \geq N.$$

Usually N depends on ϵ.

EXERCISES 2-2

For each sequence $\{S_n\}$ given below, guess its limit L. For the given ϵ, find an integer N such that $|S_n - L| < \epsilon$ whenever $n \geq N$.

1. $S_n = 2 + \frac{1}{n}$, $\epsilon = 0.01$

2. $S_n = \frac{2n + 3}{n}$, $\epsilon = 0.5$

3. $S_n = \frac{n + \sin n}{2n + 1}$, $\epsilon = 0.1$

4. $S_n = 1 - \frac{1}{3} + \frac{1}{9} - \frac{1}{27} + \cdots + (-1)^n \frac{1}{3^n}$, $\epsilon = 0.001$

5. Prove that, if $\lim_{n \to \infty} S_n = L$, and $T_n = S_{n+1}$, then $\lim_{n \to \infty} T_n = L$.

2-3 SOME PROPERTIES OF LIMITS

One of the goals we set for ourselves was to develop the theory of limits far enough to test the validity of such laws as

$$\lim_{n \to \infty} (S_n + T_n) = (\lim_{n \to \infty} S_n) + (\lim_{n \to \infty} T_n) \tag{1}$$

and

$$\lim_{n \to \infty} (S_n T_n) = (\lim_{n \to \infty} S_n) \cdot (\lim_{n \to \infty} T_n). \tag{2}$$

LIMITS OF SEQUENCES

Example 1. Suppose $S_n = \dfrac{n+2}{n}$ and $T_n = \dfrac{2n+1}{n+2}$. Then

$$S_n = 1 + \frac{2}{n} \quad \text{and} \quad \lim_{n \to \infty} S_n = 1,$$

$$T_n = \frac{(2n+4)-3}{n+2} = 2 - \frac{3}{n+2} \quad \text{and} \quad \lim_{n \to \infty} T_n = 2,$$

while

$$S_n + T_n = \left(1 + \frac{2}{n}\right) + \left(2 - \frac{3}{n+2}\right) = 3 + \frac{2}{n} - \frac{3}{n+2},$$

so

$$\lim_{n \to \infty} (S_n + T_n) = 3 = \lim_{n \to \infty} S_n + \lim_{n \to \infty} T_n.$$

Also,

$$S_n T_n = \left(1 + \frac{2}{n}\right)\left(2 - \frac{3}{n+2}\right)$$

$$= 2 + \frac{4}{n} - \frac{3}{n+2} - \frac{6}{n(n+2)},$$

so that

$$\lim_{n \to \infty} (S_n T_n) = 2 = (\lim_{n \to \infty} S_n) \cdot (\lim_{n \to \infty} T_n).$$

Theorem 2–1. *Suppose $\{S_n\}$ and $\{T_n\}$ are sequences with limits L_1 and L_2 respectively. Then*

$$\lim_{n \to \infty} (S_n + T_n) = L_1 + L_2. \tag{3}$$

Proof. By hypothesis,

$$\lim_{n \to \infty} S_n = L_1 \quad \text{and} \quad \lim_{n \to \infty} T_n = L_2. \tag{4}$$

We must show that to any positive number ϵ there corresponds a positive integer N such that

$$|(S_n + T_n) - (L_1 + L_2)| < \epsilon \quad \text{when} \quad n \geq N. \tag{5}$$

We now apply (2) of Section 1–6, with $a = S_n - L_1$ and $b = T_n - L_2$, to rewrite the left side of the desired inequality (5) as follows:

$$|(S_n + T_n) - (L_1 + L_2)|$$
$$= |(S_n - L_1) + (T_n - L_2)| \leq |S_n - L_1| + |T_n - L_2|.$$

32

From the given limits in (4), and the fact that $\epsilon/2$ is positive when ϵ is, we can assert that to any $(\epsilon/2) > 0$ there correspond integers N_1 and N_2 such that

$$|S_n - L_1| < \frac{\epsilon}{2} \quad \text{whenever} \quad n \geq N_1 \tag{6a}$$

and

$$|T_n - L_2| < \frac{\epsilon}{2} \quad \text{whenever} \quad n \geq N_2. \tag{6b}$$

Now let N be the larger of N_1 and N_2. Then, whenever $n \geq N$, both (6a) and (6b) are satisfied, so that

$$|(S_n + T_n) - (L_1 + L_2)| \leq |S_n - L_1| + |T_n - L_2| < \frac{\epsilon}{2} + \frac{\epsilon}{2} = \epsilon.$$

Q.E.D.

In Example 1, $S_n = 1 + (2/n)$, $T_n = 2 - 3/(n + 2)$. If $\epsilon > 0$, then

$$|S_n - 1| = \frac{2}{n} < \frac{\epsilon}{2} \quad \text{if} \quad n > \frac{4}{\epsilon}$$

and

$$|T_n - 2| = \frac{3}{n+2} < \frac{\epsilon}{2} \quad \text{if} \quad n > \frac{6}{\epsilon} - 2.$$

If N_1 is a positive integer $\geq 4/\epsilon$ and N_2 is a positive integer $\geq (6/\epsilon) - 2$, we let N be the larger of N_1 and N_2. For instance, with $\epsilon = 0.01$, we want $N_1 \geq 400$ and $N_2 \geq 598$, so we take $N = 598$. Then, whenever $n \geq 598$, we would be sure that $|S_n + T_n - 3| < \epsilon$ because we would be sure that

$$|S_n - 1| < \frac{\epsilon}{2} \quad \text{and} \quad |T_n - 2| < \frac{\epsilon}{2}$$

and

$$|(S_n + T_n) - 3| \leq |S_n - 1| + |T_n - 2| < \epsilon.$$

Before we go on to products of sequences, we prove that every convergent sequence is bounded. The meanings of the words *convergent* and *bounded* as applied to a sequence $\{S_n\}$ are defined as follows.

Definition 3. Convergent. A sequence $\{S_n\}$ that has a limit L is said to be *convergent*, and it *converges* to L. If a sequence does not have a limit, it is said to be *divergent*, or to *diverge*.

Definition 4. Bounded. A sequence $\{S_n\}$ is said to be *bounded* if there exists a number B such that

$$|S_n| \leq B \quad \text{for all values of } n. \tag{7}$$

Example 2. The sequence given by $S_n = (-1)^n$ diverges, because $(-1)^n = -1$ when n is odd, and $(-1)^n = +1$ when n is even. Hence there is no fixed number L such that all terms S_n with sufficiently large values of the index n are very nearly equal to L. This sequence is *bounded*, however, because there is a number B such that $|S_n| \leq B$ for all values of n. For instance, we could take $B = 2$, or $B = 10$, or $B = 1$.

Example 3. For
$$S_n = \frac{n+2}{n} = 1 + \frac{2}{n}$$
we have $2/n \leq 2$, for all $n \geq 1$, so
$$|S_n| \leq 3.$$
For
$$T_n = \frac{2n+1}{n+2} = 2 - \frac{3}{n+2},$$
we have $|T_n| \leq 2$, because $3/(n+2)$ is positive and less than or equal to 1 for all $n \geq 1$.

Theorem 2–2. *Every convergent sequence is bounded.*

Proof. Let $\{S_n\}$ be a convergent sequence. Then there is a number L to which S_n converges:
$$\lim_{n \to \infty} S_n = L.$$
In the definition of limit, take $\epsilon = 1$ (any other positive number would work as well). Then we know there is an integer N such that
$$|S_n - L| < 1 \qquad \text{whenever} \qquad n \geq N.$$
This means that S_n lies between $L - 1$ and $L + 1$ when $n \geq N$, so certainly
$$|S_n| \leq |L| + 1 \qquad \text{when} \qquad n \geq N. \tag{8}$$
We may not be able to take $B_1 = |L| + 1$ as a bound for the entire sequence, because the first inequality in (8) does not necessarily apply to $|S_1|, |S_2|, \ldots, |S_{N-1}|$; we are only sure it holds for $n \geq N$. But there are only a finite number of terms with $1 \leq n \leq N - 1$ and we take B_2 to be their maximum:
$$|S_n| \leq B_2 \qquad \text{for} \qquad 1 \leq n \leq N - 1.$$
Now take B to be the larger of B_1 and B_2. Then $|S_n| \leq B$ for all n.

Example 4. Let
$$S_n = \frac{2n-5}{n+1}.$$
Then
$$S_n = \frac{(2n+2)-7}{n+1} = 2 - \frac{7}{n+1}$$
converges to $L = 2$, and
$$|S_n - 2| = \frac{7}{n+1} < 1$$
if $n \geq 7 = N$. To find B_2, we examine $|S_1|, |S_2|, \ldots, |S_6|$, and take the maximum. For this example,

$$S_1 = \frac{2-5}{1+1} = -\frac{3}{2}, \quad S_2 = \frac{4-5}{2+1} = -\frac{1}{3},$$

$$S_3 = \frac{6-5}{3+1} = \frac{1}{4}, \quad S_4 = \frac{8-5}{4+1} = \frac{3}{5},$$

$$S_5 = \frac{10-5}{5+1} = \frac{5}{6}, \quad S_6 = \frac{12-5}{6+1} = \frac{7}{7} = 1.$$

The absolute values are $\frac{3}{2}, \frac{1}{3}, \frac{1}{4}, \frac{3}{5}, \frac{5}{6}, 1$, and the maximum of these is $B_2 = \frac{3}{2}$. Since $L = 2$ and $B_1 = L + 1 = 3 = 2 + \epsilon$ with $\epsilon = 1$, we take B to be the larger of the numbers $\frac{3}{2}$ and 3; hence $B = 3$. We therefore have
$$\left|\frac{2n-5}{n+1}\right| \leq 3$$
for every positive integer n.

We are now prepared to prove a theorem about the limit of a product of two sequences.

Theorem 2–3. Let $\{S_n\}$ and $\{T_n\}$ be convergent sequences whose limits are L_1 and L_2, respectively. Then
$$\lim_{n \to \infty} (S_n T_n) = L_1 L_2. \tag{9}$$

Proof. Let ϵ be an arbitrary positive number. We want to show that there is a positive integer N such that
$$|S_n T_n - L_1 L_2| < \epsilon \quad \text{whenever} \quad n \geq N. \tag{10}$$
To this end, we rewrite the difference $S_n T_n - L_1 L_2$ as follows:
$$S_n T_n - L_1 L_2 = S_n T_n - L_1 T_n + L_1 T_n - L_1 L_2$$

and apply inequality (2) of Section 1–6, with
$$a = S_n T_n - L_1 T_n$$
and
$$b = L_1 T_n - L_1 L_2.$$
Thus
$$|S_n T_n - L_1 L_2| \le |S_n T_n - L_1 T_n| + |L_1 T_n - L_1 L_2|$$
or
$$|S_n T_n - L_1 L_2| \le |(S_n - L_1) T_n| + |L_1 (T_n - L_2)|. \tag{11}$$

Next we use (1a) of Section 1–6,
$$|ab| = |a| \cdot |b|,$$
which enables us to rewrite (11) as follows:
$$|S_n T_n - L_1 L_2| \le |S_n - L_1| \cdot |T_n| + |L_1| \cdot |T_n - L_2|. \tag{12}$$

If we compare the inequality (12) with that in (10), we see that our goal is reached if we find an integer N such that both terms on the right side of (12) are less than $\epsilon/2$ when $n \ge N$. Apply Theorem 2–2 to $\{T_n\}$: since $\{T_n\}$ converges it is bounded and there is a number B such that
$$|T_n| \le B \quad \text{for every } n,$$
so
$$|S_n - L_1| \cdot |T_n| \le |S_n - L_1| \cdot B. \tag{13}$$

Our goal is to make the right side of (13) less than $\epsilon/2$. In general, since B is not negative, this is equivalent to making
$$|S_n - L_1| < \frac{\epsilon}{2B}. \tag{14a}$$

But the right side of (14a) is meaningless if B happens to be zero. We can avoid any risk of dividing by zero if we use $1 + B$ instead of B in the denominator in (14a). Since $\{S_n\}$ converges to L_1, there is a positive integer N_1 such that
$$|S_n - L_1| < \frac{\epsilon}{2(1 + B)} \quad \text{whenever} \quad n \ge N_1. \tag{14b}$$

Hence
$$|S_n - L_1| \cdot |T_n| \le |S_n - L_1| \cdot B < \frac{\epsilon}{2} \quad \text{whenever} \quad n \ge N_1, \tag{15}$$
because $B/(1 + B) < 1$.

Likewise, there is a positive integer N_2 such that

$$|T_n - L_2| < \frac{\epsilon}{2(1 + |L_1|)} \quad \text{whenever} \quad n \geq N_2,$$

so

$$|L_1| \cdot |T_n - L_2| < \frac{\epsilon}{2} \quad \text{whenever} \quad n \geq N_2, \quad (16)$$

because $|L_1|/(1 + |L_1|) < 1$.

Now take N equal to the larger of the two integers N_1 and N_2. Then both (15) and (16) are in force whenever $n \geq N$, and we have [from (12)]

$$|S_n T_n - L_1 L_2| \leq |S_n - L_1| \cdot |T_n| + |L_1| \cdot |T_n - L_2|$$
$$< \frac{\epsilon}{2} + \frac{\epsilon}{2} = \epsilon$$

whenever $n \geq N$.

Example 5. As in Example 3, let

$$S_n = \frac{n+2}{n} \quad \text{and} \quad T_n = \frac{2n+1}{n+2} = \frac{(2n+4) - 3}{n+2}.$$

Then $L_1 = \lim_{n \to \infty} S_n = 1$ and $L_2 = \lim_{n \to \infty} T_n = 2$. In Example 3 we found $|T_n| \leq 2 = B$, and the inequality (14b) in the proof of Theorem 2–3 becomes

$$|S_n - 1| < \frac{\epsilon}{2(1+2)} = \frac{\epsilon}{6}. \quad (17)$$

Since

$$S_n - 1 = \left(1 + \frac{2}{n}\right) - 1 = \frac{2}{n},$$

inequality (17) is satisfied whenever

$$\frac{2}{n} < \frac{\epsilon}{6}, \quad \text{or} \quad n > \frac{12}{\epsilon}.$$

Thus, we take N_1 to be any integer greater than $12/\epsilon$. Now look at the inequality (16), which becomes

$$|1| \cdot |T_n - 2| < \frac{\epsilon}{2}. \quad (18)$$

Since

$$T_n = 2 - \frac{3}{n+2},$$

inequality (18) is satisfied if

$$\frac{3}{n+2} < \frac{\epsilon}{2}, \quad \text{or} \quad n > \frac{6}{\epsilon} - 2.$$

We take N_2 to be any integer greater than $(6/\epsilon) - 2$. Since $(12/\epsilon) > (6/\epsilon) - 2$, we can here take $N = N_1$ and be sure that both inequalities (17) and (18) are satisfied for every value of n that is greater than or equal to N. For example, if our opponent specifies $\epsilon = 0.003$, then $12/\epsilon = 4000$ and we can take $N = 4001$. Notice, however, that in Example 1 we found that

$$S_n T_n = 2 + \frac{4}{n} - \frac{3}{n+2} - \frac{6}{n(n+2)}$$
$$= 2 + \frac{(4n+8) - 3n - 6}{n(n+2)} = 2 + \frac{n+2}{n(n+2)} = 2 + \frac{1}{n},$$

so

$$|S_n T_n - 2| = \frac{1}{n},$$

and, if $\epsilon = 0.003$, we can make $1/n$ less than ϵ by taking $n \geq 334$. Hence $N = 334$ will also meet the requirement

$$|S_n T_n - 2| < \epsilon \quad \text{whenever} \quad n \geq N. \tag{19}$$

This illustrates the fact that the N we find by following the steps in the proof of Theorem 2–3 isn't necessarily the smallest N that will work for a given positive ϵ. However, the definition of limit doesn't require that we produce the smallest possible N, but just that we show there is *some* positive integer N that meets the challenge.

Theorems 2–2 and 2–3 take care of the question of limits of sums and products. They also have the following easy corollaries.

Corollary 2–3.1. If $\{S_n\}$ converges to L and k is any number, then $\{kS_n\}$ converges to kL.

Proof. For each n, take $T_n = k$. Then $\lim_{n \to \infty} T_n = k$ because the difference $T_n - k$ is zero for every n, and therefore, if $\epsilon > 0$, we can take $N = 1$ and be sure that

$$|T_n - k| < \epsilon \quad \text{whenever} \quad n \geq N.$$

Therefore the sequence $\{T_n\}$ converges to k, and if we take $L_1 = L$ and $L_2 = k$ in Theorem 2–3, we get

$$\lim_{n \to \infty} S_n T_n = \lim_{n \to \infty} k S_n = kL.$$

Corollary 2–3.2. If $\{S_n\}$ converges to L_1 and $\{T_n\}$ converges to L_2, then $\lim_{n \to \infty} (S_n - T_n) = L_1 - L_2$.

Proof. By Corollary 2–3.1, with $k = -1$, we have
$$\lim_{n \to \infty} (-T_n) = -L_2.$$
Now apply Theorem 2–1 to the convergent sequences $\{S_n\}$ and $\{-T_n\}$.

Example 6. Suppose
$$S_n = \frac{2n + \cos n}{n} \quad \text{and} \quad T_n = \frac{n+1}{2n+3}.$$
Then
$$S_n = 2 + \frac{\cos n}{n}, \quad |S_n - 2| = \left|\frac{\cos n}{n}\right| \leq \frac{1}{n},$$
so $\lim_{n \to \infty} S_n = 2$. Likewise,
$$T_n = \frac{1}{2}\left(\frac{2n+2}{2n+3}\right) = \frac{1}{2}\left(\frac{2n+3-1}{2n+3}\right) = \frac{1}{2} - \frac{1}{2}\left(\frac{1}{2n+3}\right),$$
so $\lim_{n \to \infty} T_n = \frac{1}{2}$. Then
$$S_n - T_n = \frac{2n + \cos n}{n} - \frac{n+1}{2n+3} = 2 + \frac{\cos n}{n} - \frac{1}{2} + \frac{1}{2}\left(\frac{1}{2n+3}\right)$$
and
$$\lim_{n \to \infty} (S_n - T_n) = 2 - \tfrac{1}{2} = \lim_{n \to \infty} S_n - \lim_{n \to \infty} T_n.$$

There remains the question of division. We handle it by first considering the reciprocal.

Example 7. If $T_n = \dfrac{n+1}{2n+3}$, then
$$\frac{1}{T_n} = \frac{2n+3}{n+1} = \frac{(2n+2) + 1}{n+1} = 2 + \frac{1}{n+1}$$
and
$$\lim_{n \to \infty} \left(\frac{1}{T_n}\right) = 2 = \frac{1}{\lim_{n \to \infty} T_n}.$$
We now prove the following theorem.

Theorem 2–4. If $\{S_n\}$ converges to a limit L that is not zero then, for sufficiently large values of n, S_n is not zero, and
$$\lim_{n \to \infty} \frac{1}{S_n} = \frac{1}{L}.$$

Proof. Since $L \neq 0$, we can take

$$\epsilon_1 = \tfrac{1}{2}|L|$$

as a positive number, and know that, because S_n converges to L, there is a positive integer N_1 such that

$$|S_n - L| < \epsilon_1 \quad \text{whenever} \quad n \geq N_1. \tag{20}$$

If we write

$$L = (L - S_n) + S_n$$

and apply

$$|a + b| \leq |a| + |b|$$

to the right side, we get

$$|L| = |(L - S_n) + S_n| \leq |L - S_n| + |S_n| < \epsilon_1 + |S_n| \tag{21}$$

for $n \geq N_1$. The first and last parts of (21) give

$$\epsilon_1 + |S_n| > |L|,$$

or

$$|S_n| > |L| - \epsilon_1 = |L| - \tfrac{1}{2}|L| = \tfrac{1}{2}|L|,$$

or

$$|S_n| > \tfrac{1}{2}|L|, \quad \text{when} \quad n \geq N_1. \tag{22}$$

Therefore, whenever $n \geq N_1$, S_n is not zero, and

$$\left|\frac{1}{S_n} - \frac{1}{L}\right| = \left|\frac{L - S_n}{L S_n}\right| = \frac{|L - S_n|}{|L|} \cdot \frac{1}{|S_n|} \leq \frac{|L - S_n|}{|L|} \cdot \frac{2}{|L|}, \tag{23}$$

where the last inequality follows from (22) and the fact that if

$$|S_n| > \frac{|L|}{2} \quad \text{then} \quad \frac{1}{|S_n|} < \frac{2}{|L|}.$$

Now let the opponent specify any positive number ϵ. We need to show that there is a positive integer N such that

$$\left|\frac{1}{S_n} - \frac{1}{L}\right| < \epsilon \quad \text{whenever} \quad n \geq N.$$

We shall first of all require that N be at least as large as N_1, and then use the fact that $\{S_n\}$ converges to L to make the last member in (23) less than ϵ by finding N_2 such that

$$|L - S_n| < \frac{\epsilon |L|^2}{2} \quad \text{whenever} \quad n \geq N_2. \tag{24}$$

Now take N to be the larger of the positive integers N_1 and N_2, so that both (23) and (24) are satisfied whenever $n \geq N$:

$$\left|\frac{1}{S_n} - \frac{1}{L}\right| \leq \frac{2|L - S_n|}{|L|^2} < \frac{2}{|L|^2} \frac{\epsilon|L|^2}{2} = \epsilon.$$

Therefore,

$$\lim_{n \to \infty} \frac{1}{S_n} = \frac{1}{L}.$$

Example 8. To illustrate the proof of Theorem 2–4, suppose

$$S_n = \frac{3(2^n) - 1}{2^n} = 3 - \frac{1}{2^n}.$$

Then

$$\lim_{n \to \infty} S_n = 3,$$

and we take $\epsilon_1 = \frac{3}{2}$ and ask how large n needs to be to make

$$|S_n - 3| < \tfrac{3}{2}. \tag{25}$$

Inequality (25) is the same as

$$\frac{1}{2^n} < \frac{3}{2} \quad \text{or} \quad 2^n > \frac{2}{3},$$

which is satisfied for every $n \geq 1$. Hence we can take $N_1 = 1$. Now suppose our opponent challenges us with $\epsilon = 0.02$ and asks us to make

$$\left|\frac{1}{S_n} - \frac{1}{3}\right| < 0.02.$$

If we follow the proof, with $L = 3$, the inequality (24) becomes

$$|3 - S_n| < \frac{(0.02)(9)}{2} = 0.09,$$

which is the same as

$$\frac{1}{2^n} < \frac{9}{100}, \quad \text{or} \quad 2^n > \frac{100}{9} = 11\tfrac{1}{9}.$$

To satisfy this, we take $N_2 = 4$, and whenever $n \geq N_2$ we have

$$2^n \geq 16 > (100/9).$$

Now take N equal to the larger of $N_1 = 1$ and $N_2 = 4$; thus $N = 4$.

LIMITS OF SEQUENCES

We can now assert that

$$\left|\frac{1}{S_n} - \frac{1}{3}\right| < 0.02 \qquad \text{whenever} \qquad n \geq 4.$$

In fact,

$$\frac{1}{S_n} = \frac{2^n}{3(2^n) - 1} = \frac{1}{3} \cdot \frac{3(2^n)}{3(2^n) - 1} = \frac{1}{3} \cdot \frac{3(2^n) - 1 + 1}{3(2^n) - 1}$$

$$= \frac{1}{3} + \frac{1}{3} \cdot \frac{1}{3(2^n) - 1},$$

so

$$\left|\frac{1}{S_n} - \frac{1}{3}\right| = \frac{1}{3} \cdot \frac{1}{3(2^n) - 1} < 0.02$$

if

$$9(2^n) - 3 > 50, \qquad \text{or} \qquad 9(2^n) > 53,$$

or

$$2^n > \tfrac{53}{9}, \qquad \text{or} \qquad n \geq 3.$$

Hence we could meet the challenge of $\epsilon = 0.02$ by taking $N = 3$, which is slightly smaller than the answer we got by following the steps of the proof of Theorem 2–4.

Corollary 2–4.1. If $\{S_n\}$ converges to L_1 and $\{T_n\}$ converges to L_2, then

$$\lim_{n \to \infty} \frac{S_n}{T_n} = \frac{L_1}{L_2}, \qquad \text{provided} \qquad L_2 \neq 0.$$

Proof. By Theorem 2–4, $\lim_{n \to \infty} (1/T_n) = 1/L_2$, and we apply Theorem 2–3 to the product $S_n \cdot (1/T_n)$ to get

$$\lim_{n \to \infty} \frac{S_n}{T_n} = (\lim_{n \to \infty} S_n) \cdot \left(\lim_{n \to \infty} \frac{1}{T_n}\right) = L_1 \cdot \frac{1}{L_2} = \frac{L_1}{L_2}.$$

Example 9

$$\lim_{n \to \infty} \frac{2 - \frac{3}{n} + \frac{4}{n^2}}{5 + \frac{2}{n} - \frac{1}{n^3}} = \frac{\lim_{n \to \infty}\left(2 - \frac{3}{n} + \frac{4}{n^2}\right)}{\lim_{n \to \infty}\left(5 + \frac{2}{n} - \frac{1}{n^3}\right)} = \frac{2}{5}.$$

Example 10. Find

$$\lim_{n \to \infty} \frac{2n^2 + 4n - 3}{3n^2 + 5n + 6}.$$

2-3 SOME PROPERTIES OF LIMITS

Solution. Divide both the numerator and the denominator of the fraction by n^2, the highest power of n that occurs, and get

$$\frac{2n^2 + 4n - 3}{3n^2 + 5n + 6} = \frac{2 + \dfrac{4}{n} - \dfrac{3}{n^2}}{3 + \dfrac{5}{n} + \dfrac{6}{n^2}} = \frac{S_n}{T_n},$$

with

$$\lim_{n\to\infty} S_n = \lim_{n\to\infty}\left(2 + \frac{4}{n} - \frac{3}{n^2}\right) = 2,$$

and

$$\lim_{n\to\infty} T_n = \lim_{n\to\infty}\left(3 + \frac{5}{n} + \frac{6}{n^2}\right) = 3.$$

Since $\lim T_n \neq 0$, Corollary 2–4.1 gives the answer

$$\lim_{n\to\infty} \frac{S_n}{T_n} = \frac{\lim S_n}{\lim T_n} = \frac{2}{3}.$$

Remark. We can apply Theorem 2–1 in computing $\lim S_n$ above, as follows:

$$\lim_{n\to\infty} S_n = \lim_{n\to\infty}\left(2 + \frac{4}{n} - \frac{3}{n^2}\right) = \lim_{n\to\infty}(2) + \lim_{n\to\infty}\left(\frac{4}{n}\right) + \lim_{n\to\infty}\left(\frac{-3}{n^2}\right)$$
$$= 2 + 0 + 0$$
$$= 2,$$

because it is easy to verify (from the definition of limit) that

$$\lim_{n\to\infty}\frac{1}{n} = 0 \quad \text{and} \quad \lim_{n\to\infty}\frac{1}{n^2} = 0.$$

We conclude this section by proving that whenever a sequence converges its limit is unique.

Theorem 2–5. Let $\{S_n\}$ be a convergent sequence. Then $\lim_{n\to\infty} S_n$ is unique.

Proof (by contradiction). Suppose L_1 and L_2 are two different numbers such that to each positive number ϵ there correspond positive integers N_1 and N_2 such that

$$|S_n - L_1| < \epsilon \quad \text{whenever} \quad n \geq N_1, \tag{26a}$$

and

$$|S_n - L_2| < \epsilon \quad \text{whenever} \quad n \geq N_2. \tag{26b}$$

LIMITS OF SEQUENCES 2-3

If N is the larger of N_1 and N_2, we deduce from (26a) and (26b) that

$$|L_1 - L_2| = |L_1 - S_n + S_n - L_2| \leq |L_1 - S_n| + |S_n - L_2| < 2\epsilon$$

$$\text{whenever} \quad n \geq N. \qquad (27)$$

If, however, $L_1 \neq L_2$, the number $|L_1 - L_2|$ is positive, and nothing prohibits our taking $\epsilon = \tfrac{1}{2}|L_1 - L_2|$ in (27) and concluding that

$$|L_1 - L_2| < |L_1 - L_2|,$$

which is false. This contradiction comes from the assumption that a convergent sequence can have two different limits, L_1 and L_2, so such an assumption is not tenable. That is, the limit of a convergent sequence is unique.

EXERCISES 2-3

1. Prove that
$$\lim_{n \to \infty} 1/n = 0.$$

[If $\epsilon > 0$ is given, how large must N be to guarantee that $|(1/n) - 0| < \epsilon$ whenever $n \geq N$?]

2. Prove that $\lim_{n \to \infty} (1/n^2) = 0$:
    a) by applying Theorem 2-3 and Exercise 1 above, and
    b) directly from the definition by stating how large N needs to be to make $|(1/n^2) - 0| < \epsilon$ whenever $n \geq N$.

3. If $S_n = \dfrac{2n - 3}{3n + 5}$ and $T_n = \dfrac{n^2 + 1}{2n^2 + 3}$, find:

    a) $\lim\limits_{n \to \infty} S_n$ b) $\lim\limits_{n \to \infty} T_n$ c) $\lim\limits_{n \to \infty} (2S_n + 3T_n)$

    d) $\lim\limits_{n \to \infty} (S_n T_n^2)$ e) $\lim\limits_{n \to \infty} \dfrac{S_n^2}{T_n}$

Evaluate the following limits:

4. $\lim\limits_{n \to \infty} \dfrac{2 + 3 \sin n}{4 + n}$ 5. $\lim\limits_{n \to \infty} \dfrac{2n^2 + 3 \sin n}{4n^2 + n}$

6. $\lim\limits_{n \to \infty} \sqrt{\dfrac{2 + 3n}{4 + n}}$ (Can you justify your answer to this problem?)

7. Prove: If $\{S_n\}$ is a *bounded* sequence and $\{T_n\}$ is a sequence that converges to *zero*, then $\lim_{n \to \infty} S_n T_n = 0$. (Since we do not assume $\{S_n\}$ converges, we cannot deduce this theorem from Theorem 2-3, but must give a direct proof.)

8. Apply Theorems 2–1 and 2–3 to prove that if $\{S_n\}$, $\{T_n\}$, and $\{U_n\}$ are convergent sequences, then

$$\lim (S_n + T_n + U_n) = \lim S_n + \lim T_n + \lim U_n,$$

and

$$\lim (S_n T_n U_n) = (\lim S_n) \cdot (\lim T_n) \cdot (\lim U_n),$$

where all limits are as $n \to \infty$. Extend the result to an arbitrary finite number of convergent sequences.

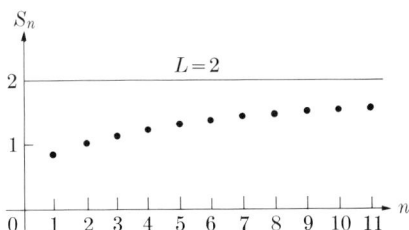

FIG. 2–2. Portion of the graph of $S_n = (2n + 3)/(n + 5) = 2 - 7/(n + 5)$.

2–4 GRAPHICAL REPRESENTATION OF SEQUENCES AND LIMITS; MONOTONE SEQUENCES

The sequence defined by

$$S_n = \frac{2n + 3}{n + 5} = \frac{(2n + 10) - 7}{n + 5}$$

can be represented by a graph consisting of the points $(n, 2 - 7/(n + 5))$ for $n = 1, 2, 3$, and so on, as in Fig. 2–2. In this example, we see that the graph steadily rises as n increases, and that the individual values of S_n are all less than 2. This exemplifies a monotonically increasing sequence that is bounded above—terms we now define.

TABLE FOR FIGURE 2–2

n	$S_n = \dfrac{2n+3}{n+5}$	n	$S_n = \dfrac{2n+3}{n+5}$
1	$\frac{5}{6} \approx 0.83$	7	$\frac{17}{12} \approx 1.42$
2	$\frac{7}{7} = 1.00$	8	$\frac{19}{13} \approx 1.46$
3	$\frac{9}{8} \approx 1.13$	9	$\frac{21}{14} = 1.50$
4	$\frac{11}{9} \approx 1.22$	10	$\frac{23}{15} \approx 1.53$
5	$\frac{13}{10} = 1.30$	11	$\frac{25}{16} \approx 1.58$
6	$\frac{15}{11} \approx 1.36$		

LIMITS OF SEQUENCES 2–4

Definition 5. Monotone Sequences. A sequence $\{S_n\}$ is said to be monotonically increasing (or monotone increasing) if $S_{n+1} \geq S_n$ for every positive integer n. Similarly, $\{S_n\}$ is monotone decreasing if $S_{n+1} \leq S_n$ for each positive integer n.

Definition 6. Bounded above, or bounded below. A sequence $\{S_n\}$ is said to be *bounded above* (or bounded from above) if there is a number M such that $S_n \leq M$ for each n. Similarly, $\{S_n\}$ is *bounded below* if there is a number K such that $S_n \geq K$ for each n.

Definition 7. If M is an upper bound of a sequence $\{S_n\}$, and no number less than M is also an upper bound of $\{S_n\}$, then M is called the least upper bound of $\{S_n\}$, abbreviated thus:

$$M = \text{lub } \{S_n\}.$$

Remark 1. A synonym for *least upper bound* is *supremum*, abbreviated

$$M = \sup \{S_n\}.$$

For the example of Fig. 2–2, where $S_n = (2n + 3)/(n + 5)$, the least upper bound is 2.

Definition 8. If K is a lower bound of the sequence $\{S_n\}$, and no number greater than K is also a lower bound, then K is called the *greatest lower bound* of $\{S_n\}$, abbreviated thus:

$$K = \text{glb } \{S_n\}.$$

Remark 2. A synonym for *greatest lower bound* is *infimum*, abbreviated

$$K = \inf \{S_n\}.$$

For the same example of Fig. 2–2, the infimum is $\frac{5}{6}$.

Remark 3. A constant sequence, one in which $S_1 = S_2 = \cdots = S_n = S_{n+1} = \cdots$ is both monotone increasing and monotone decreasing because it satisfies $S_{n+1} \geq S_n$ and $S_{n+1} \leq S_n$ for each n. Constant sequences are the only ones that are both monotone increasing and monotone decreasing.

Remark 4. From Theorem 2–2 of the previous section we know that every convergent sequence is bounded. If $|S_n| \leq B$ for each n, then $-B \leq S_n \leq B$, so the sequence is both bounded above and bounded below. It is also clear that any sequence that is bounded both from above and from below is bounded, because if $S_n \leq M$ and $S_n \geq K$, then $K \leq S_n \leq M$ and we can take B to be the larger of $|K|$ and $|M|$ and have $|S_n| \leq B$. For example, if $K = -3$ and $M = 2$, we take $B = 3$; if $K = 2$, $M = 5$, we use $B = 5$; if $K = -5$, $M = -2$, we again take $B = 5$.

Remark 5. A monotone increasing sequence is bounded from below because all its terms are at least as large as the first term, S_1.

Remark 6. Sometimes a sequence is defined in such a way that no simple formula can be found expressing S_n in terms of n. For example, suppose

$$S_1 = 1,\ S_2 = S_1 + 1,\ S_3 = S_2 + \frac{1}{2!},$$

and, for each n, $S_n = S_{n-1} + 1/(n-1)!$. Then it is easy to deduce that

$$S_n = 1 + \frac{1}{1!} + \frac{1}{2!} + \frac{1}{3!} + \cdots + \frac{1}{(n-1)!}, \quad \text{for} \quad n \geq 4. \quad (1)$$

But there is no simple way of combining these fractions to get what is called a closed form expression for the sum. It happens that a very important mathematical constant, e, used as the base of natural logarithms, is the limit of the sequence $\{S_n\}$ defined by (1). The question we now consider is this: How can we be sure that the sequence (1) has a limit when we can't find a closed form expression for S_n? The answer depends upon the theorem, which we shall soon prove, that every bounded monotone sequence is convergent. In order to prove this theorem, we use the *completeness* postulate of the system of real numbers.

Completeness Postulate. Let A be any nonempty set of real numbers that is bounded from above. Then there is a unique real number M, called the *supremum* of A, written sup A, such that sup A is the *least* upper bound for A. Likewise, if A is a nonempty set of real numbers that is bounded from below, then there is a unique real number m, called the *infimum* of A, written inf A, such that inf A is the *greatest* lower bound for A.

Example 1. Let A be the set of rational numbers greater than 1 whose squares are less than 2. Then A is nonempty because it contains 1.4, for example, and it is bounded from below by 1 and from above by 1.5. By the completeness postulate, there is a unique real number (not necessarily rational) that is the supremum of A. Because A consists of all rational numbers x such that $1 < x < \sqrt{2}$, we readily see that

$$\sup A = \sqrt{2}, \quad \text{and} \quad \inf A = 1.$$

LIMITS OF SEQUENCES   2–4

We now come to the most important theorem about bounded monotone sequences.

Theorem 2–6. Every bounded monotone sequence converges.

Proof. We shall prove the theorem for monotone increasing sequences; the proof for decreasing sequences is nearly the same. Assume that $\{S_n\}$ is a bounded monotone increasing sequence. Let A be the set of values of S_n for $n = 1, 2, 3, \ldots$ Then A is nonempty and bounded, so there is a number $L = \sup A$. We claim that $\{S_n\}$ converges to L. For suppose ϵ is any positive number. Then $L - \epsilon < L$, where L is the *least* upper bound of A, so $L - \epsilon$ is not an upper bound of A. Therefore, $S_N > L - \epsilon$ for some positive integer N. And, since the sequence is monotone increasing, $S_n \geq S_N$ whenever $n \geq N$, so

$$S_n > L - \epsilon \quad \text{for} \quad n \geq N. \tag{2}$$

But L is an upper bound, so

$$S_n \leq L \quad \text{for} \quad n \geq 1. \tag{3}$$

Combining (2) and (3) we get

$$L - \epsilon < S_n \leq L \quad \text{whenever} \quad n \geq N, \tag{4}$$

which also means that

$$|S_n - L| < \epsilon \quad \text{whenever} \quad n \geq N.$$

Therefore, $\lim_{n \to \infty} S_n = L$.

Example 2. Consider the sequence defined by

$$S_1 = 1, \quad S_2 = 1 + \frac{1}{1!}, \quad S_3 = 1 + \frac{1}{1!} + \frac{1}{2!},$$

and

$$S_n = 1 + \frac{1}{1!} + \frac{1}{2!} + \cdots + \frac{1}{(n-1)!} \quad \text{for} \quad n \geq 4.$$

Since $S_{n+1} = S_n + (1/n!)$, the sequence $\{S_n\}$ is monotone increasing. If we can prove it is also bounded above, we can conclude from Theorem 2–6 that it has a limit. To get an upper bound, we observe that for $n \geq 2$, $n! = 1 \cdot 2 \cdot 3 \cdots n$ has $n - 1$ factors that are all ≥ 2, so

$$n! \geq 2^{n-1} \quad \text{for} \quad n \geq 2.$$

Hence
$$\frac{1}{n!} \leq \frac{1}{2^{n-1}} \quad \text{for} \quad n \geq 2,$$
and
$$S_n = 1 + \frac{1}{1!} + \frac{1}{2!} + \frac{1}{3!} + \cdots + \frac{1}{(n-1)!}$$
$$\leq 1 + \frac{1}{1!} + \frac{1}{2} + \frac{1}{2^2} + \cdots + \frac{1}{2^{n-2}}. \tag{5}$$

The terms on the right side of (5), after the first term, form a geometric progression with sum
$$1 \times \left[\frac{1 - (\frac{1}{2})^{n-1}}{1 - \frac{1}{2}}\right] = 2[1 - (\tfrac{1}{2})^{n-1}] = 2 - (\tfrac{1}{2})^{n-2},$$
so the inequality (5) is the same as
$$S_n \leq 1 + 2 - (\tfrac{1}{2})^{n-2}. \tag{6}$$
Finally, since $(\tfrac{1}{2})^{n-2}$ is positive, (6) can be replaced by
$$S_n < 3. \tag{7}$$
Thus, from $S_{n+1} > S_n$ and $S_n < 3$ for all n, we conclude that $\{S_n\}$ is a bounded monotone increasing sequence, so Theorem 2–6 leads to the conclusion that the sequence converges. The limit is called e:
$$e = \lim_{n \to \infty} \left(1 + \frac{1}{1!} + \frac{1}{2!} + \cdots + \frac{1}{(n-1)!}\right), \tag{8a}$$
and this is also indicated by writing
$$e = 1 + \frac{1}{1!} + \frac{1}{2!} + \frac{1}{3!} + \cdots, \tag{8b}$$
which is abbreviated as
$$e = \sum_{k=0}^{\infty} \frac{1}{k!} = \frac{1}{0!} + \frac{1}{1!} + \frac{1}{2!} + \cdots \tag{8c}$$

The proof of Theorem 2–6 shows that $\lim_{n \to \infty} S_n$ is the least upper bound of $\{S_n\}$. Since inequality (7) gives 3 as an upper bound, we conclude that $e \leq 3$. The number e has been computed to many thousands of decimal places. To 15 places it is
$$e = 2.7\ 1828\ 1828\ 45\ 90\ 45.$$

Example 3. A sequence $\{S_n\}$ is defined by $S_1 = 1$, $S_{n+1} = 2 + \sqrt{S_n}$. Does the sequence converge and, if so, what is its limit?

Solution. $S_2 = 2 + \sqrt{1} = 3 > S_1$,
$$S_3 = 2 + \sqrt{S_2} = 2 + \sqrt{3} > S_2.$$

If, for some n, $S_n > S_{n-1}$, then $\sqrt{S_n} > \sqrt{S_{n-1}}$ and
$$S_{n+1} = 2 + \sqrt{S_n} > 2 + \sqrt{S_{n-1}} = S_n.$$

Therefore, by induction on n, $S_{n+1} > S_n$ for each positive integer n. Hence $\{S_n\}$ is a monotone increasing sequence. Is it bounded?
$$S_1 = 1, \quad S_2 = 3, \quad S_3 = 2 + \sqrt{3}$$
are all less than 4. And if any S_n is less than 4, then $S_{n+1} = 2 + \sqrt{S_n} < 2 + \sqrt{4} = 4$, so $S_{n+1} < 4$. Therefore, by induction on n,
$$S_n < 4 \quad \text{for every } n.$$

Since $\{S_n\}$ is monotone increasing and bounded, Theorem 2–6 guarantees that there is a number L such that
$$\lim_{n \to \infty} S_n = L.$$

Likewise, $\lim_{n \to \infty} S_{n+1} = L$, and* $\lim_{n \to \infty} \sqrt{S_n} = \sqrt{L}$, so we get
$$\lim_{n \to \infty} S_{n+1} = \lim_{n \to \infty} (2 + \sqrt{S_n}),$$
or
$$L = 2 + \sqrt{L}.$$

Hence
$$(L - 2)^2 = L, \quad L^2 - 5L + 4 = 0,$$
so $L = 1$ or $L = 4$. Since $S_n \geq 3$ for $n \geq 2$, the limit cannot equal 1. Therefore $L = 4$.

Caution. Neither boundedness alone nor monotonicity alone is sufficient to guarantee convergence. As we observed earlier, the sequence given by $S_n = (-1)^n$ is bounded, but it does not converge. The next example illustrates a monotone sequence that diverges.

---

* To prove that $\lim_{n \to \infty} \sqrt{S_n} = \sqrt{L}$, we can here use the fact that if $T_n = \sqrt{S_n}$, then T_n is also a bounded monotone increasing sequence, so it converges to some limit, say L'. Then, because $T_n > 0$, we also have $L' \geq 0$; and because $S_n = T_n \cdot T_n$, $\lim S_n = L = (\lim T_n)^2 = (L')^2$; so $L' = \sqrt{L}$.

Example 4. Suppose $S_n = 1 + \frac{1}{2} + \frac{1}{3} + \cdots + 1/n$. Then

$$S_{n+1} = S_n + 1/(n+1) > S_n,$$

so the sequence is monotone increasing. But

$$S_1 = 1, \quad S_2 = \tfrac{3}{2}, \quad S_4 = 1 + \tfrac{1}{2} + \tfrac{1}{3} + \tfrac{1}{4} > 1 + \tfrac{1}{2} + (\tfrac{1}{4} + \tfrac{1}{4}) = 2,$$
$$S_8 > 1 + \tfrac{1}{2} + (\tfrac{1}{4} + \tfrac{1}{4}) + (\tfrac{1}{8} + \tfrac{1}{8} + \tfrac{1}{8} + \tfrac{1}{8}) = \tfrac{5}{2},$$

and, in general, if n is a power of 2, say $n = 2^m$, then the fractions that are to be added to give S_n can be grouped as follows:

$$\tfrac{1}{2} = \tfrac{1}{2},$$
$$(\tfrac{1}{3} + \tfrac{1}{4}) > (\tfrac{1}{4} + \tfrac{1}{4}) = \tfrac{1}{2},$$
$$(\tfrac{1}{5} + \tfrac{1}{6} + \tfrac{1}{7} + \tfrac{1}{8}) > (\tfrac{1}{8} + \tfrac{1}{8} + \tfrac{1}{8} + \tfrac{1}{8}) = \tfrac{1}{2},$$
$$\left(\frac{1}{2^{m-1}+1} + \frac{1}{2^{m-1}+2} + \cdots + \frac{1}{2^m}\right) > \underbrace{\left(\frac{1}{2^m} + \frac{1}{2^m} + \cdots + \frac{1}{2^m}\right)}_{2^{m-1} \text{ terms}} = \frac{1}{2},$$

to give the result

$$S_n > \left(1 + \frac{m}{2}\right) \quad \text{if} \quad n = 2^m.$$

Therefore the sequence $\{S_n\}$ is not bounded, so it does not converge. (See Theorem 2–2.)

The following example shows how we can sometimes find the limit of a sequence by the use of inequalities.

Example 5. Find $\lim_{n \to \infty} \sqrt[n]{n}$.

Solution. Let $S_n = \sqrt[n]{n} = (n)^{1/n}$. Then $S_1 = 1$, $S_2 = \sqrt{2} \approx 1.414$, $S_3 = \sqrt[3]{3} \approx 1.442$, $S_4 = \sqrt[4]{4} = \sqrt{2} \approx 1.414$, and so on. The sequence is not monotone, nor do we have much evidence of how successive terms are related. But we can experiment some more. For

$n = 8$, $\quad S_n = (8)^{1/8} = (\sqrt{8})^{1/4} \approx (2.828)^{1/4} = (\sqrt{2.828})^{1/2}$
$\qquad \approx (1.68)^{1/2} = \sqrt{1.68} \approx 1.3;$

$n = 16$, $\quad S_n = (16)^{1/16} = (\sqrt{16})^{1/8} = (4)^{1/8} = (\sqrt{4})^{1/4} = (2)^{1/4}$
$\qquad = (\sqrt{2})^{1/2} \approx (1.414)^{1/2} \approx 1.2.$

It appears that for n equal to a power of 2, the calculation of S_n involves a sequence of square root operations yielding numbers not much larger than 1. Suppose we let h be the difference $S_n - 1$. Then

$$S_n = 1 + h = \sqrt[n]{n}, \tag{9a}$$

so

$$(1 + h)^n = n. \tag{9b}$$

By the binomial theorem,

$$n = (1 + h)^n = 1 + nh + \frac{n(n-1)}{2!} h^2 + \cdots + h^n$$

$$\geq \frac{n(n-1)h^2}{2} \quad \text{when} \quad n \geq 2,$$

so

$$h^2 \leq \frac{2n}{n(n-1)} = \frac{2}{n-1}$$

and

$$|h| \leq \sqrt{\frac{2}{n-1}} \quad \text{for} \quad n \geq 2. \tag{10}$$

When $n > 1$, $\sqrt[n]{n}$ is also greater than 1, so $h = \sqrt[n]{n} - 1$ is positive, and combining this with (10) we get

$$0 \leq h = \sqrt[n]{n} - 1 \leq \sqrt{\frac{2}{n-1}} \quad \text{for} \quad n \geq 2.$$

We now drop h from the discussion, and focus attention on

$$0 \leq \sqrt[n]{n} - 1 \leq \sqrt{\frac{2}{n-1}} \quad \text{for} \quad n \geq 2. \tag{11}$$

The leftmost term of inequality (11) is the constant 0 and the rightmost member approaches 0 as n increases without bound. We therefore find it reasonable to conclude that the middle member, $\sqrt[n]{n} - 1$, will also approach 0 as $n \to \infty$. In other words, we now claim that

$$\lim_{n \to \infty} \sqrt[n]{n} = 1. \tag{12}$$

For suppose $\epsilon > 0$. Then, from (11), we have

$$|\sqrt[n]{n} - 1| \leq \sqrt{\frac{2}{n-1}}$$

and this is less than ϵ if

$$\frac{2}{n-1} < \epsilon^2 \quad \text{or} \quad n > \frac{2}{\epsilon^2} + 1.$$

Let N be a positive integer greater than 2 and also greater than $1 + (2/\epsilon^2)$. Then
$$|\sqrt[n]{n} - 1| < \epsilon \quad \text{whenever} \quad n \geq N.$$
This establishes Eq. (12).

The foregoing example suggests the following general theorem.

Theorem 2–7. Let $\{S_n\}$, $\{T_n\}$, $\{U_n\}$ be sequences such that, for some positive integer N_0, one has
$$S_n \leq T_n \leq U_n \quad \text{whenever} \quad n \geq N_0. \tag{13}$$
If $\lim_{n \to \infty} S_n = \lim_{n \to \infty} U_n = L$, then $\lim_{n \to \infty} T_n = L$.

Proof. By hypothesis, if $\epsilon > 0$, there is an integer N_1 such that
$$|S_n - L| < \epsilon \quad \text{whenever} \quad n \geq N_1,$$
and an integer N_2 such that
$$|U_n - L| < \epsilon \quad \text{whenever} \quad n \geq N_2.$$
Let N be the maximum of N_0, N_1, N_2. Then, if $n \geq N$,
$$L - \epsilon < S_n \leq T_n \leq U_n < L + \epsilon,$$
so
$$L - \epsilon < T_n < L + \epsilon$$
and
$$|T_n - L| < \epsilon \quad \text{whenever} \quad n \geq N.$$

EXERCISES 2–4

1. Prove Theorem 2–6 for bounded monotone decreasing sequences.
2. Suppose
$$S_n = 1 + \frac{1}{2!} + \frac{1}{4!} + \cdots + \frac{1}{[2(n-1)]!}.$$
Prove that $\{S_n\}$ converges.
3. Suppose
$$S_n = \frac{1}{1 \cdot 2} + \frac{1}{2 \cdot 3} + \frac{1}{3 \cdot 4} + \cdots + \frac{1}{n \cdot (n+1)}.$$
Show that S_n can also be written in the following form:
$$S_n = (\tfrac{1}{1} - \tfrac{1}{2}) + (\tfrac{1}{2} - \tfrac{1}{3}) + (\tfrac{1}{3} - \tfrac{1}{4}) + \cdots + \left(\frac{1}{n} - \frac{1}{n+1}\right);$$
then prove that $\{S_n\}$ converges. Find $\lim_{n \to \infty} S_n$ if you can.

4. Let
$$T_n = \frac{1}{2^2} + \frac{1}{3^2} + \cdots + \frac{1}{(n+1)^2}.$$
Compare T_n with S_n of Exercise 3. Can you prove that $\{T_n\}$ converges?

5. Prove the following theorem: If $\{S_n\}$ and $\{T_n\}$ are monotone increasing sequences such that $T_n \leq S_n$ for each positive integer n, and if $\{S_n\}$ converges, then $\{T_n\}$ converges.

6. State and prove a theorem analogous to that of Exercise 5 for monotone decreasing sequences.

7. If $2 \quad (3/n) < S_n < 2 + (5/n)$, does the sequence $\{S_n\}$ converge, and if so, what is its limit?

8. a) Prove that
$$\frac{n}{n+1} < \frac{n+1}{n+2} \quad \text{for} \quad n \geq 1.$$
b) If
$$\frac{n}{n+1} < S_n < \frac{n+1}{n+2},$$
prove that $\lim_{n \to \infty} S_n = 1$.

9. Prove: If $\{S_n\}$ is a convergent sequence and $S_n \geq 0$ for each n, then $\lim_{n \to \infty} \sqrt{S_n} = \sqrt{\lim_{n \to \infty} S_n}$. Method: Let $\lim S_n = L$. Consider cases for (1) $L = 0$, (2) $L > 0$. In the second case, use
$$\sqrt{S_n} - \sqrt{L} = (S_n - L)/(\sqrt{S_n} + \sqrt{L}).$$

3
LIMITS OF FUNCTIONS THAT ARE NOT SEQUENCES

3-1 DEFINITIONS AND THEOREMS

In the preceding chapter we defined what is meant by the limit of a sequence and proved theorems about limits of sums, products, and quotients of sequences. In this chapter we deal with limits of functions that are not sequences.

Example 1. If x is any real number that is very nearly equal to 3, then $2x - 5$ is very nearly equal to 1. In fact, we can measure the difference this way:
$$|(2x - 5) - 1| = |2x - 6| = 2|x - 3|.$$

Thus, if an opponent requires us to make $2x - 5$ differ from 1 by an amount less than some preassigned positive number ϵ, we can do so by requiring that x differ from 3 by an amount less than $\epsilon/2$:

$$|(2x - 5) - 1| = 2|x - 3| < \epsilon \quad \text{whenever} \quad |x - 3| < \frac{\epsilon}{2}.$$

We can say this another way: To each positive number ϵ there corresponds a positive number $\epsilon/2 = \delta$ (Greek *delta*) such that $2x - 5$ is within ϵ distance of 1 whenever x is within δ distance of 3. This is shown graphically in Fig. 3-1.

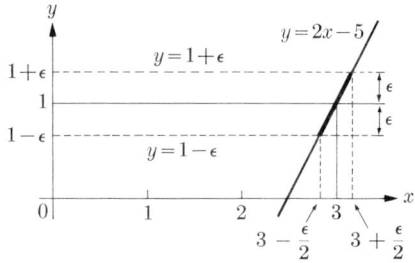

FIG. 3-1. If $\delta = \epsilon/2$, then the graph of $y = 2x - 5$ lies between the lines $y = 1 - \epsilon$ and $y = 1 + \epsilon$ when x is between $3 - \delta$ and $3 + \delta$.

Example 2. As another example, suppose x is different from 2 but very close to 2. Is there a number L such that $(5x^2 - 20)/(x - 2)$ is very close to L? To answer the question, let us rewrite the expression as follows:

$$\frac{5x^2 - 20}{x - 2} = \frac{5(x^2 - 4)}{x - 2} = 5(x + 2) = 5x + 10, \quad \text{if} \quad x \neq 2.$$

Now observe that, for any $x \neq 2$, the given expression has the same value as $5x + 10$, and hence should be very close to $L = 20$ when x is near 2. Suppose an opponent challenges us with some positive number ϵ and requires that we find a positive number δ such that

$$\left|\frac{5x^2 - 20}{x - 2} - 20\right| < \epsilon \quad \text{whenever} \quad x \neq 2 \quad \text{and} \quad |x - 2| < \delta.$$

The first requirement is the same as

$$|(5x + 10) - 20| = |5x - 10| = 5|x - 2| < \epsilon,$$

and this will be satisfied if

$$|x - 2| < \frac{\epsilon}{5}.$$

So for this example we could take $\delta = \epsilon/5$ and have a positive number that meets the challenge.

In the first example, the function $f(x) = 2x + 5$ is defined for all values of x, including $x = 3$. In the second example, $f(x) = (5x^2 - 20)/(x - 2)$ is defined for all values of x except 2. In general, if we study a function $f(x)$ for values of x close to some fixed number c we shall require that the function be defined throughout some *deleted neighborhood* of c (which we now define).

Definition 1. Deleted Neighborhood. A *neighborhood* of a real number c is an interval of real numbers x satisfying $|x - c| < h$ for some positive number h. A *deleted* neighborhood of c is a neighborhood of c from which c itself is removed. Alternatively, a deleted neighborhood of c consists of all $x \neq c$ such that $|x - c| < h$ for some positive number h; or all x such that $0 < |x - c| < h$ for some positive h. (See Fig. 3–2.)

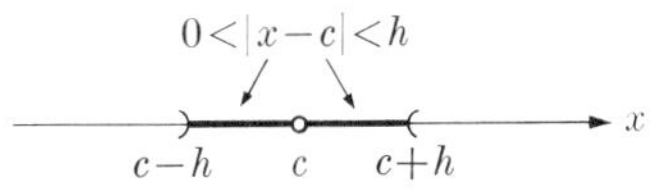

FIG. 3–2. Deleted neighborhood $0 < |x - c| < h$ is the union of the open intervals $c - h < x < c$ and $c < x < c + h$.

Thus some neighborhoods of 3 are: the interval $2 < x < 4$, or $2.9 < x < 3.1$; and some deleted neighborhoods of 2 are: the union of the intervals $1 < x < 2$ and $2 < x < 3$, or the union of the intervals $1.8 < x < 2$ and $2 < x < 2.2$. The first of these deleted neighborhoods of 2 can also be expressed as those x that satisfy $0 < |x - 2| < 1$, and the second as $0 < |x - 2| < 0.2$.

Definition 2. The Limit of a Function. Suppose $f(x)$ is defined for all x in some deleted neighborhood of c, say $0 < |x - c| < h$. If there is a number L such that to each positive number ϵ there corresponds a positive number δ, $\delta \leq h$, such that

$$|f(x) - L| < \epsilon \quad \text{when} \quad 0 < |x - c| < \delta, \qquad (1)$$

we say that $f(x)$ *converges* to L as x approaches c, and write

$$\lim_{x \to c} f(x) = L. \qquad (2)$$

Thus from Examples 1 and 2, we have

$$\lim_{x \to 3} (2x - 5) = 1 \quad \text{and} \quad \lim_{x \to 2} \frac{5x^2 - 20}{x - 2} = 20.$$

Example 3. Find the limit of $f(x) = x^2$ as x approaches 4.

Solution. Naturally, we guess that the answer is $L = 16$, because x^2 should be close to 16 if x is close to 4. Now, can we prove it? Suppose $\epsilon > 0$. We want to satisfy

$$|x^2 - 16| < \epsilon,$$

or

$$|x + 4| \cdot |x - 4| < \epsilon.$$

We are tempted to say, take $\delta = \epsilon/|x + 4|$, but x is a variable, so this does not determine δ. Is there some constant K we could use instead of the variable $|x + 4|$? Of course it isn't true that $|x + 4|$ is *equal* to K for

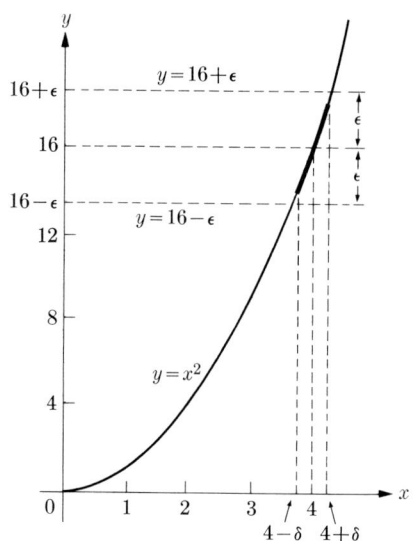

FIG. 3–3. If $\delta = \epsilon/8.5$, then the graph of $y = x^2$ lies between $y = 16 - \epsilon$ and $y = 16 + \epsilon$ when $0 < |x - 4| < \delta$. (This assumes that $\epsilon/8.5 \leq 0.5$. Otherwise, take $\delta = 0.5$.)

all x in some deleted neighborhood of 4, but perhaps we can find a *bound* for $|x + 4|$ when x is near 4. We are almost tempted to try 8 as such a bound, but this would not permit x to be greater than 4. However, if we restrict x to the deleted neighborhood $0 < |x - 4| < \frac{1}{2}$, say, then $3.5 < x < 4.5$ and $7.5 < x + 4 < 8.5$, so we have $|x + 4| < 8.5$ for all these x. Now take $\delta < \frac{1}{2}$ and also $\leq \epsilon/8.5$. Then

$$|x^2 - 16| = |x + 4| \cdot |x - 4| < (8.5)|x - 4| < \epsilon$$

whenever $0 < |x - 4| < \delta$. (See Fig. 3–3.)

Remark 1. Example 3 would be easier to solve if we had the following two theorems:

1) $\lim_{x \to c} x = c$, and

2) if
$$\lim_{x \to c} f(x) = L_1 \quad \text{and} \quad \lim_{x \to c} g(x) = L_2,$$
then
$$\lim_{x \to c} [f(x) \cdot g(x)] = L_1 L_2 = [\lim_{x \to c} f(x)] \cdot [\lim_{x \to c} g(x)].$$

For, given these two theorems, we could write

$$\lim_{x \to 4} x^2 = \lim_{x \to 4} (x \cdot x) = [\lim_{x \to 4} x] \cdot [\lim_{x \to 4} x] = 4 \cdot 4 = 16.$$

We now turn our attention to these and other fundamental theorems. We shall prove some of them and leave proofs of the others to the reader. Notice that the proofs are much like the proofs of the corresponding theorems for sequences but with N's replaced by δ's.

Theorem 3–1. $\lim_{x \to c} x = c$.

Proof. Let $\epsilon > 0$. Then $|x - c| < \epsilon$ whenever $0 < |x - c| < \delta$ if we take $\delta = \epsilon$. Therefore, to each $\epsilon > 0$ there corresponds $\delta > 0$, with $\delta = \epsilon$, such that the conditions (1) in the definition of limit are met when $f(x) = x$ and $L = c$.

Theorem 3–2. If $\lim_{x \to c} f(x) = L$, then $|f(x)|$ is bounded in some deleted neighborhood of c.

Proof. Since $1 > 0$, we can take $\epsilon = 1$ in the definition of limit and conclude that there is a positive number δ such that $|f(x) - L| < 1$

whenever $0 < |x - c| < \delta$. This means, in particular, that

$$|f(x)| < |L| + 1 \quad \text{whenever} \quad 0 < |x - c| < \delta,$$

so $|f(x)|$ is bounded in this deleted neighborhood of c.

Example 4. For $x \neq 2$, let $f(x) = (5x^2 - 20)/(x - 2)$, as in Example 2. There we found $L = 20$ and $\delta = \epsilon/5$, so with $\epsilon = 1$ and $\delta = \frac{1}{5}$ we can say

$$\left| \frac{5x^2 - 20}{x - 2} \right| < 21 \quad \text{whenever} \quad 0 < |x - 2| < \tfrac{1}{5}.$$

As a check, we find by direct calculation, for $x \neq 2$,

$$\frac{5x^2 - 20}{x - 2} = 5x + 10 < 21 \quad \text{if} \quad 5x < 11,$$

or

$$x < 2.2, \quad \text{and} \quad x \neq 2.$$

Theorem 3-3. If $\lim f(x) = L_1$ and $\lim g(x) = L_2$, then

a) $\lim [f(x) \pm g(x)] = [\lim f(x)] \pm [\lim g(x)] = L_1 \pm L_2$,

b) $\lim kf(x) = k \lim f(x) = kL_1$,

c) $\lim [f(x) \cdot g(x)] = [\lim f(x)] \cdot [\lim g(x)] = L_1 L_2$, and

d) $\lim \dfrac{f(x)}{g(x)} = \dfrac{\lim f(x)}{\lim g(x)} = \dfrac{L_1}{L_2}$, if $L_2 \neq 0$,

where all limits are taken as $x \to c$.

Proof. We omit proofs of everything except part (c), because we need only repeat the proofs of the corresponding theorems for sequences with minor changes. To illustrate those changes, we give the proof of part (c).

Suppose $\epsilon > 0$. We want to show that there is a positive number δ such that

$$|f(x)g(x) - L_1 L_2| < \epsilon \quad \text{whenever} \quad 0 < |x - c| < \delta.$$

We rewrite the first difference this way:

$$f(x)g(x) - L_1 L_2 = f(x)g(x) - L_1 g(x) + L_1 g(x) - L_1 L_2$$

and

$$\begin{aligned}
|f(x)g(x) - L_1 L_2| &= |(f(x)g(x) - L_1 g(x)) + (L_1 g(x) - L_1 L_2)| \\
&\leq |f(x)g(x) - L_1 g(x)| + |L_1 g(x) - L_1 L_2| \\
&\leq |f(x) - L_1| \cdot |g(x)| + |L_1| \cdot |g(x) - L_2|.
\end{aligned}$$

By hypothesis, $\lim g(x) = L_2$, so $|g(x)| < 1 + |L_2| = B$ in some deleted neighborhood of c, by Theorem 2–2. Let this deleted neighborhood be $0 < |x - c| < \delta_1$. By hypothesis, $\lim f(x) = L_1$, so given the positive number $\epsilon/(2B)$, there is a corresponding positive number δ_2 such that

$$|f(x) - L_1| < \epsilon/(2B) \quad \text{whenever} \quad 0 < |x - c| < \delta_2.$$

Likewise, there is a positive number δ_3 such that

$$|g(x) - L_2| < \frac{\epsilon}{2|L_1| + 1} \quad \text{whenever} \quad 0 < |x - c| < \delta_3.$$

Now let δ be the minimum of δ_1, δ_2, and δ_3. Then

$$|f(x)g(x) - L_1 L_2| \le |f(x) - L_1| \cdot |g(x)| + |L_1| \cdot |g(x) - L_2|$$

$$< \frac{\epsilon}{2B} \cdot B + |L_1| \cdot \frac{\epsilon}{2|L_1| + 1} < \epsilon$$

whenever

$$0 < |x - c| < \delta.$$

Corollary 3–3.1. If $f(x) = a_0 + a_1 x + a_2 x^2 + \cdots + a_n x^n$ is a polynomial and c is any real number, then

$$\lim_{x \to c} f(x) = f(c).$$

Proof. Apply Theorems 3–1 and 3–3(c), to deduce that

$$\lim_{x \to c} x^2 = (\lim x) \cdot (\lim x) = c^2,$$

and, if $\lim_{x \to c} x^{n-1} = c^{n-1}$, then

$$\lim_{x \to c} x^n = (\lim_{x \to c} x^{n-1}) \cdot (\lim_{x \to c} x) = c^{n-1} \cdot c = c^n.$$

Hence, by induction on n, $\lim_{x \to c} x^n = c^n$. Likewise, by Theorem 3–3(b), $\lim_{x \to c} a_k x^k = a_k c^k$, and

$$\lim_{x \to c} (a_0 + a_1 x + \cdots + a_n x^n) = \lim_{x \to c} a_0 + \lim_{x \to c} a_1 x + \cdots + \lim_{x \to c} a_n x^n$$

$$= a_0 + a_1 c + \cdots + a_n c^n = f(c).$$

Example 5. $\lim_{x \to 2} (x^2 + 5x - 6) = 4 + 10 - 6 = 8.$

Corollary 3–3.2. If $f(x) = a_0 + a_1 x + \cdots + a_n x^n$ and $g(x) = b_0 + b_1 x + \cdots + b_m x^m$, and $g(c) \ne 0$, then

$$\lim_{x \to c} \frac{f(x)}{g(x)} = \frac{f(c)}{g(c)}.$$

Proof. Apply Theorem 3–3(d) and Corollary 3–3.1.

Example 6
$$\lim_{x \to 2} \frac{x^3 - 8}{x^2 - 5x + 6} = \lim_{x \to 2} \frac{(x - 2)(x^2 + 2x + 4)}{(x - 2)(x - 3)}$$
$$= \lim_{x \to 2} \frac{x^2 + 2x + 4}{x - 3} = \frac{4 + 4 + 4}{2 - 3}$$
$$= -12.$$

Theorem 3–4. If $f(x) \leq g(x) \leq h(x)$ for all x in some deleted neighborhood of c and
$$\lim_{x \to c} f(x) = \lim_{x \to c} h(x) = L,$$
then
$$\lim_{x \to c} g(x) = L.$$

Proof. Suppose $\epsilon > 0$. By hypothesis, there are positive numbers δ_0, δ_1, δ_2 such that

a) $f(x) \leq g(x) \leq h(x)$ for $0 < |x - c| < \delta_0$,
b) $|f(x) - L| < \epsilon$ for $0 < |x - c| < \delta_1$,
c) $|h(x) - L| < \epsilon$ for $0 < |x - c| < \delta_2$.

Let δ be the minimum of δ_0, δ_1, δ_2. Then
$$L - \epsilon < f(x) \leq g(x) \leq h(x) < L + \epsilon \quad \text{for} \quad 0 < |x - c| < \delta,$$
so
$$|g(x) - L| < \epsilon \quad \text{whenever} \quad 0 < |x - c| < \delta.$$

Example 7. As an application of Theorem 3–4, we prove that
$$\lim_{\theta \to 0} \sin \theta = 0, \tag{3a}$$
$$\lim_{\theta \to 0} \cos \theta = 1, \tag{3b}$$
and
$$\lim_{\theta \to 0} \frac{\sin \theta}{\theta} = 1, \tag{3c}$$

provided θ is measured in radians. (Recall that the radian measure of an angle can be defined by
$$\theta = s/r, \tag{4}$$

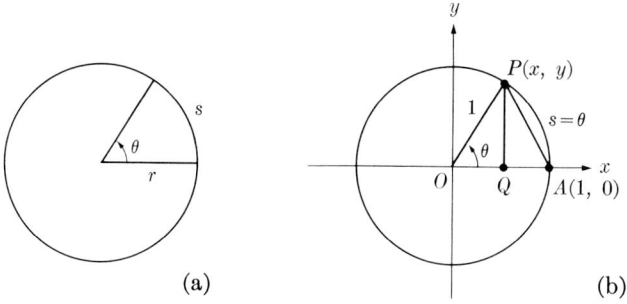

FIG. 3-4. (a) Radian measure: $\theta = s/r$; $s = r\theta$. (b) On a unit circle: $r = 1$, $s = \theta$.

where s is the length of arc the angle intercepts on a circle of radius r when the center of the circle is the vertex of the angle. Figure 3-4(a) illustrates this definition.)

In Fig. 3-4(b), O is the center of a unit circle, θ is the radian measure of an acute angle AOP, and $\triangle APQ$ is a right triangle with legs of length

$$QP = \sin\theta, \quad AQ = 1 - \cos\theta.$$

From the theorem of Pythagoras, and the fact that $AP < \theta$, we get

$$\sin^2\theta + (1 - \cos\theta)^2 = (AP)^2 < \theta^2. \tag{5}$$

Both terms on the left side of inequality (5) are positive, so each is smaller than their sum, and hence is less than θ^2:

$$\sin^2\theta < \theta^2, \tag{6a}$$

$$(1 - \cos\theta)^2 < \theta^2. \tag{6b}$$

The inequalities (6a) and (6b) also imply

$$|\sin\theta| < |\theta|, \tag{7a}$$

$$|1 - \cos\theta| < |\theta|. \tag{7b}$$

If ϵ is any positive number we can take $\delta = \epsilon$ and conclude that

$$|\sin\theta - 0| < \epsilon$$

and

$$|1 - \cos\theta| < \epsilon$$

whenever $0 < |\theta| < \delta$. Therefore,

$$\lim_{\theta \to 0} \sin\theta = 0 \quad \text{and} \quad \lim_{\theta \to 0} \cos\theta = 1.$$

LIMITS OF FUNCTIONS THAT ARE NOT SEQUENCES 3–1

In order to establish Eq. (3c), let us assume that θ is positive and less than $\pi/2$ in Fig. 3–5. We compare the areas of $\triangle AOP$, sector AOP, and $\triangle AOT$. The area of sector AOP is the same fraction of the area of a circle of radius 1 that its arc is of the total circumference*:

$$\frac{\text{area sector } AOP}{\text{area circle}} = \frac{\text{arc } AP}{\text{circumference}},$$

so

$$\frac{\text{area sector } AOP}{\pi r^2} = \frac{r\theta}{2\pi r} = \frac{\theta}{2\pi},$$

so

area sector $AOP = \tfrac{1}{2}r^2\theta$ (8a)

$\qquad\qquad\quad = \tfrac{1}{2}\theta,$ when $r = 1.$ (8b)

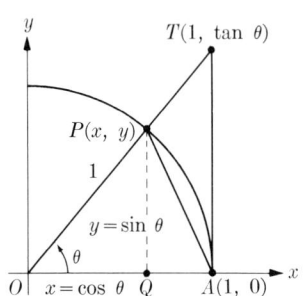

FIG. 3–5. Area $\triangle AOP <$ area sector $AOP <$ area $\triangle AOT$.

The altitude of $\triangle AOP$, with base $OA = 1$, is $y = QP = \sin\theta$. Hence

$$\text{area } \triangle AOP = \tfrac{1}{2}\sin\theta. \tag{9a}$$

The altitude of $\triangle AOT$, with base $OA = 1$, is $AT = \tan\theta$, so

$$\text{area } \triangle AOT = \tfrac{1}{2}\tan\theta. \tag{9b}$$

But

$$\text{area } \triangle AOP < \text{area sector } AOP < \text{area } \triangle AOT,$$

so

$$\tfrac{1}{2}\sin\theta < \tfrac{1}{2}\theta < \tfrac{1}{2}\tan\theta, \quad \text{if} \quad 0 < \theta < \frac{\pi}{2}. \tag{10}$$

Dividing the three terms of inequality (10) by $\tfrac{1}{2}\sin\theta$, we get

$$1 < \frac{\theta}{\sin\theta} < \frac{1}{\cos\theta}, \quad \text{if} \quad 0 < \theta < \frac{\pi}{2}. \tag{11a}$$

Recall that $\sin(-\theta) = -\sin\theta$, and $\cos(-\theta) = \cos\theta$, so (11a) also implies that

$$1 < \frac{-\theta}{\sin(-\theta)} < \frac{1}{\cos(-\theta)}, \quad \text{if} \quad 0 < -\theta < \frac{\pi}{2}$$

or

$$1 < \frac{\theta}{\sin\theta} < \frac{1}{\cos\theta}, \quad \text{if} \quad 0 > \theta > -\frac{\pi}{2}. \tag{11b}$$

---

* We assume the results $A = \pi r^2$ and $C = 2\pi r$, for the area and circumference of a circle of radius r. These do, in fact, depend on limits. See Section 2–1.

We can combine (11a) and (11b) to say

$$1 < \frac{\theta}{\sin\theta} < \frac{1}{\cos\theta}, \quad \text{if} \quad 0 < |\theta| < \frac{\pi}{2}. \tag{12}$$

Inequality (12) tells us about $\theta/(\sin\theta)$. We convert this to information about $(\sin\theta)/\theta$ by taking reciprocals and reversing the inequality signs, because

if a and b are positive and $a > b$, then $(1/a) < (1/b)$.

Therefore,

$$1 > \frac{\sin\theta}{\theta} > \cos\theta \quad \text{whenever} \quad 0 < |\theta| < \frac{\pi}{2}.$$

Now apply Theorem 3–4:

$$\lim_{\theta \to 0} 1 = 1, \quad \lim_{\theta \to 0} \cos\theta = 1,$$

so

$$\lim_{\theta \to 0} \frac{\sin\theta}{\theta} = 1.$$

Example 8

$$\lim_{x \to 0} \frac{\sin 3x}{x} = \lim_{3x \to 0} \frac{3 \sin 3x}{3x}$$

$$= 3 \lim_{\theta \to 0} \left(\frac{\sin\theta}{\theta}\right) = 3.$$

Example 9

$$\lim_{x \to 0} \frac{\tan x}{x} = \lim_{x \to 0} \left(\frac{\sin x}{x} \cdot \frac{1}{\cos x}\right)$$

$$= \left[\lim_{x \to 0} \left(\frac{\sin x}{x}\right)\right] \cdot \left[\lim_{x \to 0} \left(\frac{1}{\cos x}\right)\right]$$

$$= 1 \cdot 1 = 1.$$

Example 10

$$\lim_{x \to \infty} x \sin \frac{1}{x} = \lim_{y \to 0} \frac{\sin y}{y} \quad \left(y = \frac{1}{x}\right)$$

$$= 1.$$

In Example 10, a meaning needs to be assigned to the limit of a function, in this example the function $f(x) = x \sin(1/x)$, as x tends to infinity. When we assert that the limit is 1, we mean that when x is large, $f(x)$ is

LIMITS OF FUNCTIONS THAT ARE NOT SEQUENCES

close to 1. We take our cue from the definition of the limit of a sequence:

$\lim_{n \to \infty} S_n = L$ if and only if to each $\epsilon > 0$ there corresponds an integer N such that $|S_n - L| < \epsilon$ whenever $n \geq N$.

The only difference in defining $\lim_{x \to \infty} f(x) = L$ is that we no longer restrict x to be an integer, so we say:

$\lim_{x \to \infty} f(x) = L$ if and only if to each $\epsilon > 0$ there corresponds a number M such that $|f(x) - L| < \epsilon$ whenever $x \geq M$.

Theorems about limits of sums, products, and quotients of functions $f(x)$ and $g(x)$, when $x \to \infty$, are like the corresponding theorems for sequences as $n \to \infty$.

Theorem 3–5. If $\lim_{x \to \infty} f(x) = L_1$ and $\lim_{x \to \infty} g(x) = L_2$, then

a) $\lim_{x \to \infty} [f(x) \pm g(x)] = L_1 \pm L_2,$

b) $\lim_{x \to \infty} f(x) \cdot g(x) = L_1 L_2,$

c) $\lim_{x \to \infty} f(x)/g(x) = L_1/L_2,$ provided $L_2 \neq 0.$

The proofs are omitted because they are almost identical with the corresponding proofs for sequences.

The trick of substituting $y = 1/x$ and then saying $y \to 0$ as $x \to \infty$ is another way of handling limit problems like the next example.

Example 11

$$\lim_{x \to \infty} \frac{3x^2 - 2x + 8}{5x^2 + 7x + 1} = \lim_{x \to \infty} \frac{3 - (2/x) + (8/x^2)}{5 + (7/x) + (1/x^2)}$$

$$= \frac{3 - 0 + 0}{5 + 0 + 0} = 3/5$$

$$= \lim_{y \to 0} \frac{3 - 2y + 8y^2}{5 + 7y + y^2} \quad \left(y = \frac{1}{x}\right).$$

EXERCISES 3–1

Evaluate the following limits.

1. $\lim_{x \to 1} \dfrac{x+1}{x+2}$

2. $\lim_{x \to 1} \dfrac{x-1}{x+2}$

3. $\lim_{x \to 2} \dfrac{x^2+1}{x^2+2}$

4. $\lim_{x \to 2} \dfrac{x^2-x-2}{x^2-4}$

66

5. $\lim_{x \to 2} \dfrac{x^2 - 4}{x^3 - 8}$

6. $\lim_{x \to 0} \dfrac{x^2 - 4}{x^3 - 8}$

7. $\lim_{x \to 0} \dfrac{\sin 3x}{5x}$

8. $\lim_{h \to 0} \dfrac{\sin 3h}{\sin 5h}$

9. $\lim_{\theta \to 0} \dfrac{\sin^2 \theta}{\theta}$

10. $\lim_{\theta \to 0} \dfrac{\sin^2 \theta}{\theta^2}$

11. $\lim_{\theta \to 0} \dfrac{1 - \cos \theta}{\theta^2}$ [*Hint:* Multiply and divide by $1 + \cos \theta$.]

12. $\lim_{\theta \to 0} \dfrac{1 - \cos \theta}{\theta}$

13. Expand and simplify the left side of Eq. (5) and prove that $(1 - \cos \theta) < \theta^2/2$ for $0 < \theta < \pi/2$. Is $(1 - \cos \theta) < \theta^2/2$ true for $\theta = 0$? For $\theta = \pi/2$? For $0 < |\theta| < \pi/2$? For all $\theta \neq 0$?

Determine which of the following indicated limits exist and evaluate those that do.

14. $\lim_{x \to \infty} \dfrac{2x + 3}{5x + 7}$

15. $\lim_{z \to \infty} \dfrac{4z^3 + 3z^2 + 2}{z^4 + 5z^2 + 1}$

16. $\lim_{x \to \infty} \dfrac{\sin x}{x}$

17. $\lim_{x \to \infty} (1 - \cos x)$

18. $\lim_{x \to \infty} \dfrac{1 - \cos x}{x}$

19. $\lim_{x \to \infty} \sqrt{\dfrac{2 + 3x}{1 + 5x}}$

20. $\lim_{x \to \infty} \sqrt{\dfrac{2 + 3x}{1 - 5x}}$

21. $\lim_{x \to \infty} x^x$

22. $\lim_{x \to \infty} (x)^{1/x}$ (Try to guess an answer and argue for it even if you cannot prove it is correct.)

3–2 PROBLEMS OF THE DIFFERENTIAL AND INTEGRAL CALCULUS

The problems of the differential calculus were stated in Chapter 1. In the preceding pages an introduction to the theory of limits was given with a few examples, such as $\lim_{n \to \infty} \sqrt[n]{n}$ and $\lim_{\theta \to 0} [(\sin \theta)/\theta]$, to illustrate how theorems about limits can be used to answer fairly difficult questions. The following typical problems, of which 1 and 2 were introduced before and approached intuitively, are now presented to show how the theory of limits is used in the differential calculus; Example 3 is a typical problem of the integral calculus.

Example 1 (*Slope of a curve*). Find the slope of the tangent to the curve $y = 1/x$ at the point $P(2, \frac{1}{2})$.

Solution. The graph of a portion of the curve, near the point P, is shown in Fig. 3–6. Consider another point $Q(x, 1/x)$ near P. The line $\overleftrightarrow{PQ}$ is called a secant line of the curve. Let us call its slope m_{sec}. Then

$$m_{\text{sec}} = \frac{\text{rise}}{\text{run}} = \frac{(1/x) - (1/2)}{x - 2}. \tag{1}$$

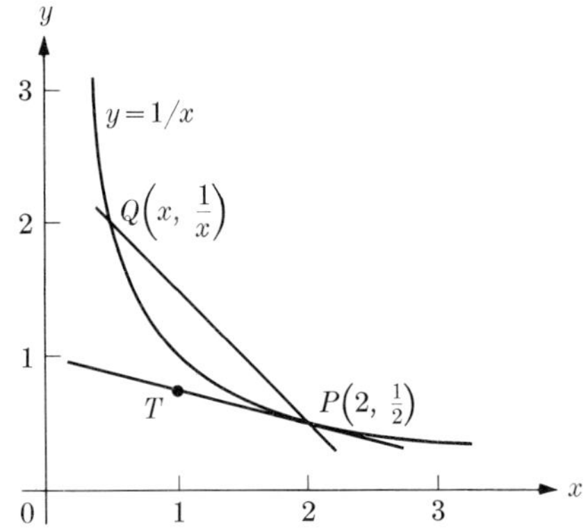

FIG. 3–6. Portion of the graph of $y = 1/x$.

If we hold P fixed and let Q approach P along the curve, with $Q \neq P$, then each position of Q determines a secant line, and a slope m_{sec}. If the slopes of these secant lines approach a *limit* as Q approaches P, we call that limit m_{tan}, the *slope of the tangent* to the curve at P. From Eq. (1) we get

$$m_{\text{sec}} = \frac{2x}{2x} \cdot \frac{(1/x) - (1/2)}{x - 2} = \frac{2 - x}{2x(x - 2)} = \frac{-1}{2x}.$$

Therefore, by Theorem 3–3

$$m_{\text{tan}} = \lim_{Q \to P} (m_{\text{sec}}) = \lim_{x \to 2} \left(\frac{-1}{2x}\right) = \frac{-1}{4}.$$

Thus the slope of the tangent to the curve $y = 1/x$ at $P(2, \frac{1}{2})$ is $-\frac{1}{4}$. To construct a line tangent to the curve at P we can locate the point $T(x_1, y_1)$ with $x_1 = 2 - 1$ and $y_1 = \frac{1}{2} + \frac{1}{4}$, and draw PT.

Example 2 (*Velocity of a moving object*). Suppose an object moves along a straight line in such a way that s, its coordinate relative to a fixed point O on the line, is given by

$$s = 5t^2 + 7t, \quad \text{for} \quad t \geq 0. \tag{2}$$

In Eq. (2), t is time measured in seconds and s is distance along the line of motion measured in feet away from O. How fast is the object moving when $t = 5$? When $t = 3$? When $t = t_1$?

Solution. Figure 3–7 represents the line of motion and the position of the object at different times.

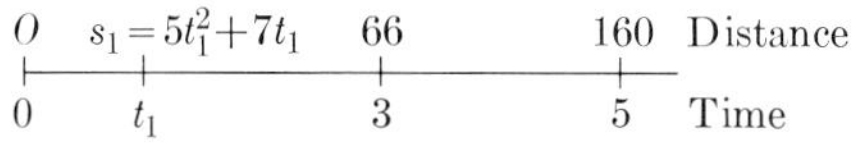

```
0   s₁=5t₁²+7t₁   66              160   Distance
├────┼────────────┼────────────────┼─
0        t₁        3                5    Time
```

FIG. 3–7. Motion on a line, with $s = 5t^2 + 7t$.

a) Consider the motion near $t = 5$, $s = 160$. For a value of t slightly greater than 5, the distance s away from O is slightly more than 160. The *average* velocity during this part of the motion is given by

$$v_{\text{ave}} = \frac{s - 160}{t - 5} = \frac{5t^2 + 7t - 160}{t - 5}. \qquad (3)$$

This average velocity is the ratio of the distance traveled to the elapsed time, during the short interval of time from 5 to t. If this average velocity has a *limit* as t approaches 5 we call that limit the *instantaneous velocity* at time $t = 5$:

$$(\text{instantaneous velocity at } t = 5) = \lim_{t \to 5}(v_{\text{ave}}). \qquad (4)$$

The numerator in the last fraction in Eq. (3) factors into $5t^2 + 7t - 160 = (t - 5)(5t + 32)$. When this is divided by $(t - 5)$ we get

$$v_{\text{ave}} = 5t + 32 = (\text{average velocity from 5 to } t \neq 5),$$

and this clearly has a limit as t approaches 5. We call this limit v, and have

$$v = \lim_{t \to 5}(5t + 32) = 57.$$

The same algebra would apply for values of t slightly less than 5, with the first fraction in Eq. (3) replaced by $(160 - s)/(5 - t)$, and we would arrive at the same limit if we let t approach 5 from below. Thus the instantaneous velocity at $t = 5$ is 57 ft/sec.

b) Next, suppose t_1 is any positive number. The corresponding position is

$$s_1 = 5t_1^2 + 7t_1. \qquad (5)$$

Let t represent an instant of time near t_1 but not equal to t_1. Then the *average* velocity over the corresponding portion of the motion is

$$v_{\text{ave}} = \frac{s - s_1}{t - t_1} = \frac{(5t^2 + 7t) - (5t_1^2 + 7t_1)}{(t - t_1)}$$

$$= \frac{5(t^2 - t_1^2) + 7(t - t_1)}{(t - t_1)}$$

$$= 5(t + t_1) + 7, \qquad \text{for} \quad t \neq t_1.$$

Now keep t_1 fixed and let t approach t_1 (from either side). The average velocity has a limit, and we call this limit the instantaneous velocity v_1 at time t_1:
$$v_1 = \lim_{t \to t_1} (5(t + t_1) + 7) = 10t_1 + 7.$$

The formula
$$v_1 = 10t_1 + 7 \tag{6}$$

enables us to find the velocity at any instant. We can check our answer of part (a) by taking $t_1 = 5$ and getting 57 as the corresponding velocity. When $t_1 = 3$, the velocity is 37.

In calculus one learns how to write down the formula for the velocity directly from the equation of motion by applying rules for finding derivatives. With the aid of those rules we can go directly from Eq. (2) to Eq. (6). The same rules are used in finding the slope of the tangent to a curve, from its equation.

Mathematical Induction

In the next example, we need to know that the sum of the squares of the first n positive integers is given by the formula

$$1^2 + 2^2 + 3^2 + \cdots + n^2 = \frac{n(n+1)(2n+1)}{6}, \tag{7}$$

for any positive integer n. For the first few positive integers, we get from Eq. (7)

$$1^2 = \frac{1(1+1)(2+1)}{6} \quad \text{or} \quad 1 = \frac{6}{6},$$

$$1^2 + 2^2 = \frac{2(2+1)(4+1)}{6} \quad \text{or} \quad 5 = \frac{30}{6},$$

$$1^2 + 2^2 + 3^2 = \frac{3(3+1)(6+1)}{6} \quad \text{or} \quad 14 = \frac{84}{6},$$

all of which are correct. No amount of such verification will establish the formula for all positive integers n, but the general method of *mathematical induction* will.

Principle of mathematical induction. Suppose that to each positive integer n there corresponds a proposition (or theorem, or formula) P_n which is either true or false. If P_1 is true, and for every positive integer

k the truth of proposition P_k implies the truth of the following proposition, P_{k+1}, then all of the propositions

$$P_1, P_2, P_3, \ldots, P_n, \ldots$$

are true. That is, P_n is true for every positive integer n.

For a detailed discussion, see R. Courant and H. Robbins, *What is Mathematics?*, Oxford University Press, 1941, pp. 9–16, or E. P. Vance, *An Introduction to Modern Mathematics*, Addison-Wesley, 1963, pp. 246–251. To apply this principle, we need, *first*, to show that P_1 is true, and, *second*, to show that $P_k \Rightarrow P_{k+1}$. (The symbol $\Rightarrow$ is read "implies.")

In the present application, let P_n be the proposition that Eq. (7) is correct for the integer n. We have verified the truth of P_1, P_2, and P_3 above. Such verification is necessary only for P_1, but the additional checking was easy and not harmful. Now we see if we can show that $P_k \Rightarrow P_{k+1}$ for any positive integer k. To test whether this is so, we assume that k is an integer for which P_k is true:

$$1^2 + 2^2 + \cdots + k^2 = \frac{k(k+1)(2k+1)}{6}.$$

If we add $(k+1)^2$ to both sides of this equation, we get

$$1^2 + 2^2 + \cdots + k^2 + (k+1)^2 = \frac{k(k+1)(2k+1)}{6} + (k+1)^2$$

$$= \frac{(k+1)}{6}[k(2k+1) + 6(k+1)]$$

$$= \frac{(k+1)}{6}(2k^2 + 7k + 6). \quad (8a)$$

Now P_{k+1} is the proposition that the sum of the squares on the left above should equal $n(n+1)(2n+1)/6$ with $n = k+1$; in other words, $(k+1)(k+2)(2k+3)/6$. Hence the truth of P_{k+1} follows from P_k provided it is true that

$$\frac{(k+1)}{6}(2k^2 + 7k + 6) = \frac{(k+1)(k+2)(2k+3)}{6}. \quad (8b)$$

This equation is easily seen to be true because $(k+2)(2k+3) = 2k^2 + 7k + 6$. Thus the two steps needed to establish Eq. (7) for all positive integers n are complete. We shall apply the result in finding an area in the next example.

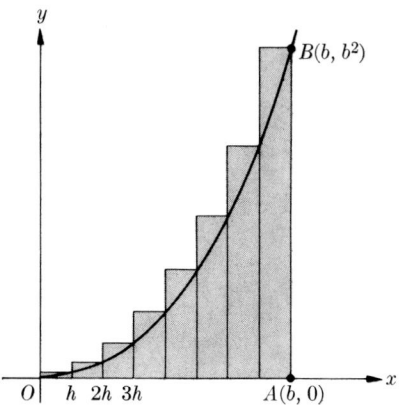

FIG. 3-8. Graph of $y = x^2$, $0 \leq x \leq b$, and circumscribed rectangles, $n = 8$.

Example 3 (*Area*). Find the area of the region AOB bounded above by the arc OB of the curve $y = x^2$, below by the segment OA of the x-axis, and on the right by the segment AB through the points $A(b, 0)$ and $B(b, b^2)$, $b > 0$ (Fig. 3-8).

Solution. Let n be a positive integer and divide the line segment OA into n subintervals, each of length $h = b/n$. In Fig. 3-8 we have used $n = 8$, but in the algebraic formulas below n can be any positive integer. With each subinterval we associate a rectangle having its base on that subinterval and its upper right vertex on the curve $y = x^2$. We number these rectangles $1, 2, \ldots, n$ in order from left to right. Thus

rectangle number	has its base on the subinterval	and has altitude equal to
1	$0 \leq x \leq h$	h^2
2	$h \leq x \leq 2h$	$(2h)^2$
3	$2h \leq x \leq 3h$	$(3h)^2$
$\vdots$	$\vdots$	$\vdots$
n	$(n-1)h \leq x \leq nh$	$(nh)^2$

Let S_n denote the sum of the areas of these n circumscribed rectangles. Then

$$S_n = h(h)^2 + h(2h)^2 + h(3h)^2 + \cdots + h(nh)^2$$
$$= h^3[1^2 + 2^2 + 3^2 + \cdots + n^2]. \tag{9}$$

But $h = b/n$, and the sum of squares in brackets in Eq. (9) is equal to $n(n+1)(2n+1)/6$, so we can also write S_n in the form

$$S_n = \frac{b^3}{n^3} \times \frac{n(n+1)(2n+1)}{6} = \frac{b^3}{6}\left(\frac{n}{n}\right)\left(\frac{n+1}{n}\right)\left(\frac{2n+1}{n}\right). \quad (10)$$

If S_n has a limit as $n \to \infty$, we define that limit to be the area of the region AOB. The last three quotients in Eq. (10) have the following limits:

$$\lim_{n \to \infty}\left(\frac{n}{n}\right) = 1, \quad \lim_{n \to \infty}\left(\frac{n+1}{n}\right) = 1, \quad \lim_{n \to \infty}\left(\frac{2n+1}{n}\right) = 2,$$

so

$$\text{area of } AOB = \lim S_n = \frac{b^3}{6}(1)(1)(2) = \frac{b^3}{3}. \quad (11)$$

The area of the parabolic region AOB is therefore $b^3/3$. Note that the area inside the triangle AOB is $b^3/2$. Thus the parabolic area is $\frac{2}{3}$ that of the corresponding triangle.

In calculus, the area of the parabolic region of Example 3 is represented by $\int_0^b x^2\, dx$, which is read "the integral from 0 to b of $x^2\, dx$." Our result could therefore be expressed as follows:

$$\int_0^b x^2\, dx = \frac{b^3}{3}. \quad (12)$$

Calculus shows how to evaluate this integral, and many others, almost instantly.

EXERCISES 3–2

By the method of mathematical induction, prove the validity of statements 1 through 4, for all positive integers n:

1. $1 + 2 + 3 + \cdots + n = \dfrac{n(n+1)}{2}$

2. $1^3 + 2^3 + 3^3 + \cdots + n^3 = \dfrac{n^2(n+1)^2}{4}$

3. $1 \cdot 2 + 3 \cdot 4 + 5 \cdot 6 + \cdots + (2n-1)(2n) = \dfrac{n(n+1)(4n-1)}{3}$

4. $|a_1 + a_2 + \cdots + a_n| \leq |a_1| + |a_2| + \cdots + |a_n|$ (This is the generalization of (2) in Section 1–6.)

5. Find a formula for $\sum_{i=1}^{n}(3i-1)$ and prove that it is correct.

6. Find a formula for $\sum_{i=1}^{n}(i^2 + i + 1)$ and prove it.

LIMITS OF FUNCTIONS THAT ARE NOT SEQUENCES   3–2

In Exercises 7 through 16, use the method of Example 1 to find the slope of the tangent to the given curve at the given point P.

7. $y = 3x + 5$, $P(0, 5)$
8. $y = 3x^2 + 5$, $P(0, 5)$
9. $y = 3x^2 + 5$, $P(x_1, y_1)$
10. $y = 2/x$, $P(1, 2)$
11. $y = 2/x$, $P(x_1, y_1)$
12. $y = \sqrt{x}$, $P(9, 3)$
13. $y = \sqrt{x}$, $P(x_1, \sqrt{x_1})$
14. $y = 2x^2 + 4x - 6$, $P(1, 0)$
15. $y = ax^2 + bx + c$, $P(x_1, y_1)$
16. $y = x^3$, $P(x_1, x_1^3)$

In Exercises 17 through 22, use the method of Example 2 to find the instantaneous velocity for the given equation of motion, at the given time.

17. $s = t^2$, $t = 2$
18. $s = t^2$, $t = t_1$
19. $s = at^2 + bt + c$, $t = t_1$
20. $s = 1/(t + 1)$, $t = 3$
21. $s = 1/(t + 1)$, $t = t_1$
22. $s = \sqrt{2t + 1}$, $t = t_1$

23. Find the area of the region AOB of Example 3, using *inscribed* rectangles instead of circumscribed rectangles. [*Check:* You should get $s_n = h^3[0^2 + 1^2 + 2^2 + \cdots + (n - 1)^2]$ in place of Eq. (9).]

24. Let S_n represent the sum of the areas of the n circumscribed rectangles used in Example 3, and let s_n denote the sum of the areas of the corresponding inscribed rectangles used in Exercise 23 above. Show that

$$S_n - s_n = h^3 n^2 = \frac{b^3}{n}.$$

Given b and $\epsilon > 0$, how large should N be in order to guarantee that $|S_n - s_n| < \epsilon$ whenever $n \geq N$? Would this also guarantee that both S_n and s_n differ from the area of the parabolic region AOB by no more than ϵ? Why?

25. Using the notation of Exercise 24, show that

$$s_n < \frac{b^3}{3} < S_n,$$

for every positive integer n.

26. Using the notation of Exercise 24, show that $\{s_n\}$ is a monotonically increasing sequence and that $\{S_n\}$ is a monotonically decreasing sequence. What is the geometric significance of these facts?

27. Prove that $n^3 - n$ is divisible by 3 for every positive integer n. [*Hint:* Use $(k + 1)^3 - (k + 1) = (k^3 - k) + 3(k^2 + k)$.] (*Note:* This is a special case of a theorem of Fermat that states that $n^p - n$ is divisible by p, if p is a prime.)

4
THEORY OF LIMITS APPLIED TO FORMAL DIFFERENTIATION

4-1 DEFINITION OF DERIVATIVE AND EXAMPLE

Let us recall that the derivative of a function f at an interior point x_1 of the domain of f is defined to be

$$f'(x_1) = \lim_{x_2 \to x_1} \frac{f(x_2) - f(x_1)}{x_2 - x_1} \qquad (1)$$

provided this limit exists. There are alternative ways of writing Eq. (1). For example, if we denote the difference $x_2 - x_1$ by Δx,

$$\Delta x = x_2 - x_1, \qquad (2a)$$

then

$$x_2 = x_1 + \Delta x, \qquad (2b)$$

and $x_2 \to x_1$ if and only if $\Delta x \to 0$. Hence Eq. (1) can be rewritten in the form

$$f'(x_1) = \lim_{\Delta x \to 0} \frac{f(x_1 + \Delta x) - f(x_1)}{\Delta x}. \qquad (3)$$

Equation (3) sometimes leads to simpler computations than Eq. (1). Before we illustrate these computations, let us note that there is nothing sacred about the use of x_1 in Eq. (3). We can equally well write

$$f'(c) = \lim_{\Delta x \to 0} \frac{f(c + \Delta x) - f(c)}{\Delta x} \qquad (4a)$$

or

$$f'(x) = \lim_{\Delta x \to 0} \frac{f(x + \Delta x) - f(x)}{\Delta x}. \qquad (4b)$$

In Eq. (4a), c is a real number interior to the domain of f, and the same is true of x in Eq. (4b). Observe in Eqs. (3), (4a), and (4b) that the variable that approaches zero during the limit process is Δx, and not x_1, nor c, nor x.

Example 1. If $f(x) = x^3 - 2x + 1$, find $f'(x)$ for any real number x.

Solution. Let x be any real number, and let $\Delta x \neq 0$. Then

$$\begin{aligned} f(x + \Delta x) &= (x + \Delta x)^3 - 2(x + \Delta x) + 1 \\ &= x^3 + 3x^2\,\Delta x + 3x(\Delta x)^2 + (\Delta x)^3 - 2x - 2\,\Delta x + 1, \end{aligned}$$

and

$$f(x) = x^3 - 2x + 1,$$

so that

$$f(x + \Delta x) - f(x) = 3x^2\,\Delta x + 3x(\Delta x)^2 + (\Delta x)^3 - 2\,\Delta x$$

and

$$\frac{f(x + \Delta x) - f(x)}{\Delta x} = (3x^2 - 2) + 3x\,\Delta x + (\Delta x)^2.$$

Hold x fixed and let Δx approach zero. Then

$$\lim_{\Delta x \to 0} 3x\, \Delta x = 0 \quad \text{and} \quad \lim_{\Delta x \to 0} (\Delta x)^2 = 0,$$

so

$$f'(x) = \lim_{\Delta x \to 0} \frac{f(x + \Delta x) - f(x)}{\Delta x} = 3x^2 - 2.$$

Observe that we have

$$f'(x) = 3x^2 - 2, \quad \text{for any real } x.$$

Thus the mapping

$$f' : x \to f'(x) = 3x^2 - 2, \quad x \text{ any real number},$$

defines a new function f', called the *derived function* of the original function $f : x \to f(x) = x^3 - 2x + 1$. In this example, both f and f' are polynomials, and their domains are the same, namely, the set of all real numbers.

EXERCISES 4–1

Find the derivative of each of the following using the definition.

1. $f(x) = 3x + 2$
2. $f(x) = 3x^2 + 2x - 1$
3. $f(x) = \sqrt{x + 1}, \quad x > -1$
4. $f(x) = x^{3/2}, \quad x > 0$
5. $f(x) = (4x^2 + 7)(3x - 2)$
6. $f(x) = ax^3(bx + c)$
7. $f(x) = \dfrac{x + 3}{x - 5}, \quad x \neq 5$
8. $f(x) = \dfrac{1}{x^2 + 3x + 5}$
9. $f(x) = \dfrac{1}{ax^2 + bx + c}$, wherever $ax^2 + bx + c \neq 0$
10. $f(x) = \dfrac{1}{\sqrt{x + 1}}, \quad x > -1$

4–2 THEOREMS

The following theorems lead to simple rules for differentiating polynomials and other rational functions. They also apply to other types of differentiable functions to be taken up later. Since the results may be well known to some readers, we first state seven theorems without proofs. Short proofs are then given following the statements of the theorems and some interpolated remarks.

THEORY OF LIMITS APPLIED TO FORMAL DIFFERENTIATION 4–2

Theorem 4–1. Constants. If f is a constant function $x \to f(x) = c$, then $f'(x) = 0$ for every real x.

Theorem 4–2. Powers. If n is a positive integer and $f(x) = x^n$, then $f'(x) = nx^{n-1}$ for every real x.*

Theorem 4–3. If u is a function, differentiable at x, and c is a number, and $f = cu$, then $f'(x) = cu'(x)$.

Theorem 4–4. Sums. If u and v are functions differentiable at x and $f = u + v$, then $f'(x) = u'(x) + v'(x)$.

Corollary 4–4.1. If each of the n functions $u_1, u_2, \ldots, u_n$ is differentiable at x, and if $f = \sum_{i=1}^{n} u_i$, then

$$f'(x) = \sum_{i=1}^{n} u'_i(x).$$

In words, the derivative of a sum of a finite number of functions is the sum of their derivatives. The proof is by induction on n. If the corollary is true for $n = k$, then when $n = k + 1$, let

$$u = u_1 + \cdots + u_k, \qquad v = u_{k+1},$$

and apply Theorem 4–4 to $f = u + v$. This leads to

$$f'(x) = u'(x) + v'(x) = \sum_{i=1}^{k} u'_i(x) + u'_{k+1}(x)$$

$$= \sum_{i=1}^{k+1} u'_i(x).$$

Thus Theorem 4–4 and Corollary 4–4.1 for $n = k$ imply the truth of the corollary for $n = k + 1$.

Theorem 4–5. Products. If u and v are functions differentiable at x and $f = uv$, then $f'(x) = u(x)v'(x) + u'(x)v(x)$.

Corollary 4–5.1. If n is an integer ≥ 2 and $u_1, u_2, \ldots, u_n$ are functions differentiable at x and

$$f = u_1 u_2 u_3 \cdots u_n,$$

then

$$f'(x) = u'_1(x) u_2(x) u_3(x) \cdots u_n(x) + u_1(x) u'_2(x) u_3(x) \cdots u_n(x)$$
$$+ \cdots + u_1(x) u_2(x) u_3(x) \cdots u'_n(x).$$

---

* For $n = 1$, $f'(x) \equiv 1$, and we have to modify the statement slightly when $x = 0$, or abuse the language, because 0^0 is undefined.

In words, if f is the product of the n functions $u_1, u_2, \ldots, u_n$, then $f'(x)$ is the

sum of n terms,

each of which is a

product of n terms consisting of the derivative of one of the functions and the $(n-1)$ other original function values.

The notation $\prod_{i=1}^{n}$ is an abbreviation for a product just as $\sum_{i=1}^{n}$ is an abbreviation for a sum. And, in the present context,

$$\prod_{\substack{i=1 \\ i \neq j}}^{n} u_i$$

means the same as

$$\left(\prod_{i=1}^{j-1} u_i\right)\left(\prod_{i=j+1}^{n} u_i\right).$$

Thus the formula could also be written as

$$\left(\prod_{i=1}^{n} u_i\right)'(x) = \sum_{j=1}^{n}\left(u_j'(x) \prod_{\substack{i=1 \\ i \neq j}}^{n} u_i(x)\right).$$

Corollary 4–5.2. If u is a function differentiable at x and n is a positive integer, then the derivative at x of u^n is $n[u(x)]^{n-1} u'(x)$.

Remark 1. Corollary 4–5.1 follows from Theorem 4–5 by induction on n. Theorem 4–5 is the case $n = 2$. If we assume that the corollary is true for $n = k$, then we can deduce its truth for $n = k + 1$ by applying Theorem 4–5 to the function $f = gh$, with

$$g = u_1 u_2 \cdots u_k, \qquad h = u_{k+1}.$$

By Theorem 4–5, $f'(x) = g(x)h'(x) + g'(x)h(x)$. And, by Corollary 4–5.1 for the case $n = k$,

$$g'(x) = \sum_{j=1}^{k}\left(u_j'(x) \prod_{\substack{i=1 \\ i \neq j}}^{k} u_i(x)\right),$$

so that

$$g'(x)h(x) = \sum_{j=1}^{k}\left(u_j'(x) \prod_{\substack{i=1 \\ i \neq j}}^{k+1} u_i(x)\right).$$

Moreover,

$$g(x)h'(x) = h'(x)g(x) = u_{k+1}'(x) \prod_{i=1}^{k} u_i(x).$$

This latter is the same as

$$u'_{k+1}(x) \prod_{\substack{i=1 \\ i \neq k+1}}^{k+1} u_i(x),$$

which combines with the earlier sum to give the desired result:

$$f'(x) = \sum_{j=1}^{k+1} \left(u'_j(x) \prod_{\substack{i=1 \\ i \neq j}}^{k+1} u_i(x) \right).$$

Hence Corollary 4–5.1 is true for $n = k + 1$ whenever it is true for the positive integer k ($k \geq 2$). The principle of mathematical induction shows that Corollary 4–5.1 follows from Theorem 4–5.

Remark 2. For $n \geq 2$, Corollary 4–5.2 follows from Corollary 4–5.1 by taking

$$u_1 = u_2 = \cdots = u_n = u.$$

For $n = 1$, we interpret $n[u(x)]^{n-1}$ as 1.

Theorem 4–6. Quotients. If u and v are functions differentiable at x, and $v(x) \neq 0$, and $f = u/v$, then

$$f'(x) = \frac{v(x)u'(x) - u(x)v'(x)}{[v(x)]^2}.$$

Theorem 4–7. Polynomials. If f is a polynomial function,

$$f(x) = \sum_{k=0}^{n} a_k x^k = a_0 + a_1 x + a_2 x^2 + \cdots + a_n x^n,$$

then

$$f'(x) = \sum_{k=1}^{n} k a_k x^{k-1} = a_1 + 2a_2 x + \cdots + n a_n x^{n-1}.$$

Remark 3. Theorem 4–7 is a corollary of Theorems 4–1 through 4–4. First, Theorem 4–4 says that the derivative of a sum of two functions is the sum of their derivatives, and this can easily be extended (Corollary 4–4.1) to the sum of any finite number of functions. Theorems 4–2 and 4–3 together say that the derivative of $a_k x^k$ is $k a_k x^{k-1}$, for $k \geq 1$; and Theorem 4–1 says that the derivative of a_0 is zero. Combining these results, we get Theorem 4–7.

Most of the proofs of these theorems are abbreviated.

Theorem 4–7 says that polynomials are everywhere differentiable. Combining this result with Theorem 4–6, we find that a rational function

$P(x)/Q(x)$ is differentiable wherever it is defined, that is, wherever the denominator $Q(x)$ is not zero.

Proof of Theorem 4–1. If f is constant, then
$$f(x + \Delta x) = f(x) = c$$
and
$$\frac{f(x + \Delta x) - f(x)}{\Delta x} = \frac{c - c}{\Delta x} = \frac{0}{\Delta x} = 0,$$
so
$$f'(x) = \lim_{\Delta x \to 0} (0) = 0.$$

Proof of Theorem 4–2. If n is a positive integer and $f(x) = x^n$, then, by the binomial theorem,
$$f(x + \Delta x) = (x + \Delta x)^n = x^n + \sum_{k=1}^{n} \binom{n}{k} x^{n-k} (\Delta x)^k$$
and
$$\frac{f(x + \Delta x) - f(x)}{\Delta x} = \sum_{k=1}^{n} \binom{n}{k} x^{n-k} (\Delta x)^{k-1}. \tag{1}$$

The term for $k = 1$ on the right-hand side of Eq. (1) is
$$\binom{n}{1} x^{n-1} (\Delta x)^0 = nx^{n-1},$$
and it remains fixed as $\Delta x \to 0$. If $n = 1$, there are no other terms in the summation; but if $n \geq 2$, the terms on the right-hand side of Eq. (1) for $k > 1$ are:
$$\binom{n}{2} x^{n-2} \Delta x, \quad \binom{n}{3} x^{n-3} (\Delta x)^2, \quad \ldots, \quad \binom{n}{n} (\Delta x)^{n-1},$$
and each of these has zero for its limit when $\Delta x \to 0$. Therefore, since the limit of the sum of a finite number of terms is the sum of the limits,
$$f'(x) = \lim_{\Delta x \to 0} \frac{f(x + \Delta x) - f(x)}{\Delta x} = nx^{n-1}.$$

Proof of Theorem 4–3. If $f = cu$, then
$$f(x + \Delta x) = c\,u(x + \Delta x),$$
$$f(x) = c\,u(x),$$
and
$$\frac{f(x + \Delta x) - f(x)}{\Delta x} = c\,\frac{u(x + \Delta x) - u(x)}{\Delta x}. \tag{2}$$

THEORY OF LIMITS APPLIED TO FORMAL DIFFERENTIATION   4–2

By hypothesis, $u'(x)$ exists, and is

$$u'(x) = \lim_{\Delta x \to 0} \frac{u(x + \Delta x) - u(x)}{\Delta x}.$$

Therefore, taking limits in Eq. (2), we get

$$f'(x) = cu'(x).$$

Proof of Theorem 4–4. Let $f = u + v$. Then

$$f(x + \Delta x) = u(x + \Delta x) + v(x + \Delta x),$$
$$f(x) = u(x) + v(x),$$

and

$$\frac{f(x + \Delta x) - f(x)}{\Delta x} = \frac{u(x + \Delta x) - u(x) + v(x + \Delta x) - v(x)}{\Delta x}$$
$$= \frac{u(x + \Delta x) - u(x)}{\Delta x} + \frac{v(x + \Delta x) - v(x)}{\Delta x}. \quad (3)$$

By hypothesis, $u'(x)$ and $v'(x)$ exist and are

$$u'(x) = \lim_{\Delta x \to 0} \frac{u(x + \Delta x) - u(x)}{\Delta x}, \quad v'(x) = \lim_{\Delta x \to 0} \frac{v(x + \Delta x) - v(x)}{\Delta x}.$$

Since each difference quotient in the last line of Eq. (3) has a limit, so has their sum, and taking limits as $\Delta x \to 0$ in Eq. (3), we get

$$f'(x) = u'(x) + v'(x).$$

Proof of Theorem 4–5. To simplify the writing, let $u(x + \Delta x) - u(x) = \Delta u$ and $v(x + \Delta x) - v(x) = \Delta v$. Then, if $f = uv$, we have

$$f(x + \Delta x) = (u(x) + \Delta u) \cdot (v(x) + \Delta v)$$
$$= u(x)v(x) + u(x)\,\Delta v + v(x)\,\Delta u + \Delta u\,\Delta v$$

and

$$\frac{f(x + \Delta x) - f(x)}{\Delta x} = u(x)\frac{\Delta v}{\Delta x} + v(x)\frac{\Delta u}{\Delta x} + \Delta u \cdot \frac{\Delta v}{\Delta x}. \quad (4)$$

By hypothesis, the following limits exist:

$$u'(x) = \lim_{\Delta x \to 0} \frac{u(x + \Delta x) - u(x)}{\Delta x} = \lim_{\Delta x \to 0} \frac{\Delta u}{\Delta x}, \quad v'(x) = \lim_{\Delta x \to 0} \frac{\Delta v}{\Delta x},$$

and

$$\lim_{\Delta x \to 0} (\Delta u) = \lim_{\Delta x \to 0} \left(\Delta x \cdot \frac{\Delta u}{\Delta x} \right) = (\lim \Delta x) \cdot \lim \left(\frac{\Delta u}{\Delta x} \right) = 0 \cdot u'(x) = 0.$$

Therefore, the right-hand side of Eq. (4) has a limit when $\Delta x \to 0$, and that limit is
$$f'(x) = u(x)v'(x) + v(x)u'(x) + 0.$$

Proof of Theorem 4–6. Using the same notation, with $f = u/v$, we have (see Exercise 11 for discussion of a technical point)

$$f(x + \Delta x) - f(x) = \frac{u(x) + \Delta u}{v(x) + \Delta v} - \frac{u(x)}{v(x)}$$
$$= \frac{v(x)u(x) + v(x) \cdot \Delta u - u(x)v(x) - u(x) \cdot \Delta v}{v(x)[v(x) + \Delta v]}.$$

Therefore,

$$\frac{f(x + \Delta x) - f(x)}{\Delta x} = \frac{v(x)\frac{\Delta u}{\Delta x} - u(x)\frac{\Delta v}{\Delta x}}{v(x)[v(x) + \Delta v]}. \tag{5}$$

By hypothesis, the following limits exist:

$$u'(x) = \lim \frac{\Delta u}{\Delta x}, \qquad v'(x) = \lim \frac{\Delta v}{\Delta x},$$

and
$$0 = \lim \Delta v.$$

The major denominator on the right-hand side of Eq. (5) has the limit $[v(x)]^2$, which is different from zero if $v(x) \neq 0$. Hence, taking limits as $\Delta x \to 0$ in Eq. (5), we get

$$f'(x) = \frac{v(x)u'(x) - u(x)v'(x)}{[v(x)]^2}.$$

EXERCISES 4–2

In Exercises 1 through 10, use the theorems of this section to find the derivative of as many exercises of Section 4–1, repeated below, as possible. If the theorems are not applicable, indicate why.

1. $f(x) = 3x + 2$
2. $f(x) = 3x^2 + 2x - 1$
3. $f(x) = \sqrt{x + 1}, \quad x \geq -1$
4. $f(x) = x^{3/2}, \quad x > 0$
5. $f(x) = (4x^2 + 7)(3x - 2)$
6. $f(x) = ax^3(bx + c)$
7. $f(x) = \dfrac{x + 3}{x - 5}, \quad x \neq 5$
8. $f(x) = \dfrac{1}{x^2 + 3x + 5}$
9. $f(x) = \dfrac{1}{ax^2 + bx + c}$, wherever $ax^2 + bx + c \neq 0$
10. $f(x) = \dfrac{1}{\sqrt{x + 1}}, \quad x \geq -1$

11. In the proof of Theorem 4–6, we tacitly assumed that $v(x + \Delta x)$, or $v(x) + \Delta v$, is not zero. Prove that, if x_1 is a real number in the domain of v at which $v(x_1) \neq 0$ and $v'(x_1)$ exists, then there exists an $h > 0$ such that $v(x_1 + \Delta x) \neq 0$ if $0 < |\Delta x| < h$. The proof of Theorem 4–6 goes through if we so restrict Δx. [*Hint:* Let $\epsilon = \frac{1}{2}|v(x_1)|$. Then $\epsilon > 0$. Now work with this ϵ and use

$$\lim_{\Delta x \to 0} v(x_1 + \Delta x) = \lim_{\Delta x \to 0} \left(v(x_1) + \frac{\Delta v}{\Delta x} \Delta x \right)$$

to establish the existence of $\delta > 0$ such that

$$|v(x_1 + \Delta x) - v(x_1)| < \epsilon \quad \text{when} \quad |\Delta x| < \delta].$$

Find $f'(x)$ in each of the following:

12. $f(x) = (3x^2 - 2)^3$
13. $f(x) = (3x^2 - 2)^n$, n any positive integer
14. $f(x) = (x^2 + 5x + 1)^{56}$
15. $f(x) = (3x^5 - 5x^3 + 2)^2 + (x^3 - 3x - 2)^3$
16. $f(x) = (x^m + a)^n + (x^m + a)^m$, m and n positive integers
17. $f(x) = (x^2 - 4)(x^3 + 5)$
18. $f(x) = (2x^2 - x + 6)(3x^2 + x - 6)$
19. $f(x) = (x^3 - 2)^2(x^2 + 3)^3$
20. $f(x) = (x^m - a)^n(x^n - a)^m$, m and n positive integers
21. $f(x) = \dfrac{x}{x-2}$, $x \neq 2$
22. $f(x) = (3x - 2)(2x + 3)(x - 5)$
23. $f(x) = \dfrac{(x+3)^2}{x-5}$, $x \neq 5$
24. $f(x) = \dfrac{(x+3)^3}{(x-5)^2}$, $x \neq 5$
25. $f(x) = \dfrac{3x^2}{x^3 - 2}$, $x \neq \sqrt[3]{2}$
26. $f(x) = \dfrac{3x^2}{(x^3 - 2)^2}$, $x \neq \sqrt[3]{2}$
27. $f(x) = \dfrac{(3x^2 + 2)^3}{(2x^3 - 5)^2}$, $x \neq \frac{1}{2}\sqrt[3]{5}$
28. $f(x) = \dfrac{x+a}{x-a}$, $x \neq a$

4–3 DERIVATIVES OF THE TRIGONOMETRIC FUNCTIONS

The theorems of Section 4–2 have so far been applied only to polynomials and other rational functions. We can, however, easily apply the results to the trigonometric functions tangent, cotangent, secant, and cosecant after we have established the existence of, and formulas for, the derivatives of the sine and cosine functions. We do this in the next two theorems.

Theorem 4–8. *The sine function is everywhere differentiable, and for every real number x, $(\sin)'(x) = \cos x$.*

Proof. Let $f(x) = \sin x$. Then
$$f(x + \Delta x) = \sin(x + \Delta x) = \sin x \cos \Delta x + \cos x \sin \Delta x$$
and
$$f(x + \Delta x) - f(x) = (\sin x)(-1 + \cos \Delta x) + \cos x \sin \Delta x.$$
Therefore,
$$\frac{f(x + \Delta x) - f(x)}{\Delta x} = (-\sin x) \cdot \frac{1 - \cos \Delta x}{\Delta x} + \cos x \cdot \frac{\sin \Delta x}{\Delta x}. \tag{1}$$

When $\Delta x \to 0$, we have the following limits:
$$\lim_{\Delta x \to 0} \frac{\sin \Delta x}{\Delta x} = 1, \tag{2a}$$

$$\lim_{\Delta x \to 0} \frac{1 - \cos \Delta x}{\Delta x} = \lim_{\Delta x \to 0} \frac{(1 - \cos \Delta x)(1 + \cos \Delta x)}{\Delta x \cdot (1 + \cos \Delta x)}$$
$$= \lim_{\Delta x \to 0} \frac{\sin \Delta x}{\Delta x} \cdot \frac{\sin \Delta x}{1 + \cos \Delta x}$$
$$= \lim_{\Delta x \to 0} \frac{\sin \Delta x}{\Delta x} \cdot \lim_{\Delta x \to 0} \frac{\sin \Delta x}{1 + \cos \Delta x}$$
$$= 1 \cdot \frac{0}{2} = 0. \tag{2b}$$

Now let $\Delta x \to 0$ in Eq. (1). The terms involving Δx on the right-hand side of (1) have limits given by Eqs. (2a) and (2b). Therefore,
$$\lim_{\Delta x \to 0} \frac{f(x + \Delta x) - f(x)}{\Delta x} = f'(x) = (-\sin x) \cdot 0 + \cos x \cdot 1 = \cos x.$$

Since $f(x) = \sin x$, this means that the sine function is everywhere differentiable and
$$(\sin)'(x) = \cos x. \tag{3}$$

Remark 1. Equation (3) says that the slope of the tangent to the sine curve, at x, is equal to the ordinate of the cosine curve, at x. Thus, if we sketch both a sine curve and a cosine curve on the same coordinate plane as in Fig. 4–1, we can relate Eq. (3) to the following features of the sine curve (Fig. 4–1). At $x = 0$, $\sin 0 = 0$, $\cos 0 = 1$, so the sine curve passes through the origin with slope $= +1$. Therefore, the sine curve is tangent to the line $y = x$ at the origin. But as x increases through small positive values, $\cos x$ decreases, so the slope of the sine curve decreases. At $x = \pi/2$, the sine has reached its maximum value, $\sin(\pi/2) = +1$, and the cosine is zero, $\cos(\pi/2) = 0$. Thus the tangent to the sine curve is horizontal at the maximum. Between $x = \pi/2$ and $x = \pi$, the cosine is negative, so the sine curve has negative slope and its graph runs downhill

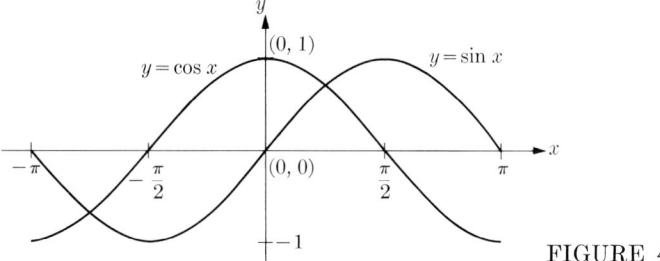

FIGURE 4–1

from $(\pi/2, 1)$ to $(\pi, 0)$. The rate of this decline increases and the curve is at its steepest descent when $x = \pi$, $\cos x = -1$, and $\sin x = 0$. Because $\sin^2 x + \cos^2 x = 1$, we know that whenever $\sin x = 0$, we have $\cos x = +1$ or -1, and these are the extreme values of the cosine. In other words, whenever the sine curve crosses the x-axis, it does so at a slope of $+1$, or -1, and these represent the steepest places on the sine curve. On the other hand, when $\sin x$ is at an extreme value of either $+1$ or -1, then $\cos x = 0$, and the sine curve has a horizontal tangent. Some of these properties are shared by graphs of other functions. In fact, the differential calculus is often used to find maximum or minimum values of a function by first locating the places where the derivative of the function is zero. The further study of maximum and minimum values of a function (existence, location, and techniques for finding) will be continued in Chapters 5 and 6.

Remark 2. The notation $(\sin)'(x)$ is not standard. It is like the notation $f'(x)$ when f is the sine function. The more familiar notations are:

$$\frac{d(\sin x)}{dx} = \cos x, \tag{4a}$$

or

$$D_x(\sin x) = \cos x. \tag{4b}$$

In Eq. (4b), the symbol D_x is read "derivative, with respect to x, of ..." In Eq. (4a), this same meaning is attached to the symbols d/dx. Later, we shall introduce the definition of the *differential* and see that a derivative is also the ratio of two differentials. Equation (4a) then has the interpretation

$$\frac{\text{(differential of sine) at } x}{\text{differential of } x} = \cos x.$$

Remark 3. If we examine the two graphs in Fig. 4–1, we see that the cosine graph is just the sine graph shifted $\pi/2$ units to the left. This means that the slope of the cosine graph at a point $(x, \cos x)$ is the same as the slope of the sine graph at the point $(x + \pi/2, \sin(x + \pi/2))$, which is $\cos(x + \pi/2)$, by Theorem 4-8. But

$$\cos\left(x + \frac{\pi}{2}\right) = \cos x \cos\left(\frac{\pi}{2}\right) - \sin x \sin\left(\frac{\pi}{2}\right) = 0 - \sin x.$$

This leads us to the following theorem.

Theorem 4-9. The cosine function is everywhere differentiable, and for every real number x,

$$(\cos)'(x) = -\sin x. \tag{5}$$

Remark 4. The trigonometric identities

$$\tan x = \frac{\sin x}{\cos x} \quad \text{and} \quad \sec x = \frac{1}{\cos x}$$

can now be used, with Theorem 4-6, to obtain formulas for the derivatives of $\tan x$ and $\sec x$ (Exercises 3 and 8). Similar techniques apply to

$$\cot x = \frac{\cos x}{\sin x} \quad \text{and} \quad \csc x = \frac{1}{\sin x}.$$

EXERCISES 4-3

1. Using the methods of Theorem 4-8, prove Theorem 4-9.
2. Use the formula for $\sin A - \sin B$ to prove Theorem 4-8.
3. If x is not an odd multiple of $\pi/2$, so that $\cos x \neq 0$, prove that

$$\tan'(x) = \sec^2 x.$$

4. Use the definition for the derivative to find $\sin' x$ at $x = 0$.
5. Find the derivative of $\sin(x + \pi/2)$ in two ways.
6. Find the derivative of $\cos(x + \pi/2)$.
7. Derive a formula for the derivative of $\cot x$, for $\sin x \neq 0$.
8. Derive a formula for the derivative of $\sec x$, for $\cos x \neq 0$.
9. Given the function $y = |\sin x|$, draw the graph for $|x| \leq 2\pi$. For what values of x does it appear that y' fails to exist? Why? Find y' (use the concept of periodicity). Draw the graph of y'.
10. Given the function $y = \sin |x|$, sketch the graph for $-2\pi \leq x \leq 2\pi$. [*Hint:* The curve is symmetric with respect to what?] Does y' exist for all x in D_f? If not, where does it fail to exist? Find y' for all other x in D_f.
11. Using Theorems 4-8 and 4-9, together with trigonometric identities, prove by mathematical induction that, for n any positive integer,

$$\frac{d(\sin nx)}{dx} = n \cos nx \quad \text{and} \quad \frac{d(\cos nx)}{dx} = -n \sin nx.$$

5
MAXIMA AND MINIMA

5-1 EXAMPLES AND TERMINOLOGY

We introduce some terminology, and illustrate some of the ideas, in the following examples.

Example 1. The function, $f: x \to f(x) = 4 - x^2$, $-\infty < x < \infty$, takes only values that are ≤ 4. Since the function actually assumes the value 4 (at $x = 0$), we say that the *maximum* of f is 4. On the other hand, the range of this function is not bounded below, because no matter how large the number M may be, $4 - x^2 < -M$ whenever $x^2 > 4 + M$. Thus, f has no minimum. (See Fig. 5–1(a).)

Example 2. Consider the function, $g: x \to g(x) = 4 - x^2$, $-3 \leq x \leq 3$ (Fig. 5–1(b)). This is not the same as the function f in Example 1, because its domain is different. And, since the domain is different, its range is also different. In fact, the range of g is the set $\{y \mid -5 \leq y \leq 4\}$, because

$$-3 \leq x \leq 3 \Rightarrow 0 \leq |x| \leq 3 \Rightarrow 0 \leq x^2 \leq 9$$
$$\Rightarrow 0 \geq -x^2 \geq -9 \Rightarrow 4 \geq 4 - x^2 \geq -5.$$

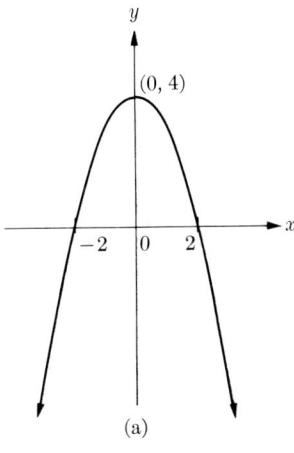

(a)

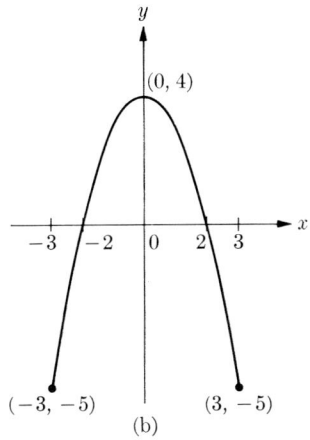

(b)

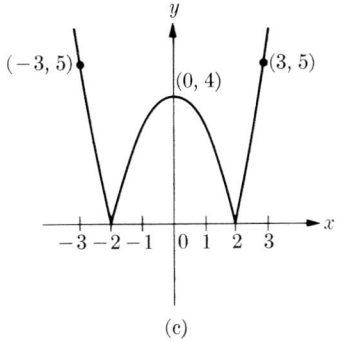

(c)

FIG. 5–1. (a) Portion of the graph of $y = 4 - x^2$, $-\infty < x < \infty$.
Range of f: $-\infty < y \leq 4$.
(b) Graph of $y = 4 - x^2$, $-3 \leq x \leq 3$.
Range of g: $-5 \leq y \leq 4$.
(c) Portion of the graph of $y = |4 - x^2|$, $-\infty < x < +\infty$.

The function g has a *maximum* at $x = 0$ and a *minimum* at $x = -3$ and at $x = +3$.

Example 3. The function
$$h : x \to h(x) = |4 - x^2|, \quad -\infty < x < \infty,$$
has a graph consisting of the points (x, y) such that
$$y = 4 - x^2, \quad x^2 \leq 4,$$
$$y = x^2 - 4, \quad x^2 > 4.$$

See Fig. 5–1(c). The range of h is the set of all $y \geq 0$. The function h has a minimum, 0, at $x = -2$ and at $x = +2$. It has no maximum, but it does have a *relative* maximum, 4, at $x = 0$.

Example 4. The function
$$f : x \to f(x) = x^3, \quad -\infty < x < \infty,$$
has neither a maximum nor a minimum. As $|x|$ increases without limit, so does $|f(x)|$. Its domain and its range both consist of the set of all real numbers. However, it does have a point at which the value of the derivative is zero, namely $(0, 0)$.

What do these examples suggest about maxima and minima of functions? Does a function always have a maximum? A minimum? A relative maximum, or a relative minimum? If it has a relative maximum or minimum, is its derivative zero there? What if the derivative fails to exist? Does a zero derivative imply a maximum or a minimum?

In order to answer these questions, we need more precise definitions. Before continuing with Section 5–2, the reader should examine the derivative for each of the four preceding examples, especially at the critical points.

5–2 DEFINITIONS: MAXIMA AND MINIMA

Let f be a function whose domain D_f and range R_f are sets of real numbers. If for some c in the domain of f there exists an $h > 0$ such that

$$f(x) \leq f(c) \quad \text{whenever } x \in D_f \text{ and } |x - c| < h, \qquad (1)$$

then f is said to have a *relative maximum*, or *local maximum*, at c. If

$$f(x) \geq f(c) \quad \text{whenever } x \in D_f \text{ and } |x - c| < h, \qquad (2)$$

then f is said to have a *relative minimum*, or *local minimum*, at c. If

$$f(x) \leq f(c) \quad \text{for all } x \text{ in } D_f, \tag{3}$$

then f has a *maximum* (an *absolute maximum*) at c; and if

$$f(x) \geq f(c) \quad \text{for all } x \text{ in } D_f, \tag{4}$$

then f has a *minimum* (an *absolute minimum*) at c.

Remark 1. An absolute maximum or minimum is necessarily also a local maximum or minimum, but not conversely. If we think in terms of a competition, the function f has a *local* maximum at c if $f(c)$ is greater than or equal to competing values $f(x)$, for x's *near* c. There are no nearby higher points on the graph of f. For an *absolute* maximum, the competition is extended to include *all* x's in the domain of f; no points on the graph of f are higher than the point $(c, f(c))$.

Remark 2. If the domain of f has an isolated point, then that point is automatically both a relative maximum and a relative minimum for f. For example, the equation

$$f(x) = \sqrt{x^2(x^2 - 4)} \tag{5}$$

yields a real value for $f(x)$ only if $x = 0$ or $|x| \geq 2$. Thus, if we take $c = 0$, the only value of x in the domain of f and close to c is $x = 0$. Therefore, we have both

$$f(x) \leq f(c) \quad \text{and} \quad f(x) \geq f(c)$$

for all values of x in the domain of f, and less than 2 units away from 0. Here, $c = 0$ wins the competition by default: there just aren't any other x's close to 0 to enter the competition. There is also a local minimum at $x = 2$ and at $x = -2$. Part of the graph is shown in Fig. 5–2.

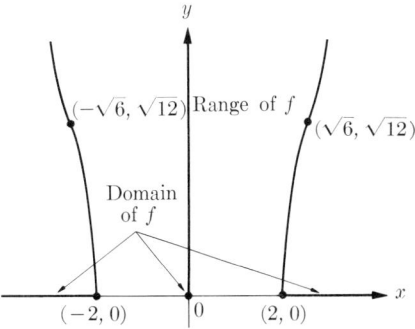

FIG. 5–2. Portion of the graph of $y = \sqrt{x^2(x^2 - 4)}$.

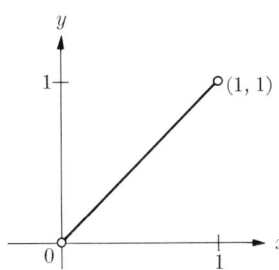

FIG. 5–3. Graph of $y = x$, $0 < x < 1$. Open circles indicate excluded points.

MAXIMA AND MINIMA

Remark 3. If f is a constant, then $f(x) = f(c)$ for every x, and every c, in the domain of f. Thus both

$$f(x) \leq f(c) \quad \text{and} \quad f(x) \geq f(c),$$

and there is an absolute maximum and an absolute minimum at each point in the domain of f.

Remark 4. In contrast with the last example, a function may be bounded and have no relative maxima or minima. For example, suppose the domain of f is the open interval $0 < x < 1$, and f is the identity mapping $f(x) = x$. Then for any c in the domain, $f(x) < f(c)$ when $0 < x < c$, and $f(x) > f(c)$ when $c < x < 1$. That is, the point $(c, f(c))$ is higher than some nearby points and lower than others, so f has neither a local maximum nor a local minimum at c. (See Fig. 5-3.)

EXERCISES 5-2

For each of the following, state whether or not there exists a maximum (relative or absolute), a minimum (relative or absolute), and give the values. Sketch the graph in each case.

1. $y = -|x|$
2. $y = |x| - 4$
3. $y = |x^2 - 4|$
4. $y = x^2 - x - 6$
5. $y = |x^2 - x - 6|$
6. $y = |x|^2 - |x| - 6$
7. $y = \begin{cases} \dfrac{x^2 - 4}{x - 2}, & \text{if } x \neq 2 \\ 5, & \text{if } x = 2 \end{cases}$
8. $y = \begin{cases} \dfrac{x^2 - 4}{x - 2}, & \text{if } x \neq 2 \\ 4, & \text{if } x = 2 \end{cases}$
9. $y = \begin{cases} \dfrac{x^2 - 4}{x - 2}, & \text{if } x \neq 2 \\ 3, & \text{if } x = 2 \end{cases}$
10. $y = \sqrt{x} - 4, \quad x \geq 0$
11. $y = 4 - \sqrt{x}, \quad x \geq 0$
12. $y = \sqrt{x} - 4, \quad x \geq 4$
13. $y = \sqrt{4 - x}, \quad x \leq 4$
14. $y = |\sin x| + |x|$
15. $y = \left| \dfrac{\sin x}{1 + \cos x} \right|, \quad -\pi < x < \pi$
16. $y^{2/3} + x^{2/3} = a^{2/3}, \quad y \geq 0$

5-3 CRITICAL VALUES

Let us now turn our attention to a point $(c, f(c))$ that is a relative maximum or minimum on the graph of a function f, and suppose that f has a derivative at c. The point $(0, 4)$ on the curves in Fig. 5-1(a), (b), or (c) is a good example. When we assume that $f'(c)$ exists, we mean that the

difference quotient
$$\frac{f(c + \Delta x) - f(c)}{\Delta x} \tag{1a}$$

is meaningful for sufficiently small $|\Delta x| \neq 0$, and that this quotient has a limit when Δx approaches zero. In particular,

a) the domain of f must include $c + \Delta x$ for both positive and negative values of Δx, when $|\Delta x|$ is sufficiently small, and

b) the limit of the quotient in (1a) is the same whether Δx approaches zero through positive or negative values.

To be specific, suppose f has a relative maximum at c. Then
$$f(c + \Delta x) \leq f(c), \quad \text{for small } |\Delta x|.$$

Thus, for some positive number h,
$$f(c + \Delta x) - f(c) \leq 0, \quad \text{when} \quad |\Delta x| < h.$$

Therefore,
$$\frac{f(c + \Delta x) - f(c)}{\Delta x} \leq 0, \quad \text{if} \quad 0 < \Delta x < h, \tag{1b}$$

and
$$\frac{f(c + \Delta x) - f(c)}{\Delta x} \geq 0, \quad \text{if} \quad -h < \Delta x < 0. \tag{1c}$$

First, let $\Delta x \to 0$ through positive values. The limit of the difference quotient in (1b) is $f'(c)$, and the inequality also applies in the limit. Therefore,
$$f'(c) = \lim_{\Delta x \to 0^+} \frac{f(c + \Delta x) - f(c)}{\Delta x} \leq 0. \tag{2a}$$

On the other hand, if we let $\Delta x \to 0$ through negative values, inequality (1c) implies that
$$f'(c) = \lim_{\Delta x \to 0^-} \frac{f(c + \Delta x) - f(c)}{\Delta x} \geq 0. \tag{2b}$$

Since the limit, $f'(c)$, is the same in both cases, (2a) and (2b) both hold, so that
$$0 \leq f'(c) \leq 0.$$

Hence
$$f'(c) = 0. \tag{3}$$

The argument for a relative minimum at c would be modified only by reversing the inequalities in (1b), (1c), and in (2a), (2b); and the new

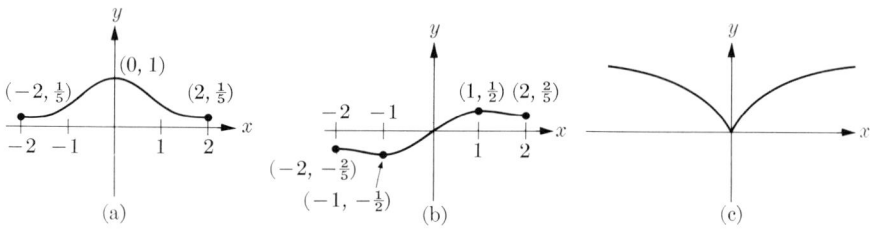
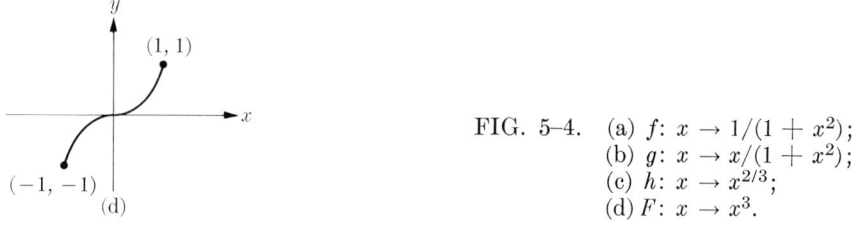

FIG. 5–4. (a) $f: x \to 1/(1+x^2)$;
(b) $g: x \to x/(1+x^2)$;
(c) $h: x \to x^{2/3}$;
(d) $F: x \to x^3$.

inequalities would again lead to Eq. (3). Thus we have proved the following theorem.

Theorem 5–1. Let f be a function from D_f to R_f, where D_f and R_f are sets of real numbers. If f has a relative maximum or a relative minimum at c and $f'(c)$ exists, then $f'(c) = 0$.

Remark. At the risk of belaboring the obvious, we remind the reader that $f'(c) = 0$ does not imply that f has a maximum or a minimum at $(c, f(c))$, as we saw in the example $f(x) = x^3$ with $c = 0$. But if an extremum occurs at $x = c$, then either $f'(c) = 0$ or else the function is not differentiable at c. In Fig. 5–2 the function has minima at $x = -2$, $x = 0$, and $x = 2$, but is not differentiable at any of these points. To summarize: in order to find the points at which a function f assumes a maximum or a minimum for $a \leq x \leq b$, examine

1. those points for which $f'(c) = 0$, $a < c < b$,
2. those points for which $f'(x)$ fails to exist,
3. $f(a)$ and $f(b)$, the values of f at the endpoints of the interval.

We consider some examples.

Example 1. Let $f(x) = 1/(1+x^2)$, $-2 \leq x \leq 2$. Then

$$f'(x) = -\frac{2x}{(1+x^2)^2}.$$

The function f has an absolute maximum at $x = 0$; $f'(0)$ exists and equals 0; f has absolute minima at $x = -2$ and $x = 2$, the endpoints of the interval.

Both $f'(-2)$ and $f'(2)$ fail to exist, because the domain of f is restricted to $[-2, 2]$. (See Fig. 5-4(a).)

Example 2. Let $g(x) = x/(1 + x^2)$, $-2 \leq x \leq 2$. Then

$$g'(x) = \frac{1 - x^2}{(1 + x^2)^2}.$$

This function has an absolute maximum at $x = 1$ and a relative maximum at $x = -2$. (See Fig. 5-4(b).) It has an absolute minimum at $x = -1$ and a relative minimum at $x = 2$.

Example 3. Let $h(x) = x^{2/3}$. Then $h'(x) = \frac{2}{3}x^{-1/3}$, and h has an absolute minimum at $x = 0$, but $h'(0)$ fails to exist. [Note cusp at $(0, 0)$; compare the graph in Fig. 5-4(c) with the graph of $f(x) = |x|$.]

Example 4. Let $F(x) = x^3$ over the interval $[-1, 1]$. Then $F'(x) = 3x^2$. The function F has an absolute maximum at $x = 1$ and an absolute minimum at $x = -1$. There is neither a maximum nor a minimum at $x = 0$, but $F'(0) = 0$ (Fig. 5-4(d)).

Note. If, in Example 4, the domain of definition is the open interval $(-1, 1)$, then F has neither a maximum nor a minimum.

Theorem 5–1 is the key to the solution of a wide variety of applied maxima and minima problems. To apply the theorem, we represent the quantity that is to be a maximum or a minimum as the value of a function of a real variable x, over some appropriate domain. Any place c in the domain where either

$$\text{a)} \quad f'(c) = 0 \tag{4a}$$

or

$$\text{b)} \quad \text{the derivative fails to exist} \tag{4b}$$

is considered to be a *critical* value of x. We then examine the values of f at these critical values of x to learn which, if any, correspond to local maxima, or minima, or neither. By Theorem 5–1, if the derivative exists at a local maximum or minimum, then (4a) must hold; otherwise (4b) applies. Hence the set of points, in the domain, at which f has a local maximum or minimum is a subset of the set of critical points. Thus, in Example 3 of Section 5–1, Fig. 5–1(c), the critical points are: $0, -2, +2$. At 0, the derivative exists and is zero. At ± 2, the derivative does not exist.*

---

* The theoretical questions about existence of extreme values will be dealt with in Chapter 6.

MAXIMA AND MINIMA

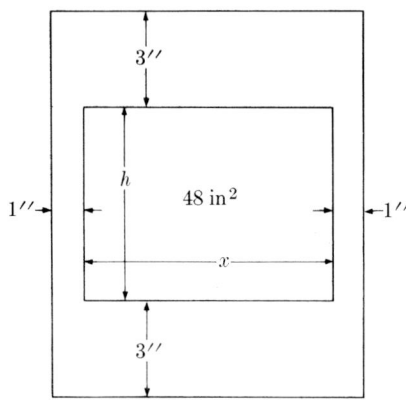
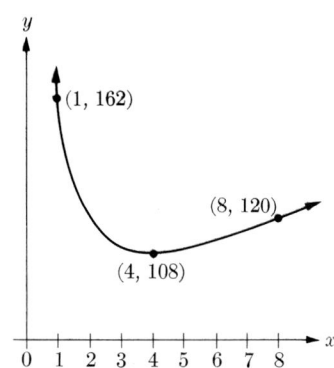

FIG. 5–5. Poster for Example 5.

FIG. 5–6. Portion of the graph of $A = 6x + (96/x) + 60,\ x > 0$.

Example 5. A rectangular poster is to have a margin of 3 inches at the top, 3 inches at the bottom, and 1 inch at each side, as shown in Fig. 5–5. The region inside these margins is to have an area of 48 square inches. What are the dimensions of the poster if its total area is to be a minimum?

Solution. Let the dimensions of the region inside the borders be x inches for the width and h inches for the height. The area of the poster is

$$A = (x + 2)(h + 6) = xh + 6x + 2h + 12$$
$$= 48 + 6x + \frac{96}{x} + 12$$
$$= 6x + \frac{96}{x} + 60. \qquad (5)$$

For what values of x is Eq. (5) meaningful? Evidently, for $x > 0$. That is

$$A = f(x) = 6x + \frac{96}{x} + 60, \qquad x > 0.$$

This function f has a derivative at every $x > 0$, and

$$f'(x) = 6 - \frac{96}{x^2}.$$

The critical values are the solutions of

$$f'(c) = 0,$$

that is

$$6 - \frac{96}{c^2} = 0, \quad \text{or} \quad c^2 = \frac{96}{6} = 16, \quad \text{or} \quad c = 4.$$

96

The root $c = -4$ is not in the domain of f, so has no meaning for our problem. Therefore, if there is a poster of minimal area, its dimensions are

$$2 + c = 6, \quad \text{width in inches,}$$

and

$$6 + \frac{48}{c} = 18, \quad \text{height in inches.}$$

The minimal area is

$$f(c) = 6c + \frac{96}{c} + 60 = 108 \text{ in}^2.$$

Can we prove that this actually is the minimum? We compute

$$f(x) - f(c) = 6x + \frac{96}{x} + 60 - 108$$

$$= 6x + \frac{96}{x} - 48$$

$$= 6\left(x + \frac{16}{x} - 8\right)$$

$$= 6\left(\sqrt{x} - \frac{4}{\sqrt{x}}\right)^2.$$

Since $6(\sqrt{x} - 4/\sqrt{x})^2 \geq 0$ for all $x > 0$, we conclude that

$$f(x) \geq f(c), \quad \text{for all } x > 0.$$

Thus f has an absolute minimum at $c = 4$, $f(c) = 108$. A portion of the graph of f is shown in Fig. 5–6. The lowest point on the graph is $(c, f(c)) = (4, 108)$. Incidentally, had we been clever enough to recognize that

$$f(x) = 6\left(\sqrt{x} - \frac{4}{\sqrt{x}}\right)^2 + 108, \quad x > 0,$$

we would not have needed calculus to see that $f(x) \geq 108$, and that $f(x) = 108$ if and only if $\sqrt{x} - 4/\sqrt{x} = 0$ or $x = 4$. But hindsight is easier than foresight, here as elsewhere in life.

5–4 TESTS FOR EXTREMA

The reader is no doubt aware that there are ways of deciding whether a particular critical point corresponds to a relative maximum, or minimum, or neither. We state one such test below, but defer its complete theoretical justification to a later time (after the proof of the mean-value theorem).

Figure 5–7 illustrates some types of critical points. We assume that the domain of the function is the interval $c_1 \leq x \leq c_7$. At the endpoints c_1 and c_7, the derivative fails to exist, because

$$\lim_{\Delta x \to 0} \frac{f(x + \Delta x) - f(x)}{\Delta x}$$

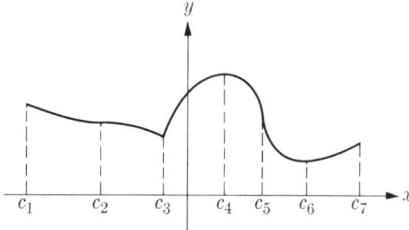

FIG. 5–7. Critical points of a function f, $c_1 \leq x \leq c_7$.

doesn't exist, since Δx cannot be negative if $x = c_1$, and Δx cannot be positive if $x = c_7$. The derivative also fails to exist at c_3 and c_5, although the curve is smooth enough to have a vertical tangent at $(c_5, f(c_5))$. The derivative exists and is zero at c_2, c_4, and c_6. The *critical values* are

c_1, c_7	endpoints,
c_2, c_4, c_6	points where derivative is zero,
c_3, c_5	interior points of the domain where the derivative does not exist.

But c_2 and c_5 correspond to neither maxima nor minima. In the open intervals between critical points, the derivative exists, and its sign can be used to provide information about the shape of the graph.

Interval	Sign of $f'(x)$	Remark
$c_1 < x < c_2$	−	max at c_1
$c_2 < x < c_3$	−	min at c_3
$c_3 < x < c_4$	+	max at c_4
$c_4 < x < c_5$	−	curve slopes downward
$c_5 < x < c_6$	−	min at c_6
$c_6 < x < c_7$	+	max at c_7

Starting from the left-hand endpoint of the domain, the negative derivative means that the curve slopes downward to the right; that is, the function decreases as we move away from $x = c_1$, so there is a relative maximum at c_1. But the slope is negative on both sides of c_2, so that in spite of the zero slope at c_2 there is no maximum or minimum there. The negative slope before and after c_5 rules out an extremum at that point too. The curve has negative slope to the left of c_3 and positive slope to the right of c_3, which indicates a relative minimum at c_3. The slope is positive as we come toward c_4 from the left, and negative as we go beyond c_4 to the right, so there is a relative maximum at c_4. The behavior near c_6 and c_7

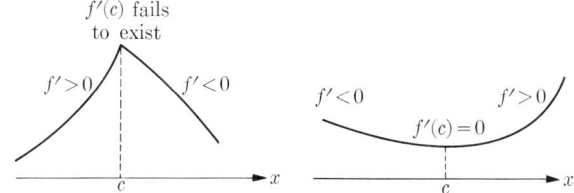

FIG. 5–8. Change of sign of $f'(x)$ as x increases through c.

can be analyzed the same way. More generally, the following test applies (see Fig. 5–8).

First-derivative Test. Let c be a critical value for the function f and an interior point of the domain of f. Suppose f is continuous at c and $f'(x)$ exists at least throughout some deleted neighborhood of c: $0 < |x - c| < h$. If the sign of $f'(x)$ changes from

$$+ \text{ for } x < c \quad \text{to} \quad - \text{ for } x > c,$$

then f has a relative maximum at c. If the sign of $f'(x)$ changes from

$$- \text{ for } x < c \quad \text{to} \quad + \text{ for } x > c,$$

then f has a relative minimum at c.

The test applies whether the derivative is zero at c or fails to exist there. However, the test does not apply to a discontinuous function such as

$$f(x) = \begin{cases} 1/x^2 & \text{for} \quad x \neq 0, \\ 0 & \text{for} \quad x = 0, \end{cases}$$

for which

$$f'(x) = -2/x^3 \quad \text{for} \quad x \neq 0.$$

Here the slope is positive for $x < 0$ and negative for $x > 0$, but the function (as defined) has a minimum (and not a maximum) at the critical point $x = 0$. (See Fig. 5–9.)

If the domain of the given function is a closed interval, or is closed at one end, the endpoint(s) should always be considered, as is seen in the following example.

Example 1. A straight fence 80 yards long stands on a ranch. It is to be left standing, and a part or all of it is to be used in forming a rectangular corral. A roll of fencing 240 yards long will be used for the other three sides and only for these additional sides, the fourth side being composed only of fencing already standing. Find the maximum area that can be so enclosed.

MAXIMA AND MINIMA

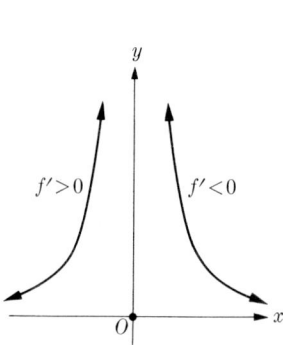

FIG. 5-9. Portion of the graph of
$$y = \begin{cases} 1/x^2 & \text{for } x \neq 0, \\ 0 & \text{for } x = 0. \end{cases}$$

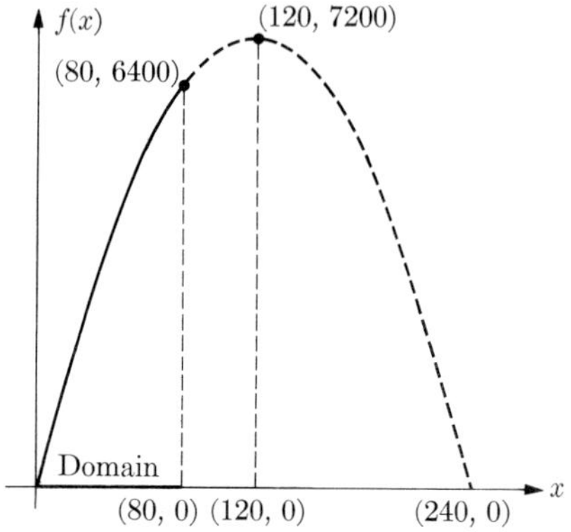

FIG. 5-10. Portion of the graph of $f(x) = 120x - x^2/2$.

Solution. Let $x =$ length of fence parallel to existing fence. Then the length of fence perpendicular to this is $(240 - x)/2$, and the enclosed area is

$$A = f(x) = x \left(\frac{240 - x}{2} \right) = 120x - \frac{x^2}{2}.$$

We have

$$f'(x) = 120 - x, \quad \text{and} \quad f'(x) = 0 \Rightarrow x = 120.$$

Since the graph of $f(x)$ is a parabola opening downward, it is apparent that $f'(120) = 0$ implies a maximum value exists at $x = 120$, but the use of 120 yards of fencing parallel to the 80 yards standing doesn't make sense. Is there a maximum area? Examination of the graph (Fig. 5–10) will make the result clear. In our problem, the domain of x is $0 < x \leq 80$, and obviously the maximum value occurs when $x = 80$, the right-hand endpoint of the domain.

EXERCISES 5-4

In Exercises 1 through 6, examine the functions for extrema and state the values of the extrema.

1. a) $f(x) = x^2 - 8$, $-\infty < x < \infty$ b) $f(x) = x^2 - 8$, $|x| \leq 2$
   c) $f(x) = x^2 - 8$, $|x| < 2$ d) $f(x) = |x^2 - 8|$, $-\infty < x < \infty$
2. a) $f(x) = x^3 + x^2 - x - 1$, $-\infty < x < \infty$
   b) $f(x) = x^3 + x^2 - x - 1$, $-2 \leq x \leq \frac{3}{2}$
   c) $f(x) = |x^3 + x^2 - x - 1|$, $x \geq 0$
3. $f(x) = 1/x + x$, $x \neq 0$ and $f(0) = 0$
4. $y^3 = 2x^2 + 1$

5. $f(x) = x - [x]$, where $[x]$ denotes the greatest integer in x.
6. $f(x) = x^2 + 4\cos x$
7. Suppose that in Example 1, the new fencing material may be used to form three sides and also a part of the fourth side. What is the maximum area thus enclosed? How does it compare with that of Example 1?
8. The diagram at the right shows the wire frame for a rectangular box with a square base and a volume V. What dimensions will require a minimum amount of wire? Carry out the computation for $V = 88$ in^3.

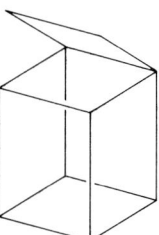

9. For the box in Exercise 8, what dimensions would yield a maximum volume if the total length of the wire is L in. per box? Compute for $L = 132$ in.
10. Repeat Exercise 9 for L units of wire used for twelve edges only (open box, no cover). Compute for $L = 72$ in.

6
CONTINUOUS FUNCTIONS. EXISTENCE OF MAXIMUM AND MINIMUM VALUES

6-1 INTRODUCTION, DEFINITIONS, AND EXAMPLES

What is a continuous function? Can a function be continuous at just one point? Can a function be continuous at just the rationals, or just the irrationals? These are some of the questions we study in this chapter. But these are only descriptive features and there are more important matters.

Much of elementary calculus is concerned with functions like polynomials, rational functions, trigonometric functions and their inverses, exponential and logarithmic functions, and combinations of these. In particular, in calculus we study these functions to discover where they have derivatives and to provide easy rules for computing those derivatives. Because existence of the derivative at a point implies continuity at that point, we shall not enter into a separate program of proving the continuity of these functions. But such a program would be relatively easy for polynomials and other rational functions, and would involve the following sequence of theorems.

Theorem 6–1. A constant function, $f(x) = c$, is everywhere continuous.

Theorem 6–2. The identity function $f(x) = x$, is everywhere continuous.

Theorem 6–3. If f and g are functions that are continuous at a, then $f + g$ and $f \cdot g$ are continuous at a.

Theorem 6–4. Polynomials are everywhere continuous.

Theorem 6–5. If f is continuous at a, and $f(a) \neq 0$, then $1/f$ is continuous at a.

Theorem 6–6. If f and g are polynomials, the rational function f/g is continuous at every value of x such that $g(x) \neq 0$.

Proofs of these theorems will be postponed. Theorem 6–4 follows directly from Theorems 6–1 through 6–3, and Theorem 6–6 from Theorems 6–4 and 6–5. The proofs require only moderate skill in handling inequalities and absolute values in addition to an understanding of the definition of continuity. There are, however, three other theorems that are used in elementary calculus whose proofs are much more difficult. A substantial portion of this chapter is devoted to studying and proving the first of the following three theorems. Unfamiliar terms that appear in the theorems will be explained later, and Theorems C-II and C-III will be proved in Chapter 7.

Theorems C-I, C-II, and C-III. Let f be a function whose domain D is the closed bounded interval $a \leq x \leq b$, and suppose f is continuous

CONTINUOUS FUNCTIONS.   MAXIMUM AND MINIMUM VALUES

at each point of D. Then:

C-I. f has a maximum and a minimum on D.

C-II. If c is any number between $f(a)$ and $f(b)$, then there exists at least one value of x in D such that $f(x) = c$.

C-III. f is uniformly continuous on D.

Let us now return to the opening question: What is a continuous function?

Continuity

Continuity of a function is a local property. Roughly, a function f is continuous at a point a in its domain provided f maps x's near a onto y's near $f(a)$. Nearness is measured by absolute values, and the following ϵ, δ (epsilon, delta) definition is widely used.

> **Definition 1.** Let f be a real-valued function of a real variable with domain D. Let a belong to D. Then f is said to be *continuous* at a provided that to each $\epsilon > 0$ there corresponds $\delta > 0$ such that
>
> $$|f(x) - f(a)| < \epsilon \quad \text{whenever} \quad |x - a| < \delta \text{ and } x \in D. \qquad (1)$$

Remark 1. The δ of inequality (1) may depend, and usually does depend, upon the particular function f, upon a, and upon ϵ. Thus we do not set out to find δ until the function f, the point a, and $\epsilon > 0$ are given. But when these are given, we can represent the domain of f on the x-axis and the range on the y-axis and try to find a δ-neighborhood of a such that when x is in this neighborhood, and in the domain of f, the image $f(x)$ is in the ϵ-neighborhood of $f(a)$. (See Fig. 6–1.)

Example 1. (This example will include proofs of Theorems 6–1 and 6–2 as special cases.) Suppose f is a linear function:

$$f(x) = mx + b, \quad -\infty < x < \infty. \qquad (2)$$

Then $f(x) - f(a) = (mx + b) - (ma + b) = m(x - a)$. Therefore, if $\epsilon > 0$, we have

$$|f(x) - f(a)| = |m(x - a)| = |m|\,|x - a| < \epsilon,$$

provided that either

$$\text{i) } m = 0 \qquad (3a)$$

or

$$\text{ii) } |x - a| < \frac{\epsilon}{|m|} \quad \text{if} \quad m \neq 0. \qquad (3b)$$

6–1 INTRODUCTION, DEFINITIONS, AND EXAMPLES

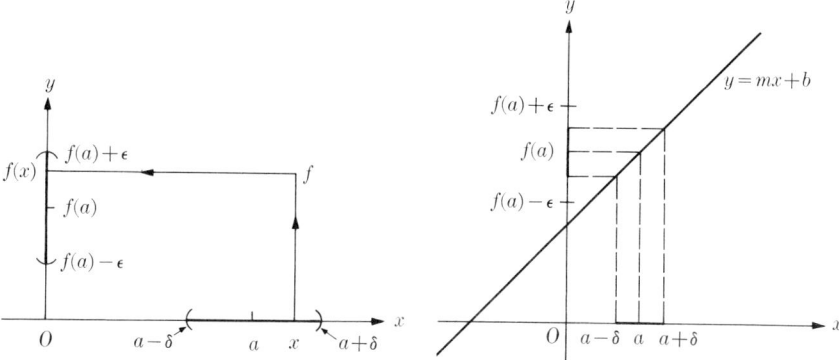

FIG. 6–1. If $x \in N_\delta(a)$ then $f(x) \in N_\epsilon(f(a))$.

FIG. 6–2. If $f(x) = mx + b$, and $\epsilon > 0$, take $\delta = \epsilon/(1 + |m|)$. (Here $m = 1$, $\delta = \epsilon/2$.)

In the first case any positive δ will do. But we can absorb this special case and satisfy (3b) as well by taking

$$\delta = \frac{\epsilon}{1 + |m|}. \qquad (4)$$

Thus, for the function f of Eq. (2), if a is any real number and ϵ is positive, Eq. (4) yields a positive number δ such that

$$|f(x) - f(a)| < \epsilon \quad \text{whenever} \quad |x - a| < \delta.$$

Therefore, the linear function (2) is continuous at a for every a. More briefly, we say "f is continuous everywhere." Figure 6–2 illustrates the relation (4) between δ and ϵ for $m = 1$.

Observe that Theorem 6–1 is the special case $m = 0$, and Theorem 6–2 the case $m = 1$, $b = 0$.

Example 2. Suppose f is the greatest-integer function

$$f(x) = [x]. \qquad (5)$$

There are two cases to consider:

i) $a = n$, an integer;
ii) a not an integer.

In the first case, $f(a) = n$ and

$$f(x) = \begin{cases} n - 1 & \text{for } x < a, \ |x - a| < 1, \\ n & \text{for } x \geq a, \ |x - a| < 1. \end{cases}$$

Therefore, no matter how small the positive δ might be, there would be values of x within δ-distance of a for which $|f(x) - f(a)| = 1$, and con-

CONTINUOUS FUNCTIONS. MAXIMUM AND MINIMUM VALUES

dition (1) would fail if $\epsilon < 1$. Hence, f is *discontinuous* at $a =$ an integer. (See Fig. 1–15.)

In the second case, a not an integer, let $n = [a]$. Then $n < a < n + 1$, and the two numbers

$$\delta_1 = a - n, \qquad \delta_2 = (n + 1) - a \tag{6a}$$

are both positive. Let

$$\delta = \min(\delta_1, \delta_2) = \min(a - [a], [a] + 1 - a). \tag{6b}$$

Then $\delta > 0$, and if $|x - a| < \delta$, then x also lies between $[a]$ and $[a] + 1$, so that

$$f(x) = [x] = [a] = f(a) \quad \text{and} \quad |f(x) - f(a)| = 0.$$

Therefore, the greatest-integer function is *continuous* at every $a \neq$ an integer, because for any such a and $\epsilon > 0$, Eq. (6b) yields $\delta > 0$ such that

$$|f(x) - f(a)| = 0 < \epsilon \quad \text{whenever} \quad |x - a| < \delta.$$

Example 3. Let $f(x) = \sqrt{x}$, $x \geq 0$. We consider the cases $a = 0$ and $a > 0$ separately.

If $a = 0$, we have $|f(x) - f(a)| = |\sqrt{x} - \sqrt{0}| = \sqrt{x}$, and if $\epsilon > 0$, then $\sqrt{x} < \epsilon$ provided $0 \leq x < \epsilon^2$. Therefore, with $a = 0$ and $\epsilon > 0$, let us take $\delta = \epsilon^2$ to satisfy the continuity requirement. The function is continuous at $a = 0$.

If $a > 0$, we have

$$|f(x) - f(a)| = |\sqrt{x} - \sqrt{a}| = \frac{|x - a|}{\sqrt{x} + \sqrt{a}} \leq \frac{|x - a|}{\sqrt{a}}.$$

If $\epsilon > 0$, and $\delta = \epsilon\sqrt{a}$, then $\delta > 0$, and

$$|x - a| < \delta \Rightarrow \frac{|x - a|}{\sqrt{a}} < \frac{\epsilon\sqrt{a}}{\sqrt{a}} = \epsilon$$
$$\Rightarrow |f(x) - f(a)| < \epsilon.$$

Hence, the square-root function is continuous everywhere on its domain.

Example 4. Let

$$f(x) = \begin{cases} 0 & \text{if } x \text{ is rational,} \\ 1 & \text{if } x \text{ is irrational.} \end{cases}$$

Since rational and irrational numbers occur in every interval of real numbers, for any real a and any $h > 0$, the neighborhood $N_h(a)$ contains both rational and irrational numbers. Hence no matter how small h may be, we can arrange to have x rational or irrational according as a is the opposite. Given any real number a and any positive $\epsilon < 1$, $|f(x) - f(a)| = 1 > \epsilon$ for some x in the δ-neighborhood of a, for any $\delta > 0$. This function is everywhere discontinuous.

Example 5. *Ruler function.* Let

$$f(x) = \begin{cases} 0 & \text{if } x \text{ is irrational,} \\ 1/q & \text{if } x \text{ is the rational number } p/q \text{ in lowest terms, } q > 0. \end{cases}$$

Thus

$$f(0) = f(1) = 1, \quad f(\tfrac{1}{2}) = f(\tfrac{3}{2}) = \tfrac{1}{2}, \quad f(\tfrac{1}{3}) = f(\tfrac{2}{3}) = \tfrac{1}{3},$$
$$f(\tfrac{1}{4}) = f(\tfrac{3}{4}) = \tfrac{1}{4}, \quad f(\tfrac{1}{8}) = f(\tfrac{3}{8}) = f(\tfrac{5}{8}) = f(\tfrac{7}{8}) = \tfrac{1}{8},$$

and so on. If we plot some of the points with rational coordinates between 0 and 1 and draw a line segment of length $f(x)$ up from x, the resulting graph (Fig. 6–3) resembles the marks on a ruler, hence the name *ruler function*. (Shift this portion to the right and to the left by $1, 2, 3, \ldots$ units to get the rest of the graph.)

Where is the ruler function continuous? Certainly not at any rational value of a. For, if a is rational and $f(a) = 1/q$, then $\epsilon = 1/(2q)$ is a positive number for which no corresponding δ exists, because there are irrational values of x arbitrarily near a.

Next consider an irrational value of a. Then $f(a) = 0$, and if $\epsilon > 0$, the condition

$$|f(x) - f(a)| < \epsilon$$

becomes

$$|f(x)| < \epsilon. \tag{7}$$

Inequality (7) holds for all irrational values of x, because $f(x) = 0$ for them. If x is the rational number p/q in lowest terms, then (7) is the same as

$$\frac{1}{q} < \epsilon \quad \text{or} \quad q > \frac{1}{\epsilon}. \tag{8}$$

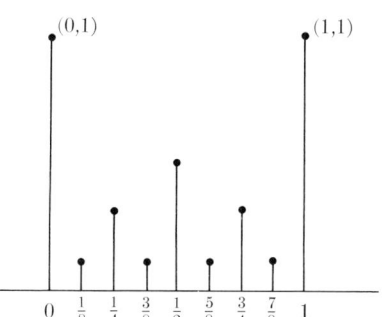

FIG. 6–3. Portion of the graph of the ruler function. (The vertical line segments are only to guide the eye; they are not part of the graph, except for their upper endpoints.)

Inequality (8) is satisfied when q is large enough. Can we find $\delta > 0$ so as to guarantee that q must be that large when x is within δ-distance of a? Yes, in the following way. Consider the positive integers

$$1, 2, \ldots, [1/\epsilon] \tag{9}$$

CONTINUOUS FUNCTIONS. MAXIMUM AND MINIMUM VALUES

that do *not* satisfy (8). (If by chance the given ϵ is greater than 1, skip the list (9) and take $\delta = 1$, because (7) will be satisfied for every x.) There is only a finite number of integers in the list (9). If q is any of these, there are at most q choices for p that will satisfy

$$a - \frac{1}{2} < \frac{p}{q} < a + \frac{1}{2}.$$

Thus, if we restrict attention to a neighborhood of radius $\frac{1}{2}$ centered at a, we find only a finite number (at most n^2, where $n = [1/\epsilon]$) of rational numbers p/q which have denominators that *fail* to satisfy inequality (8). If we represent these particular rational numbers by $r_1, r_2, \ldots, r_N$, $N \leq n^2$, then

$$|a - r_1|, |a - r_2|, \ldots, |a - r_N| \qquad (10)$$

are all positive, because the r's are rational and a is irrational. Let δ be the *minimum* of these numbers* (10). Then $\delta > 0$, and any rational number p/q that is *less* than δ-distance from a cannot have its denominator in the list (9), so $q > 1/\epsilon$. Therefore,

$$|x - a| < \delta \Rightarrow \begin{cases} x \text{ irrational,} & \text{or} \\ x \text{ rational,} & p/q, \text{ and } q > 1/\epsilon, \end{cases}$$

$$\Rightarrow |f(x) - f(a)| = \begin{cases} 0, & \text{or} \\ 1/q \end{cases} < \epsilon.$$

Hence the ruler function is continuous at every irrational a.

Remark. The idea behind the argument following (8) is just this: if we confine our attention to those rational numbers whose denominators are so small that they fail to satisfy the continuity requirement, there is a minimum positive distance from a to any of them. Rationals that are less than this minimum distance from a must have large denominators, and these satisfy the requirement. Since we are only interested in the behavior of the function near a, we lose nothing by restricting our attention to the interval from $a - \frac{1}{2}$ to $a + \frac{1}{2}$.

EXERCISES 6–1

1. Let f be the function such that

$$f(x) = \begin{cases} x & \text{if } x \text{ is rational,} \\ 0 & \text{if } x \text{ is irrational.} \end{cases}$$

Where is f continuous? Prove your assertion.

---

* See Exercises 8 and 9 regarding the existence of the maximum or minimum of a finite set.

2. Let f be the function
$$f(x) = \begin{cases} 2x & \text{for } 0 \le x \le 1, \\ c - 2x & \text{for } 1 < x \le 2. \end{cases}$$
For what value(s) of c is f continuous at 1? Sketch the graph of f for such a value of c, and for one other value of c.

3. Draw the graph of $y = -2|x - 1| + 2$ for $0 \le x \le 2$. Compare with Exercise 2.

4. Draw the graph of the neighborhood or deleted neighborhood of x for which the following inequalities hold.
   a) $|x - 2| < \frac{1}{2}$
   b) $0 < |x - 2| < c/2$
   c) $|x + c| < c/2$
   d) $0 < |x + c| < c/2$

5. Let the function f be defined by $f(x) = 2x + 1$.
   a) For what values of x will $f(x)$ differ by no more than $\frac{1}{2}$ from $f(1)$?
   b) Find the values of $f(0.5), f(0.6), f(0.7), f(0.75), f(0.8), f(1.1), f(1.2), f(1.25), f(1.3), f(1.5)$.
   c) Draw a graph to show the result of (b).
   d) Find a neighborhood, centered at $c = 2$, such that for any $\epsilon > 0$, $|f(x) - f(2)| < \epsilon$.
   e) Repeat (d) above if $f: x \to (2x^2 - 3x - 2)/(x - 2)$.

6. In the discussion of the ruler function, Example 5, suppose we replace $f(x) = 1/q$ by $f(x) = 1/q^2$ when $x = p/q$ in the lowest terms. How does this change affect the formulas (8), (9), (10)? Where is this new function continuous, and where is it discontinuous?

7. Sketch graphs of the following functions:
   a) $f(x) = x - [x]$
   b) $f(x) = [2x - 3]$
   c) $f(x) = 2[x] - 3$

8. A set S of real numbers is said to have a *maximum* if S contains a number c such that $x \le c$ for all x in S. It is obvious that the maximum of a set S that contains only one number is that number. Prove that a set S that contains n numbers has a maximum if n is any positive integer.

9. By analogy with Exercise 8, define *minimum* for a set S, and prove that if S has n elements, where n is any positive integer, then it has a minimum.

10. Give an example of a nonempty set of real numbers that has neither a maximum nor a minimum.

11. Let D be the open interval $0 < x < 1$. For any a in D show that there is a neighborhood $N_h(a)$ that is completely contained in D. Illustrate with a sketch.

12. Let $f(x) = 1/x$, $x \ne 0$. Show that f is continuous at every $a \ne 0$.

13. Prove part 1 of Theorem 6–3 (the sum of two continuous functions is continuous).

14. Prove part 2 of Theorem 6–3 (the product of two continuous functions is continuous).

15. Prove Theorem 6–5 (at a point where a continuous function is not zero, its reciprocal is continuous).

6-2 EXISTENCE OF A MAXIMUM OR MINIMUM OF A FUNCTION. EXAMPLES

As previously mentioned, the functions we consider are to be real-valued functions of a real variable. If D is the domain of such a function and c is a number in D such that

$$f(x) \leq f(c) \qquad \text{for all} \quad x \text{ in } D, \quad (1)$$

we say that f has a *maximum* at c. Similarly, if d is a number in D such that

$$f(x) \geq f(d) \qquad \text{for all} \quad x \text{ in } D, \quad (2)$$

then f has a *minimum* at d. (See Fig. 6-4.)

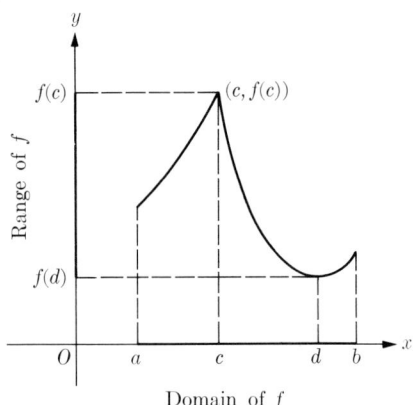

FIG. 6-4. Graph shows a maximum of f at c and a minimum at d.

There are both theoretical and applied problems in calculus that deal with the existence and location of the maximum or minimum of a function. In the previous section we stated Theorem C-I, which says that if the domain of f is a closed bounded interval and f is everywhere continuous on that domain, then f *has* both a maximum and a minimum. Before we begin to prove this theorem let us look at some examples where the conclusion does not apply, because the domain is not closed, or not bounded, or the function is not everywhere continuous on its domain.

Example 1. Let

$$f_1(x) = \begin{cases} \frac{1}{2} & \text{if } x = 0, \\ 1/x & \text{if } 0 < x \leq 1. \end{cases}$$

The domain of f_1 is the closed interval $0 \leq x \leq 1$, but the function is discontinuous at $x = 0$ (see Fig. 6-5) and it has no maximum. The minimum is $f_1(0) = \frac{1}{2}$.

Example 2. Let

$$f_2(x) = x, \qquad 0 < x < 1.$$

The domain is the open interval $0 < x < 1$. This function is everywhere continuous on its domain. But it has neither a maximum nor a minimum, because for any value attained by the function, there is both a larger value for a slightly larger x and a smaller value for a smaller x. The graph is shown in Fig. 6-6.

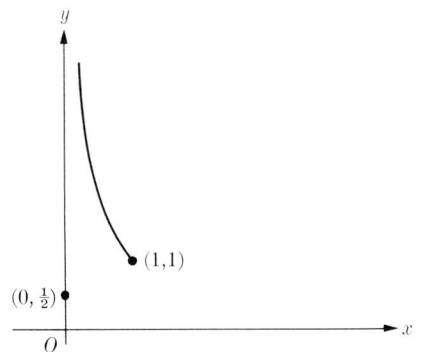

FIG. 6–5. Portion of the graph of Example 1.

FIG. 6–6. Neither the origin nor the point (1, 1) belongs to the graph.

Example 3. Let
$$f_3(x) = \begin{cases} \frac{3}{2} & \text{if } x = 0, \\ x & \text{if } 0 < x < 1, \\ \frac{1}{2} & \text{if } x = 1. \end{cases}$$

The domain of f_3 is the closed interval $0 \leq x \leq 1$, but the function is discontinuous at $x = 0$ and at $x = 1$. The maximum is $f_3(0) = \frac{3}{2}$; but there is no minimum. The graph is shown in Fig. 6–7. By changing to a value $-\frac{1}{2}$, say, at $x = 1$, we could get a discontinuous function that has both a maximum and a minimum, even though it is discontinuous at the endpoints of its domain.

Example 4. Let
$$f_4(x) = x^2, \quad -\infty < x < \infty.$$

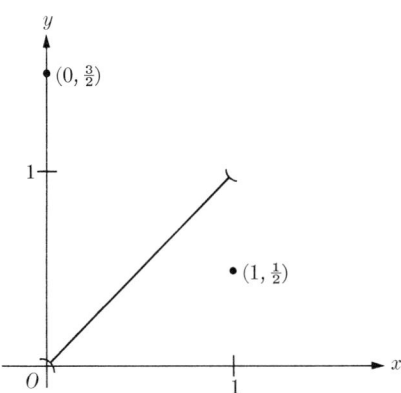

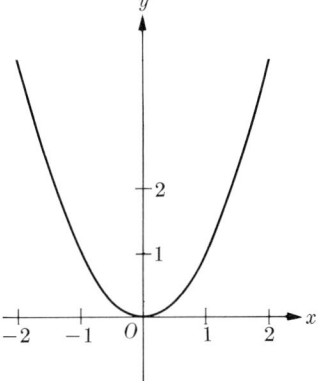

FIG. 6–7. $f_3(0) = \frac{3}{2}$, $f_3(1) = \frac{1}{2}$, $f_3(x) = x$ if $0 < x < 1$.

FIG. 6–8. Portion of the graph of $f_4(x) = x^2$.

CONTINUOUS FUNCTIONS. MAXIMUM AND MINIMUM VALUES   6–3

This function is everywhere continuous. It has a minimum value $f_4(0) = 0$, but no maximum. (See Fig. 6–8.)

EXERCISES 6–2

1. Give an example of a function with domain $0 \leq x \leq 1$ which:
   a) Is discontinuous only at $x = \frac{1}{2}$, has a maximum at $x = 0$ and a minimum at $x = 1$. Sketch its graph.
   b) Is discontinuous only at $x = \frac{1}{2}$ and has neither a maximum nor a minimum. Sketch its graph.
2. Give an example of a function with domain $-\infty < x < \infty$ which:
   a) Is everywhere continuous and has both a maximum and a minimum. Sketch its graph.
   b) Is everywhere continuous and has a maximum at $x = 0$, but has no minimum. Sketch its graph.
   c) Is everywhere discontinuous and has both a maximum and a minimum. Describe its graph.
   d) Is everywhere discontinuous and has neither a maximum nor a minimum. Describe its graph.

6–3 BOUNDEDNESS AND EXISTENCE OF MAXIMUM AND MINIMUM. UPPER AND LOWER BOUNDS

As with sequences, a set S of real numbers is said to be *bounded above*, or bounded from above, if there exists a number M such that

$$x \leq M \text{ for every } x \text{ in } S.$$

Such a number M is called an *upper bound* for S. Thus for $f_2(x)$ as in Fig. 6–6, the range is the set

$$S = \{y : 0 < y < 1\}.$$

There are many upper bounds for this set: 1, $\sqrt{2}$, 2, π, 73. The *least upper bound* is 1. More generally, L is a *least upper bound* for a set S if and only if:

a) L is an upper bound for S, and
b) no number less than L is an upper bound for S.

Condition (b) implies that there cannot be more than one least upper bound for a given set. Thus when a least upper bound L of a set S exists, it is *the* unique least upper bound for S, and we write

$$L = \text{lub}(S).$$

6–3 BOUNDEDNESS AND EXISTENCE OF MAXIMUM AND MINIMUM

Similarly, a number l is a *lower bound* for a set S provided that

$$l \leq x \text{ for every } x \text{ in } S.$$

If l is a lower bound for S, then every number less than l is also a lower bound and there is no least lower bound. However, l is a *greatest lower bound* for S provided:

a) l is a lower bound for S, and
b) no number greater than l is a lower bound for S.

When a greatest lower bound exists, it is unique and is designated by

$$l = \text{glb}(S).$$

A set that is bounded both from above and from below is said to be *bounded*. A fundamental property of the system of real numbers is *completeness*, as set forth in the completeness postulate introduced in Section 2–4 and repeated here for convenience.

Completeness Postulate. If S is a bounded nonempty set of real numbers, there exist real numbers l and L such that

$$l = \text{glb}(S) \quad \text{and} \quad L = \text{lub}(S).$$

Remark 1. We sometimes paraphrase this by saying, "a bounded nonempty set of real numbers has a greatest lower bound and a least upper bound," but in this statement the verb "has" should not be interpreted as implying that the numbers glb (S) and lub (S) are members of S. For example, if S is an open interval $a < x < b$, then glb $(S) = a$ and lub $(S) = b$, and neither is a member of S.

Remark 2. The *rational* number system is not complete. For example, suppose that S is the set of all positive rational numbers whose squares are less than 2:

$$S = \{x : x > 0, x \text{ rational}, x^2 < 2\}.$$

Then one can prove (with a little effort) that no *rational* number is a least upper bound for S. In fact, lub $(S) = \sqrt{2}$, and it is well known that $\sqrt{2}$ is irrational.

Maximum and Minimum

If a function has a maximum, then that maximum is the lub of the range of the function. Thus for $f_3(x)$ as in Fig. 6–7, the range is

$$\{\tfrac{3}{2}\} \cup \{y : 0 < y < 1\},$$

and its least upper bound is $\frac{3}{2}$. If the function has a *minimum*, then its range is bounded from below and the *greatest lower bound* of the range is the minimum value of the function. Therefore, the range of a function that has both a maximum and a minimum is bounded:

(existence of max and min) $\Rightarrow$ (boundedness).

In other words, *boundedness* is a necessary condition for the existence of a maximum and a minimum.

Theorem C-I of Section 6–1 states: Let f be a function whose domain D is the closed bounded interval $a \leq x \leq b$, and suppose f is continuous at each point of D. Then f has a maximum and a minimum on D. The proof follows this pattern: (i) we show that under the hypotheses of that theorem, the range of f is bounded; and (ii) the least upper bound and greatest lower bound of the range are in it. When we have proved (i) and (ii), we shall have proved the theorem.

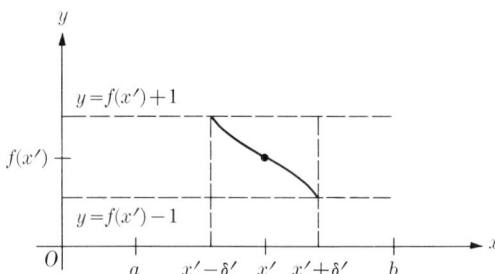

FIGURE 6–9

From now on, until we have completed the proof of Theorem C-I, we assume that a and b are real numbers, $a < b$, and f is continuous on $a \leq x \leq b$. To prove that the range of f is bounded (which we abbreviate to "f is bounded") on the entire domain, we might start at a, get an interval from a to $a + \delta_1$ in which $f(x)$ stays within 1 unit of $f(a)$; choose an x_1 between a and $a + \delta_1$ and get an interval from x_1 to $x_1 + \delta_2$ in which $f(x)$ stays within 1 unit of $f(x_1)$; and keep working toward the right, choosing an x_{k+1} between x_k and $x_k + \delta_{k+1}$, then getting an interval x_{k+1} to $x_{k+1} + \delta_{k+2}$ in which $f(x)$ stays within 1 unit of $f(x_{k+1})$. Proceeding in this fashion, we get overlapping intervals covering that part of the domain from a to $x_k + \delta_{k+1}$ in $(k + 1)$ steps, and for all x in this part of the domain, $f(x)$ stays within $k + 1$ units of $f(a)$. If we could be certain that the process would carry us all the way to the right endpoint b, in a finite number of steps, say M steps, we could conclude that

$$f(a) - M \leq f(x) \leq f(a) + M \quad \text{for all } x \text{ in } a \leq x \leq b,$$

and boundedness of the range of f would thereby be established. This technique could be applied to a specific function f, but it is not quite adequate for a general proof. Part of the trouble lies in the vagueness of the choice of $x_1, x_2, \ldots$ and of the associated δ's.

A modification of the procedure just described is more successful. Instead of trying to choose a finite number of points $a, x_1, x_2, \ldots$ and associating δ's in the way described, we construct a δ-neighborhood for *every* x in the domain of f, as follows:

By the definition of continuity, with $\epsilon = 1$, we know that for each x' in the domain $a \leq x' \leq b$, there exists a positive number $\delta' = \delta(x')$ such that

$$|f(x) - f(x')| < 1 \quad \text{if} \quad |x - x'| < \delta' \quad \text{and} \quad a \leq x \leq b. \tag{1}$$

(See Fig. 6–9.) The open interval $x' - \delta' < x < x' + \delta'$ is centered at x', so it certainly covers x', because $\delta' > 0$. Now imagine *all* of the points x' from a to b, each covered with such an associated open interval. Even the endpoints a and b are covered with their associated intervals

$$a - \delta(a) < x < a + \delta(a) \quad \text{and} \quad b - \delta(b) < x < b + \delta(b).$$

In the former of these, $f(x)$ differs from $f(a)$ by less than 1 unit whenever $|x - a| < \delta(a)$ and x is in the domain; that is,

$$\text{if} \quad a \leq x < a + \delta(a), \quad \text{then} \quad |f(x) - f(a)| < 1. \tag{2a}$$

Similarly,

$$\text{if} \quad b - \delta(b) < x \leq b, \quad \text{then} \quad |f(x) - f(b)| < 1. \tag{2b}$$

The union of *all* these subintervals

$$x' - \delta(x') < x < x' + \delta(x'), \quad a \leq x' \leq b, \tag{3}$$

provides an "open covering" of the domain of f. Because that domain, $a \leq x \leq b$, is a *bounded, closed* interval, we can apply a remarkable theorem of Heine and Borel* (see Section 6–4), which assures the existence of a *finite number* of the given collection of intervals $x' - \delta(x') < x < x' + \delta(x')$ which also *cover* $a \leq x \leq b$. Suppose that the number of intervals in this finite covering is N, and that their centers are $x_1, x_2, \ldots, x_N$ in the order

$$a \leq x_1 < x_2 < \cdots < x_N \leq b.$$

---

* This theorem was first stated by the German Heinrich Edward Heine (1821–1881) and then generalized in 1894 by the Frenchman Emile Borel (1871–1956).

CONTINUOUS FUNCTIONS. MAXIMUM AND MINIMUM VALUES

Then, for any x, $a \leq x \leq b$, there is an x_k, $1 \leq k \leq N$, such that
$$|x - x_k| < \delta(x_k),$$
and therefore,
$$|f(x) - f(x_k)| < 1.$$

The *finite* set of numbers
$$|f(x_1)|, |f(x_2)|, \ldots, |f(x_N)|$$
has a maximum, say M. Hence
$$|f(x)| < M + 1 \qquad \text{for all} \quad x, \qquad a \leq x \leq b.$$

Thus, when we have proved the Heine-Borel theorem in the next section, we shall also have completed the proof of the following theorem.

Theorem 6–7. Let f be a real-valued function, continuous everywhere on its domain $a \leq x \leq b$, where a and b are real numbers. Then the range of f is bounded.

Assuming Theorem 6–7, let us see how it can be applied to give Theorem C-I on the existence of a maximum and a minimum.

Theorem C-I. Let f be continuous on its domain
$$D = \{x : a \leq x \leq b\}. \tag{4}$$

Then f has a maximum and a minimum on D.

Proof. (Assuming Theorem 6–7.) Let R_f denote the range of f. Then, by Theorem 6–7, R_f is bounded. Let
$$m = \text{glb } R_f \qquad \text{and} \qquad M = \text{lub } R_f. \tag{5}$$
Then, for all x in D,
$$m \leq f(x) \leq M. \tag{6}$$

We wish to show that the numbers m and M are in the range of f, or, to put it another way, there exist values of x, say x_1 and x_2, in D, such that
$$f(x_1) = m \tag{7a}$$
and
$$f(x_2) = M. \tag{7b}$$

Equations (7a, b), together with the inequalities (6), will then show that m is the minimum value of $f(x)$ and M is its maximum.

We prove (7a) indirectly. [In Exercise 11 the reader is asked to give the corresponding proof for (7b).] To this end, assume that

$$f(x) \neq m \qquad \text{for all } x \text{ in } [a, b]. \tag{8}$$

The first inequality in (6) then says that

$$f(x) > m \qquad \text{for} \qquad a \leq x \leq b. \tag{9}$$

Accordingly, the function g defined by

$$g(x) = \frac{1}{f(x) - m} \qquad \text{for} \qquad a \leq x \leq b \tag{10}$$

is continuous and positive valued. Now apply Theorem 6–7 to this function; because it is continuous, it is bounded. In particular, there exists a positive number k such that

$$g(x) < k \qquad \text{for} \qquad a \leq x \leq b. \tag{11}$$

Since $g(x)$ and k are both positive, inequality (11) implies that

$$\frac{1}{g(x)} > \frac{1}{k}. \tag{12}$$

Combining Eq. (10) and this last inequality, we get

$$f(x) > m + \frac{1}{k}. \tag{13}$$

But inequality (13) contradicts the definition of m as the *greatest* lower bound of the range of f; it says that there is a greater lower bound, namely $m + (1/k)$. Since this is false, the assumption (8) cannot be true. Therefore, $f(x) = m$ for some value of x in $[a, b]$. Q.E.D.

EXERCISES 6–3

For each of the following functions, find: (a) the range, (b) the maximum, (c) the minimum. (d) Draw the graph.

1. $f(x) = x - x^3$, $-1 \leq x \leq 1$
2. $f(x) = x + |x|$, $-1 \leq x \leq 1$
3. $f(x) = x^4 - 2x^2 - 3$, $-1 \leq x \leq 2$
4. $f(x) = |x^2 - 4|$, $-2 \leq x \leq 1$
5. $f(x) = \dfrac{x}{x^2 + 1}$, $-3 \leq x \leq 3$

CONTINUOUS FUNCTIONS, MAXIMUM AND MINIMUM VALUES

6. $f(x) = 2 + 3 \sin x$, $0 \le x \le 2\pi$
7. $f(x) = x + \sin x$, $-\pi/2 \le x \le 2\pi$
8. $f(x) = \sin x + \cos x$, $0 \le x \le \pi$
9. $f(x) = \dfrac{\sin x}{1 + \cos x}$, $|x| \le \pi/2$
10. $f(x) = x + (1/x)$, $\frac{1}{4} \le |x| \le 2$
11. Complete the proof of Theorem C-I, assuming Theorem 6–7, by establishing Eq. (7b). *Suggestion:* If $f(x) < M$ for $a \le x \le b$, the function $h(x) = 1/(M - f(x))$ is continuous and bounded on $[a, b]$.

6–4 THE HEINE-BOREL COVERING THEOREM

In Section 6–3 we referred to, and used, the Heine-Borel theorem. In this section we first look at some examples and then prove that theorem.

Example 1. Let S be the set

$$\left\{1, \frac{1}{2}, \frac{1}{3}, \ldots, \frac{1}{n}, \ldots\right\}.$$

For purposes of illustration, we consider a collection of neighborhoods $N_1, N_2, \ldots$, one neighborhood associated with each number in S as follows.

$N_1: \frac{3}{4} < x < \frac{5}{4}$, covering 1,

$N_2: \frac{1}{2} - \frac{1}{12} < x < \frac{1}{2} + \frac{1}{12}$, covering $\frac{1}{2}$,

$N_3: \frac{1}{3} - \frac{1}{24} < x < \frac{1}{3} + \frac{1}{24}$, covering $\frac{1}{3}$,

$\vdots$

$N_n: \dfrac{1}{n} - \dfrac{1}{2n(n+1)} < x < \dfrac{1}{n} + \dfrac{1}{2n(n+1)}$, covering $\dfrac{1}{n}$.

The neighborhood N_n centered at $1/n$ has radius

$$h = \frac{1}{2}\left(\frac{1}{n} - \frac{1}{n+1}\right) = \frac{1}{2n(n+1)}.\text{*}$$

It therefore contains $1/n$ but not $1/(n-1)$ or $1/(n+1)$. Now the union of all these neighborhoods,

$$N_1 \cup N_2 \cup N_3 \cup \cdots \cup N_n \cup \cdots,$$

is an open set that contains, or *covers*, S. We call the collection of sets $N_1, N_2, \ldots$ an *open covering* of S.

---

* Thus, we could write $N_n = N_h(1/n)$.

There are some technical terms to be defined.

Definition 2. Let S be a set of real numbers. The number c is said to be an *inner point* of S if some neighborhood $N_h(c)$ is a subset of S. If every point of S is an inner point of S, then S is said to be *open*.

Example 2. Let S be the real numbers $0 < x < 1$ (what we have called an *open interval*). Then S is an open set. For, if c belongs to S and

$$h = \min(c, 1 - c),$$

then $h > 0$ and $N_h(c)$ is a neighborhood of c that is completely contained in S.

Definition 3. *Open covering.* Let S be a set of real numbers. Let $\mathcal{C}$ be a collection of *open* sets $S_1, S_2, \ldots$, whose union contains S. Then $\mathcal{C}$ is called an *open covering* of S.

Remark. If S is a bounded set, with $\alpha = \text{glb}(S)$ and $\beta = \text{lub}(S)$, then the open interval $\alpha - 1 < x < \beta + 1$ is an open covering of S. Thus a bounded set always has an open covering consisting of a single open interval. Indeed, if S is an arbitrary set of real numbers, and S_1 is the set of *all* real numbers, then S_1 is an open covering of S. Or, with each rational number r, associate the open intervals $r - h < x < r + h$, where h is any of the rational numbers between 0 and 1, $0 < h < 1$. Thus for each r, we have infinitely many neighborhoods corresponding to the infinitely many possible values of h. And there are infinitely many values of r. The collection of all these neighborhoods $N_h(r)$, where r and h are rational and $0 < h < 1$, covers the set of *all* real numbers, and hence covers any set of real numbers. Thus, open coverings always exist. A given open covering may be a finite collection or an infinite collection of open sets. The Heine-Borel theorem is concerned with the question of extracting a finite open covering from a *given* open covering. In Example 1, above, if we omit any neighborhood $N_h(1/n)$, $h = 1/2n(n + 1)$, then the remaining open sets fail to cover the point $1/n$. Therefore, no finite collection from the given covering $N_1, N_2, \ldots$ will be a covering. In other words, from the given open covering in Example 1 it is not possible to extract a *finite* subcollection that still covers S.

Example 3. Let S be the set $\{0, 1, \frac{1}{2}, \ldots, 1/n, \ldots\}$. As in Example 1, cover $1/n$ with the interval $N_n = N_h(1/n)$, where $h = 1/2n(n + 1)$. The intervals $N_1, N_2, \ldots$ do not cover 0; so we include one more open set S_0 that contains 0. Then $S_0, N_1, N_2, \ldots$ form an open covering of S. From this open covering, we *can* extract a *finite* open covering, as follows. Since

0 is an *inner point* of S_0, there exists a positive number h_0 such that the interval $-h_0 < x < h_0$ is contained in S_0. This same interval also contains the numbers $1/n$ for every integer n that is greater than $1/h_0$. Let k be the greatest integer in $1/h_0$, $k = [1/h_0]$. Then $S_0, N_1, N_2, \ldots N_k$ is a *finite* open covering of S. (For example, if $h_0 = 0.003$, then $1/h_0 = \frac{1000}{3} = 333\frac{1}{3}$, $k = 333$, and the 334 sets $S_0, N_1, \ldots, N_{333}$ are an open covering of S. We get rid of all the rest of the sets $N_{334}, \ldots$ because the corresponding numbers $\frac{1}{334}, \frac{1}{335}, \ldots$ in S are already covered by the one open set S_0 which contains the interval $-0.003 < x < 0.003$.)

The reason that Example 3 differs from Example 1 is that the set S of Example 1 does not contain the limit point 0, but that of Example 3 does.

Example 4. Let S be the interval $0 \leq x \leq 1$. Any point x outside of S has a minimum distance from S, equal to $|x|$ if $x < 0$ and equal to $x - 1$ if $x > 1$. Any neighborhood about x with radius less than this minimum distance contains no points of S. Therefore, such a point x is an inner point of the complement of S. That is, the complement of S is an open set. We say that any set S is *closed* when its complement is *open*. Hence the interval $0 \leq x \leq 1$ is closed. Indeed, if $a < b$, the interval $a \leq x \leq b$ is closed.

Definition 4. A set is *closed* if its complement is open.

Theorem 6–8. Heine-Borel. Let S be a closed bounded interval $[a, b]$. Let $\mathcal{C}$ be an open covering of S. Then some *finite* collection of open sets in $\mathcal{C}$ covers S.

Proof. Imagine starting at the left-hand endpoint a and working toward the right. Say that the point x in $[a, b]$ is *accessible* if some *finite* collection of open sets from the given collection $\mathcal{C}$ covers all the points from a to x inclusive. If b is not accessible, we shall show that an impossible situation arises. This will imply that b is accessible, and the proof of the theorem will be complete.

Assume, therefore, that b is not accessible. Consider the set T of all points in $[a, b]$ that are not accessible. By hypothesis, T is nonempty, and it is bounded below by a. Hence, by the completeness postulate, there is a greatest lower bound for T, say u:

$$u = \text{glb } T. \tag{1}$$

Now u is a point in $[a, b]$ and is therefore in some open set, say G', belonging to the collection $\mathcal{C}$. Since u is an inner point of G', there is a positive number δ such that the neighborhood $N_\delta(u)$ is covered by G'. (See Fig. 6–10.) However, none of the points from a to $u - \delta/2$ is in T, because u is a lower bound for T. This means that all of those points were accessible. In particular, $u - \delta/2$ is accessible, so there exist sets $G_1, G_2, \ldots, G_m$, belonging to $\mathcal{C}$, which cover the interval $[a, u - \delta/2]$. Now add the set G'. The $m + 1$ sets $G_1, G_2, \ldots, G_m, G'$ cover the interval $[a, u + \delta]$, so all of these points are accessible. Hence, if T is nonempty, all of its points lie to the right of $u + \delta$. In other words, $u + \delta$ is a lower bound for T. But this contradicts the definition of u as the *greatest* lower bound of T. Therefore, T must be empty; all points of $[a, b]$ are accessible, including b. Q.E.D.

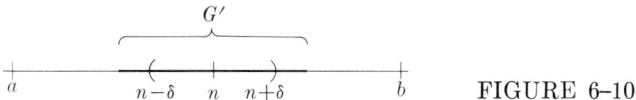

FIGURE 6–10

Example 5. Let $f(x) = x^2, 0 \leq x \leq 2$. If x_1 and x_2 are two real numbers that are nearly equal, we expect their squares also to be nearly equal. To be more explicit, suppose we fix a value of $x_1, 0 \leq x_1 \leq 2$, and try to find a neighborhood $x_1 - h < x < x_1 + h$ centered at x_1 such that $|x^2 - x_1^2| < 0.01$ for all x in that neighborhood. Factoring $x^2 - x_1^2$, we get

$$|x^2 - x_1^2| = |(x + x_1)(x - x_1)|$$
$$= |x + x_1| |x - x_1|$$
$$< 4|x - x_1|, \quad \text{if} \quad 0 \leq x, x_1 \leq 2.$$

Therefore, we shall have $|x^2 - x_1^2| < 0.01$, if

$$|x - x_1| < \frac{0.01}{4} = 0.0025, \quad 0 \leq x, \; x_1 \leq 2.$$

Thus, if $h = 0.0025$, and $0 \leq x_1 \leq 2$, we can be sure that x^2 differs by less than 0.01 from x_1^2, if x is in the neighborhood $N_h(x_1)$, and $0 \leq x \leq 2$. Let $\mathcal{C}$ be the collection of *all* these (open) neighborhoods $N_h(x_1)$, $0 \leq x_1 \leq 2$. There is one such neighborhood for each x_1 from 0 to 2, and there are infinitely many such values of x_1, so that there are infinitely many of these neighborhoods. $\mathcal{C}$ is an open covering of the closed bounded interval $S: 0 \leq x \leq 2$. By the Heine-Borel theorem there also exists a *finite* collection of the neighborhoods in $\mathcal{C}$ such that this finite collection also covers S. In this example we can easily pick out such a finite covering. For example, the 1600 neighborhoods centered at

$$\frac{h}{2}, h, \frac{3h}{2}, 2h, \frac{5h}{2}, \ldots, 800h, \quad h = 0.0025,$$

form such a finite open covering. For, the neighborhood of radius $h = 0.0025$

centered at	covers the interval
$\dfrac{h}{2}$	$-\dfrac{h}{2} < x < \dfrac{3h}{2}$
h	$0 < x < 2h$
$\dfrac{3h}{2}$	$\dfrac{h}{2} < x < \dfrac{5h}{2}$
$2h$	$h < x < 3h$
$\vdots$	$\vdots$
$800h = 2$	$799h < x < 801h$

The union of these neighborhoods completely covers S. (These neighborhoods overlap so that most points of S are actually covered twice. It is inevitable that some points will be covered twice. In any event, we have exhibited a *finite* collection, from the infinite collection $\mathcal{C}$, that covers S.)

EXERCISES 6–4

1. In Example 5 above, with $h = 0.0025$ and $r = h/2$, we used neighborhoods of radius h centered at $r, 2r, 3r, \ldots, 1600r$ to cover S. If we were to take $r = 0.9h$ (in place of $h/2$), and used neighborhoods of radius h centered at $r, 2r, \ldots, nr$, what value of n would suffice to give a complete covering of S?

2. (*Modification of Example* 5.) Suppose we want neighborhoods such that $|x^2 - x_1^2| < 0.2$ (instead of $|x^2 - x_1^2| < 0.01$), and $0 \leq x_1 \leq 5$, $0 \leq x \leq 5$. Find an appropriate value of h such that

$$|x^2 - x_1^2| < 0.2 \quad \text{if} \quad |x - x_1| < h \quad \text{and} \quad 0 \leq x, x_1 \leq 5.$$

Let $\mathcal{C}$ be the collection of all these neighborhoods $N_h(x_1)$ for $0 \leq x_1 \leq 5$. Describe a *finite* collection, from $\mathcal{C}$, that also covers the interval $0 \leq x \leq 5$. [*Hint:* Try intervals centered at $0, h, 2h, \ldots, 5.$]

7
UNIFORM CONTINUITY AND INTERMEDIATE VALUES

7–1 UNIFORM CONTINUITY

In the ϵ, δ-condition for continuity, we can sometimes find a δ that works for a given ϵ for the entire domain of the function. For example, for the linear function $f(x) = mx + b$ we found in Example 1 of Section 6–1 that for any $\epsilon > 0$ we could use

$$\delta = \frac{\epsilon}{1 + |m|}$$

and be sure that

$$|x_1 - x_2| < \delta \Rightarrow |f(x_1) - f(x_2)| < \epsilon.$$

The value of δ does not depend upon x_1 or x_2, and the same δ works throughout the entire domain of the function. This illustrates the notion of *uniform* continuity.

Definition 1. Uniform Continuity. Let f be a real-valued function defined on a domain D of real numbers. If to each positive number ϵ there corresponds a positive number δ such that

$$|f(x_1) - f(x_2)| < \epsilon \tag{1a}$$

whenever x_1 and x_2 are in D and

$$|x_1 - x_2| < \delta, \tag{1b}$$

we say that f is *uniformly* continuous on D.

Example 1. Let D be the interval

$$D: -2 \leq x \leq 2, \tag{2a}$$

and $f(x) = x^2$. Suppose $\epsilon > 0$. Condition (1a) becomes

$$|x_1^2 - x_2^2| = |x_1 + x_2| \, |x_1 - x_2| < \epsilon. \tag{2b}$$

If both x_1 and x_2 are in D, then

$$|x_1 + x_2| \leq |x_1| + |x_2| \leq 4,$$

so that

$$|x_1 + x_2| \cdot |x_1 - x_2| \leq 4|x_1 - x_2|.$$

Hence, the continuity requirement is satisfied if x_1 and x_2 are in D and

$$4|x_1 - x_2| < \epsilon,$$

or

$$|x_1 - x_2| < \epsilon/4.$$

Therefore, to any $\epsilon > 0$ there corresponds $\delta = \epsilon/4 > 0$ such that

$$|x_1^2 - x_2^2| < \epsilon$$

whenever

$$|x_1| \leq 2, \quad |x_2| \leq 2, \quad \text{and} \quad |x_1 - x_2| < \delta.$$

Thus the function

$$f(x) = x^2, \quad -2 \leq x \leq 2,$$

is *uniformly* continuous on its domain, because the same $\delta = \epsilon/4$ works for all values of x in the domain.

Example 2. Suppose $f(x) = x^2$, as in Example 1; but now let the domain be the set of all real numbers,

$$D = \{x: -\infty < x < +\infty\}.$$

Let $\epsilon > 0$ be given, and suppose we try to find $\delta > 0$ to satisfy the requirement

$$|x_1^2 - x_2^2| = |x_1 + x_2| \cdot |x_1 - x_2| < \epsilon \tag{3a}$$

whenever

$$|x_1 - x_2| < \delta. \tag{3b}$$

Since D is unbounded, the factor $|x_1 + x_2|$ is unbounded. No matter what positive number δ may be, we can find x_1 and x_2, within δ-distance of one another, but both so large that (3a) is violated. For example, take

$$x_1 - x_2 = \frac{\delta}{2}$$

and

$$x_2 = \frac{2\epsilon}{\delta}, \quad x_1 = \frac{2\epsilon}{\delta} + \frac{\delta}{2}.$$

Then

$$|x_1 - x_2| < \delta,$$

but

$$|x_1^2 - x_2^2| = \left(\frac{4\epsilon}{\delta} + \frac{\delta}{2}\right)\left(\frac{\delta}{2}\right) > \frac{4\epsilon\delta}{2\delta} = 2\epsilon > \epsilon.$$

Remark. Example 2 illustrates the difference between the concepts of "everywhere continuous" and "uniformly continuous." The function $f(x) = x^2$, $-\infty < x < \infty$, is everywhere continuous. That is, given

UNIFORM CONTINUITY AND INTERMEDIATE VALUES

any real number x_1 and any positive number ϵ, we can find a positive number δ, say

$$\delta = \min\left(1, \frac{\epsilon}{1 + 2|x_1|}\right), \tag{4a}$$

such that

$$|x_1^2 - x_2^2| < \epsilon \quad \text{whenever} \quad |x_1 - x_2| < \delta. \tag{4b}$$

But the δ given by Eq. (4a) depends upon x_1 as well as ϵ. The order of events is crucial: if we are given x_1, in addition to $\epsilon > 0$, then we can find $\delta > 0$ from (4a) to satisfy (4b). But given only $\epsilon > 0$, and no restriction on x_1 and x_2 except that they be close together, we cannot find a fixed $\delta > 0$ depending only upon ϵ, that satisfies (4b). However, if we restrict the domain of f so that $|x_1|$ is bounded in (4a), say $|x_1| \leq M$, then

$$\delta = \min\left(1, \frac{\epsilon}{1 + 2M}\right) \tag{4c}$$

can be used in place of (4a). The number given by (4c) no longer depends upon the location of x_1 (or x_2) and can be used, if the domain is restricted to $|x| \leq M$, to satisfy the condition (4b).

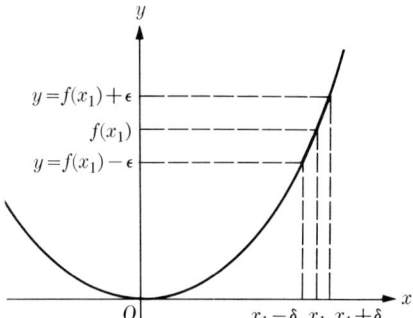

FIG. 7–1. If $f(x) = x^2$, and $\epsilon > 0$, the δ-neighborhood around x_1 required to make $|f(x_2) - f(x_1)| < \epsilon$ gets smaller as x_1 gets larger, because of the steepness of the curve.

What the above analysis amounts to is this: when the domain is appropriately restricted, among all the δ's given by (4a) there is a smallest or least favorable one—namely, the one that corresponds to the largest $|x_1|$. We can use it. But when the domain consists of all the real numbers, there is no largest $|x_1|$, and no smallest positive δ that satisfies (4b). The graph of the parabola $y = x^2$ (Fig. 7–1) rises ever more steeply as $|x|$ increases. Therefore, to keep $|\Delta y| < \epsilon$ we must make $|\Delta x|$ smaller and smaller as $|x|$ gets larger.

The property of *uniform* continuity is of great importance in the theory of integration. The following theorem, Theorem C-III of Section 6–1, applies to any function that is continuous on a closed and bounded interval.

Theorem 7–1. Let a and b be real numbers, $a < b$. Let f be a real-valued function that is continuous on the closed bounded interval

$$D: a \leq x \leq b.$$

Then f is uniformly continuous on D.

Proof. Let $\epsilon > 0$. Since f is continuous at each point of D, we can find, for each x_1 in D, a $\delta = \delta(x_1, \epsilon) > 0$ such that

$$|f(x) - f(x_1)| < \frac{\epsilon}{2} \quad \text{whenever} \quad |x - x_1| < \delta(x_1, \epsilon) \quad \text{and} \quad x \in D. \quad (5)$$

(The reason for using $\epsilon/2$, instead of ϵ, in (5) will come out in the course of the proof, along with the reason for using $\delta/2$ in place of δ in the next sentence.) The union of all the open neighborhoods

$$x_1 - \tfrac{1}{2} \delta(x_1, \epsilon) < x < x_1 + \tfrac{1}{2} \delta(x_1, \epsilon), \qquad x_1 \in D, \quad (6)$$

certainly covers D. By the Heine-Borel theorem, there exists a *finite* collection of these neighborhoods that still covers D. Let $a_1, a_2, \ldots, a_n$ be the centers of the neighborhoods in this finite covering. Finally, let

$$\delta_0 = \min \{\tfrac{1}{2} \delta(a_i, \epsilon) : i = 1, 2, \ldots, n\}. \quad (7)$$

The minimum of the *finite* collection of positive numbers in (7) exists and is a positive number.

The rest of the proof of the theorem consists in showing that

$$\text{if} \quad x' \text{ and } x'' \text{ are in } D \quad \text{and} \quad |x' - x''| < \delta_0, \quad (8a)$$

then

$$|f(x') - f(x'')| < \epsilon. \quad (8b)$$

We assume (8a). Then x' is in one of the neighborhoods of the finite covering, so that there exists one of the a's, say a_i, such that

$$|x' - a_i| < \tfrac{1}{2} \delta(a_i, \epsilon). \quad (9a)$$

Also

$$|x'' - x'| < \delta_0 \leq \tfrac{1}{2} \delta(a_i, \epsilon). \quad (9b)$$

Hence

$$|x'' - a_i| = |x'' - x' + x' - a_i|$$
$$\leq |x'' - x'| + |x' - a_i|$$
$$< \tfrac{1}{2} \delta(a_i, \epsilon) + \tfrac{1}{2} \delta(a_i, \epsilon),$$

or

$$|x'' - a_i| < \delta(a_i, \epsilon). \quad (10)$$

UNIFORM CONTINUITY AND INTERMEDIATE VALUES   7–1

The reason for the $\delta/2$ in place of δ in the inequalities (6) is clear from the inequalities leading to (10). From (9a) and (10) we see that both x' and x'' are within $\delta(a_i,\epsilon)$-distance of a_i, so that (5) applied with $x = x'$ (or x'') and $x_1 = a_i$ gives

$$|f(x') - f(a_i)| < \frac{\epsilon}{2} \tag{11a}$$

and

$$|f(x'') - f(a_i)| = |f(a_i) - f(x'')| < \frac{\epsilon}{2}. \tag{11b}$$

Therefore,

$$\begin{aligned}|f(x') - f(x'')| &= |f(x') - f(a_i) + f(a_i) - f(x'')| \\ &\leq |f(x') - f(a_i)| + |f(a_i) - f(x'')| \\ &< \frac{\epsilon}{2} + \frac{\epsilon}{2} = \epsilon. \qquad\qquad \text{Q.E.D.}\end{aligned}$$

EXERCISES 7–1

1. Let D be the open interval $0 < x < 1$, and $f(x) = 1/x$ for each x in D.
   a) Prove that f is everywhere continuous on D.
   b) Prove that f is not uniformly continuous on D.

2. a) Let D be the open interval $1 < x < 2$, and $f(x) = 1/x$ for each x in D. Prove that f is uniformly continuous on D.
   b) Replace D by the set of all $x \geq 1$, and again let $f(x) = 1/x$. Is f uniformly continuous on this new domain?

3. From the trigonometric identities

$$\sin(A + B) = \sin A \cos B + \cos A \sin B,$$
$$\sin(A - B) = \sin A \cos B - \cos A \sin B,$$

we can deduce that $\sin(A + B) - \sin(A - B) = 2\cos A \sin B$. By letting

$$A + B = x_1, \qquad A - B = x_2$$

or

$$A = \tfrac{1}{2}(x_1 + x_2), \qquad B = \tfrac{1}{2}(x_1 - x_2),$$

we get

$$\sin x_1 - \sin x_2 = 2\cos\left(\frac{x_1 + x_2}{2}\right)\sin\left(\frac{x_1 - x_2}{2}\right).$$

   a) Use this result to show that

$$|\sin x_1 - \sin x_2| \leq |x_1 - x_2| \qquad \text{for all } x_1, x_2.$$

   b) Use part (a) to prove that the function

$$f(x) = \sin x, \qquad -\infty < x < \infty,$$

   is uniformly continuous.

4. Suppose $a < b$ are real numbers and f is uniformly continuous on the domain $a < x < b$. Prove that the range of f is bounded.

For each of the following functions f on the given domains D, find $\delta > 0$ in terms of $\epsilon > 0$ such that $|f(x') - f(x'')| < \epsilon$ whenever $|x' - x''| < \delta$, where both x' and x'' are in D.

5. $f(x) = x^2$, $-2 \le x \le 3$
6. $f(x) = 2x^2$, $-3 \le x \le 2$
7. $f(x) = 1/x^2$, $1 \le x \le 1000$
8. $f(x) = 3 \sin x$, $-\infty < x < \infty$. [*Hint:* See Exercise 3 above.]
9. $f(x) = \sin(3x)$, $-\infty < x < \infty$
10. $f(x) = \sqrt{x}$, $0 \le x < \infty$. [*Hint:* Show that $|\sqrt{x+h} - \sqrt{x}| \le \sqrt{h}$, for any $x \ge 0$, $h \ge 0$.]

7-2 INTERMEDIATE VALUES

When we sketch the graph of $y = x^2$ we usually plot a few points and connect them with a smooth unbroken curve. As our pencil point moves from $(0, 0)$ to $(2, 4)$, if we do not lift it from the paper, it intersects any horizontal line between the x-axis and the line $y = 4$. That is, if c is any number between 0 and 4, there is at least one value of x between 0 and 2 where the function $f(x) = x^2$ takes on the value c. (See Fig. 7-2.) For this particular example, the appropriate value of x is $x = \sqrt{c}$. But for only slightly more complicated expressions, we may not be able to write down an explicit formula for the appropriate x.

Example 1. Suppose we want to solve the equation

$$x + 1 = 3 \sin x. \qquad (1)$$

If we let

$$f(x) = x + 1 - 3 \sin x,$$

then

$$f(0) = +1,$$

and

$$f\left(\frac{\pi}{2}\right) = \left(\frac{\pi}{2}\right) + 1 - 3 \approx 2.57 - 3 = -0.43.$$

The function f is continuous for all x. Since $f(0)$ is positive and $f(\pi/2)$ is negative, the equation $f(x) = 0$ should have a solution for some x between 0 and $\pi/2$ (Fig. 7-3). There is no simple formula that gives such an x explicitly, though there are methods (for example, Newton's method) for approximating it with any desired decimal accuracy. An iterative tech-

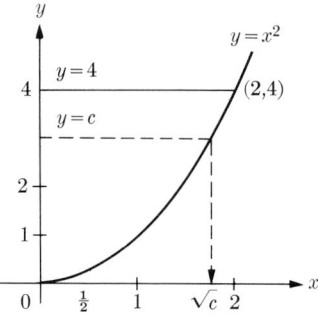
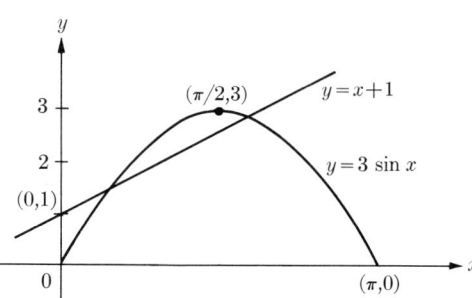

FIG. 7–2. If $0 < c < 4$, there is an x between 0 and 2 such that $x^2 = c$.

FIG. 7–3. The line $y = x + 1$ and the curve $y = 3 \sin x$ intersect at a point between $x = 0$ and $x = \pi/2$, at which $x + 1 - 3 \sin x = 0$.

nique gives a sequence $x_1, x_2, \ldots$ that converges to a root as follows: let $x_1 = 0$, and for each positive integer n, let

$$x_{n+1} = \sin^{-1}\left(\frac{1 + x_n}{3}\right). \tag{2}$$

If the sequence $\{x_n\}$ converges to a limit,

$$\lim_{n \to \infty} x_n = \lim_{n \to \infty} x_{n+1} = L,$$

then

$$L = \sin^{-1}\left(\frac{1 + L}{3}\right)$$

and

$$3 \sin L = L + 1. \tag{3}$$

The geometric interpretation of Eqs. (2) and (3) is illustrated by Fig. 7–4.

Our present aim is to illustrate and establish the *intermediate-value theorem* rather than to develop techniques for solving equations like (1).

The following theorem, Theorem C-II of Section 6–1, asserts that a function f that is continuous on a closed bounded interval $[a, b]$ takes all values in the interval $[f(a), f(b)]$.

Theorem 7–2. Intermediate-Value. Let a and b be real numbers such that $a < b$. Let f be a real-valued function continuous on $a \leq x \leq b$. If $f(a) \neq f(b)$ and c is a number between $f(a)$ and $f(b)$, then there is at least one number x_0 between a and b such that

$$f(x_0) = c. \tag{4}$$

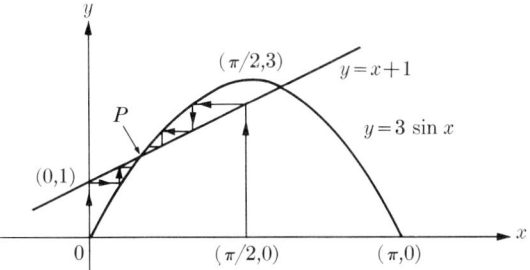

FIG. 7-4. If $x_1 = 0$ and $x_{n+1} = \sin^{-1}[(1 + x_n)/3]$ for $n \geq 1$, or $3 \sin(x_{n+1}) = 1 + x_n$, we get x_{n+1} by going horizontally from the point $(x_n, 1 + x_n)$ on the line $y = x + 1$ to the point $(x_{n+1}, 3 \sin x_{n+1})$ on the curve $y = 3 \sin x$. The staircase spiral converges to a point P, where the line and the curve intersect. (We could also start from $x_1 = \pi/2$ and move toward P from the right.)

Proof. We give the proof for

$$f(a) < c < f(b). \tag{5}$$

When the inequalities in (5) are reversed, the proof is only slightly different and will not be given in detail. Under the assumption (5) the set

$$S = \{x : f(x) < c, a \leq x \leq b\} \tag{6}$$

is nonempty because it contains a, and it is bounded above by b. By the completeness postulate for real numbers, there exists a real number x_0 such that

$$x_0 = \text{lub }(S). \tag{7}$$

Then, $a \leq x_0 \leq b$ and there are just three cases to consider for $f(x_0)$:

    i) $f(x_0) < c$,
    ii) $f(x_0) = c$,
    iii) $f(x_0) > c$.

We prove that neither (i) nor (iii) holds, and thereby establish Eq. (4).

CASE (i). If $f(x_0) < c$, the numbers

$$\tfrac{1}{2}(c - f(x_0)) \quad \text{and} \quad \tfrac{1}{2}(f(b) - c)$$

are positive. Let ϵ_0 be the smaller of these. Because $a \leq x_0 \leq b$, f is continuous at x_0 and there exists $\delta_0 > 0$ such that

$$|f(x) - f(x_0)| < \epsilon_0 \quad \text{when } |x - x_0| < \delta_0 \text{ and } a \leq x \leq b.$$

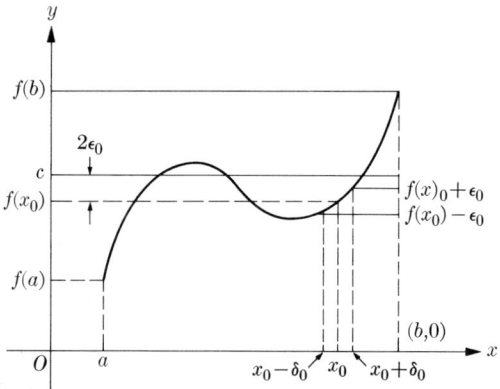

FIG. 7–5. If $f(x_0) < c$, then $x_0 \neq \text{lub } \{x : f(x) < c, a \leq x \leq b\}$.

But then (see Fig. 7–5)
$$f(x_0) - \epsilon_0 < f(x) < f(x_0) + \epsilon_0$$
implies
$$f(x) < f(x_0) + \tfrac{1}{2}(c - f(x_0)) = \tfrac{1}{2}(c + f(x_0)) < \tfrac{1}{2}(c + c) = c$$
for $x_0 < x < x_0 + \delta_0$. This means that S contains numbers $x > x_0$, which is contrary to Eq. (7). Therefore case (i) is invalid.

CASE (iii). If $f(x_0) > c$, the number $\epsilon' = \tfrac{1}{2}(f(x_0) - c)$ is positive. Also f is continuous at x_0. Therefore, there is a positive number δ' such that
$$|f(x) - f(x_0)| < \epsilon' \quad \text{when } |x - x_0| < \delta' \quad \text{and} \quad a \leq x \leq b.$$
(See Fig. 7–6). This implies that
$$f(x) > c \quad \text{for} \quad x_0 - \delta' < x < x_0. \tag{8a}$$
But also, by definition of the set S and Eq. (7),
$$f(x) \geq c \quad \text{for} \quad x_0 < x < b. \tag{8b}$$
Combining (8a, b) and $f(x_0) > c$, we see that
$$f(x) \geq c \quad \text{for} \quad x_0 - \delta' < x \leq b. \tag{8c}$$
Therefore, none of the numbers between $x_0 - \delta'$ and b belong to S, which is contrary to the fact that x_0 is the *least* upper bound for S. Hence case (iii) is also invalid.

Consequently, $f(x_0) = c$. Q.E.D.

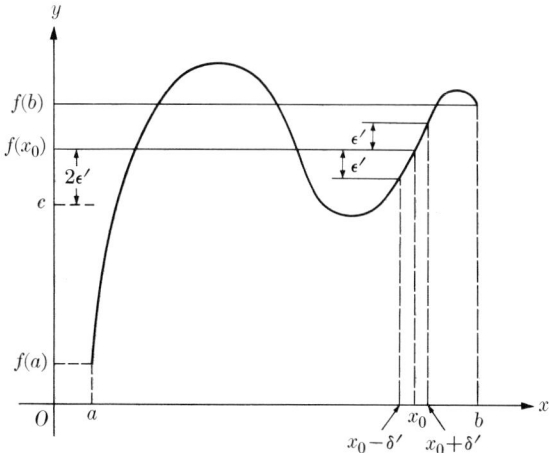

FIG. 7–6. If $f(x_0) > c$, then $x_0 \neq \text{lub}\ \{x: f(x) < c, a \leq x \leq b\}$.

Corollary 7–2.1 (a). If f is continuous on $a \leq x \leq b$ and $f(a) < 0$, $f(b) > 0$, then the equation $f(x) = 0$ has at least one root between a and b.

Proof. This is just a special case of Theorem 7–2.

Corollary 7–2.1 (b) Let n be a positive integer and let c be a positive real number. Then the equation

$$x^n = c \quad \text{or} \quad x = \sqrt[n]{c}, \tag{9}$$

has a unique positive solution.

Proof. Apply Theorem 7–2 to the continuous function

$$f(x) = x^n, \quad \text{with} \quad a = 0,\ b = 1 + c.$$

Then

$$f(a) = 0^n = 0,$$
$$f(b) = (1 + c)^n = 1 + nc + \frac{n(n-1)}{2!} c^2 + \cdots$$
$$\geq 1 + nc > c, \quad \text{because} \quad n \geq 1,$$

so that we have

$$f(a) < c < f(b).$$

Hence Theorem 7–2 guarantees that there is at least one x_0 between 0 and $1 + c$ such that

$$f(x_0) = x_0^n = c \quad \text{or} \quad x_0 = \sqrt[n]{c}.$$

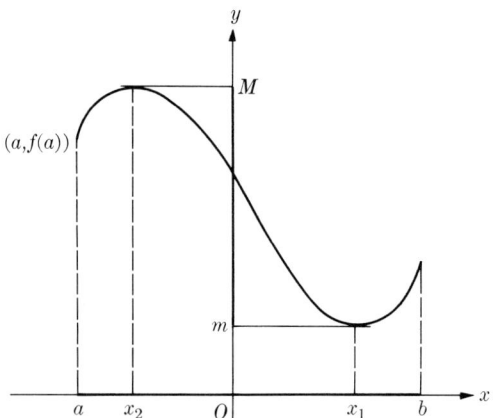

FIG. 7-7. A continuous function maps a closed bounded interval $a \leq x \leq b$ onto a closed bounded interval $m \leq y \leq M$.

To prove that there is *only* one such solution of (9) with $x > 0$, we need only observe that x^n is a strictly increasing function of x, for $x > 0$, so that it cannot take the same value for two different positive x's. Q.E.D.

Remark 1. Corollary 7-2.1(b) establishes the existence of such positive real numbers as $\sqrt{2}$, $\sqrt[3]{2}$, $\sqrt[5]{4}$, and so on.

Remark 2. We can now define a^x for any positive real number a and positive rational number $x = p/q$:

$$a^{p/q} = \sqrt[q]{a^p}; \qquad a > 0, \quad p \text{ and } q \text{ positive integers.}$$

If $x = -p/q$ is a negative rational number, we define

$$a^{-(p/q)} = \frac{1}{a^{p/q}}; \qquad a > 0, \quad p \text{ and } q \text{ positive integers.}$$

If $a > 1$, and $x = p/q > x' = p'/q'$, then

$$a^x > a^{x'},$$

so that the function $x \to a^x$, $a > 1$, is increasing on the domain of rational numbers. Finally, for x irrational and $a > 1$, we define a^x to be the least upper bound of the set of numbers $\{a^r : r < x, r \text{ rational}\}$. If $0 < a < 1$, let $b = 1/a$, then $b > 1$ and we define $a^x = b^{-x}$.

Following such a program, we could prove the fundamental law

$$a^{x_1} a^{x_2} = a^{(x_1 + x_2)}, \qquad a > 0,$$

but the proof involves substantial work with least upper bounds. We shall not pursue that program here, but deal with the exponential and logarithmic functions in Chapter 13.

Remark 3. The intermediate-value theorem, together with the theorem on maximum and minimum values, implies that if f is continuous on a closed bounded interval (the domain of f), then the range of f is also a closed bounded interval. For, if m is the minimum and M the maximum of f on $[a, b]$, then there exist x_1 and x_2 in $[a, b]$ such that

$$f(x_1) = m, \quad f(x_2) = M;$$

and if c is any number between m and M, there exists x in $[x_1, x_2]$ such that $f(x) = c$. Thus the range of f is the closed interval $m \leq y \leq M$. (See Fig. 7-7.)

EXERCISES 7-2

1. Suppose f has the graph shown in the figure. The line $y = c$ intersects the graph at points $A, B, C, D,$ and E as shown. Let

$$S = \{x : f(x) < c, \ a \leq x \leq b\}$$

and

$$T = \{x : f(x) > c, \ a \leq x \leq b\}.$$

Suppose

$$x_1 = \text{glb }(S), \quad x_2 = \text{lub }(S),$$
$$x_3 = \text{glb }(T), \quad x_4 = \text{lub }(T).$$

Which points on the graph below have as abscissas x_1? x_2? x_3? x_4?

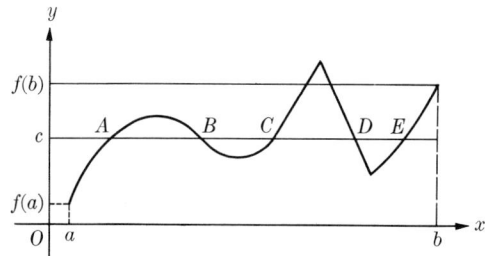

2. In the proof of Theorem 7-2, replace the inequalities in (5) by $f(a) > c > f(b)$, and let $S = \{x : f(x) > c, a \leq x \leq b\}$. If you were trying to define x_1 so that $f(x_1) = c$, would you let $x_1 = \text{lub }(S)$, or would you take $x_1 = \text{glb }(S)$? Illustrate with a variety of sketches.

3. In the proof of Theorem 7-2, with $f(a) < c < f(b)$, suppose

$$T = \{x : f(x) > c, a \leq x \leq b\},$$

and $x_2 = \text{glb }(T)$. Illustrate with graphs. Is $f(x_2) < c$? $> c$? or $= c$?

4. *The method of false position* for finding a root of an equation $f(x) = 0$ goes this way: locate an interval $[a, b]$ where $f(a)$ and $f(b)$ are of opposite signs. Bisect the interval. If $f((a + b)/2) = 0$, stop, because then $x = (a + b)/2$ is a root. Otherwise, $f((a + b)/2)$ has sign opposite to that of $f(a)$, or to that of $f(b)$. Thus, if f is continuous on $[a, b]$, we have either located a root exactly, or in an interval $[a_1, b_1]$ of length $\frac{1}{2}(b - a)$. Now repeat the process. After n times we have either located a root exactly or know that there is one in an interval $[a_n, b_n]$ of length $(\frac{1}{2})^n(b - a)$. Suppose the process continues indefinitely,

$$a \leq a_1 \leq a_2 \leq a_3 \leq \cdots \leq b$$

and

$$b \geq b_1 \geq b_2 \geq \cdots \geq a.$$

Show the following.

i) $\lim_{n \to \infty} a_n$ exists.

ii) $\lim_{n \to \infty} b_n$ exists.

iii) $\lim a_n = \lim b_n$.

iv) If $x = \lim a_n$, then $f(x) = 0$.

8
COMPOSITE FUNCTIONS AND INVERSE FUNCTIONS

8–1 COMPOSITE FUNCTIONS

Sometimes a function represents a mapping that can be achieved by performing two simpler mappings in succession. We illustrate this process of *composition of functions* with examples, and then give a general definition.

Example 1. The functions

$$f_1(x) = \sin(x^2), \qquad -\infty < x < \infty, \tag{1a}$$

$$f_2(x) = \sin^2 x, \qquad -\infty < x < \infty, \tag{1b}$$

are examples of composite functions. In Eq. (1a), we first square x, then find the sine of the result. Thus, if

$$g(x) = x^2 \tag{2a}$$

and

$$h(x) = \sin x, \tag{2b}$$

we can write Eq. (1a) in the form

$$f_1(x) = h(g(x)), \tag{3a}$$

or

$$f_1 = h \circ g, \tag{3b}$$

where (3a) is read "f_1 of x is equal to h of g of x," and (3b) is read "f_1 is the composition of h with g." The symbol $h \circ g$ is used to distinguish the composition of h with g from the ordinary product of h and g.

In the second equation, Eq. (1b), we first evaluate $\sin x$, then square the result. Using the notation of Eqs. (2a) and (2b), we can rewrite (1b) in the form

$$f_2(x) = g(h(x)) \tag{4a}$$

or, what amounts to the same thing,

$$f_2 = g \circ h. \tag{4b}$$

In both examples (1a) and (1b), the domains are the same—the set of all real numbers.

More generally, suppose we define the composition of g with h by the following rule:

$$f = g \circ h \tag{5a}$$

means that

$$f(x) = g(h(x)) \tag{5b}$$

for all values of x for which the right-hand side of Eq. (5b) is defined.

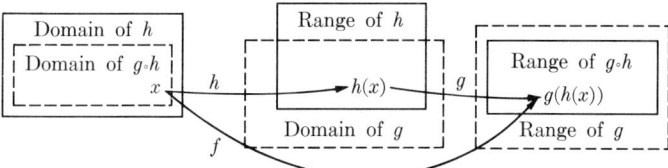

FIG. 8-1. Composition of g with h; $f(x) = g(h(x))$, $f = g \circ h$.

Here g and h represent any two functions, no longer restricted as in (2a) and (2b). To evaluate (5b), we start with a value of x in the domain of h. (See Fig. 8-1.) Then $h(x)$ is a number in the range of h. If this number $h(x)$ is also in the domain of g, then $g(h(x))$ is a number in the range of g. Thus, the domain of $g \circ h$ is a subset of the domain of h, and the range of $g \circ h$ is a subset of the range of g.

Example 2. Suppose that

$$g(x) = \sqrt{1-x}, \qquad x \leq 1, \tag{6a}$$

and

$$h(x) = \sqrt{x}, \qquad x \geq 0. \tag{6b}$$

Then

$$g(h(x)) = \sqrt{1 - h(x)}; \qquad x \geq 0, \quad h(x) \leq 1$$

$$= \sqrt{1 - \sqrt{x}}; \qquad x \geq 0, \quad \sqrt{x} \leq 1.$$

From the restrictions $x \geq 0$, $\sqrt{x} \leq 1$, we find that the domain of

$$f = g \circ h$$

is the interval $0 \leq x \leq 1$. Figure 8-2 (a), (b), (c) shows graphs of the functions h, g, $g \circ h$.

Definition 1. Composition of Functions. Let g be a real-valued function with domain D_g and range R_g, and let h be a real-valued function with domain D_h and range R_h. The composition of g with h is the function f defined by

$$f(x) = g(h(x))$$

with domain D_f consisting of those values x in D_h such that $h(x) \in D_g$.

Remark 1. Here is another way to look at the composite mapping $f = g \circ h$. Start with a number y in the intersection of D_g and R_h. Because this y is in the range of h, there is an x in D_h such that $h(x) = y$. And,

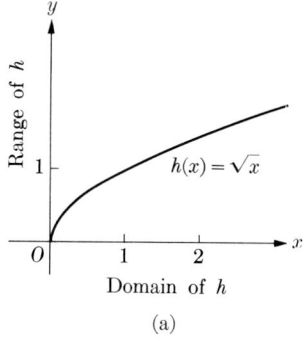
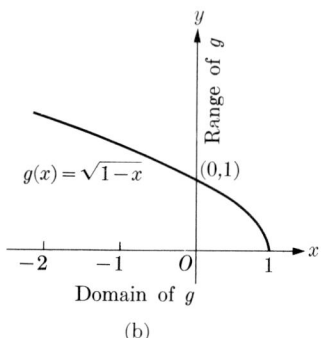
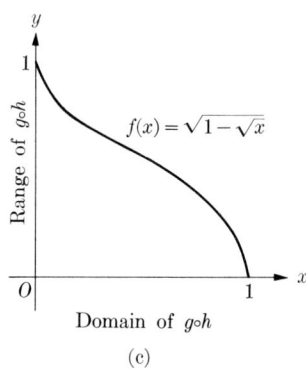

FIG. 8–2. Graphs of h, g, and $f = g \circ h$.

because y is also in the domain of g, there is a number $z = g(y)$. Combining, we get

$$z = g(h(x)) = f(x).$$

Thus,

(x, y) is an element of the function h,

(y, z) is an element of the function g,

and

(x, z) is an element of the composite function $f = g \circ h$.

Therefore the ordered pair (x, z) belongs to the composite function $f = g \circ h$ if (and only if) there is a number y such that

(x, y) belongs to h and (y, z) belongs to g.

Remark 2. By using a three-dimensional (x, y, z)-coordinate system, we can get a graphical representation of the composition:

$$y = h(x), \quad z = g(y),$$
$$f(x) = g(h(x)).$$

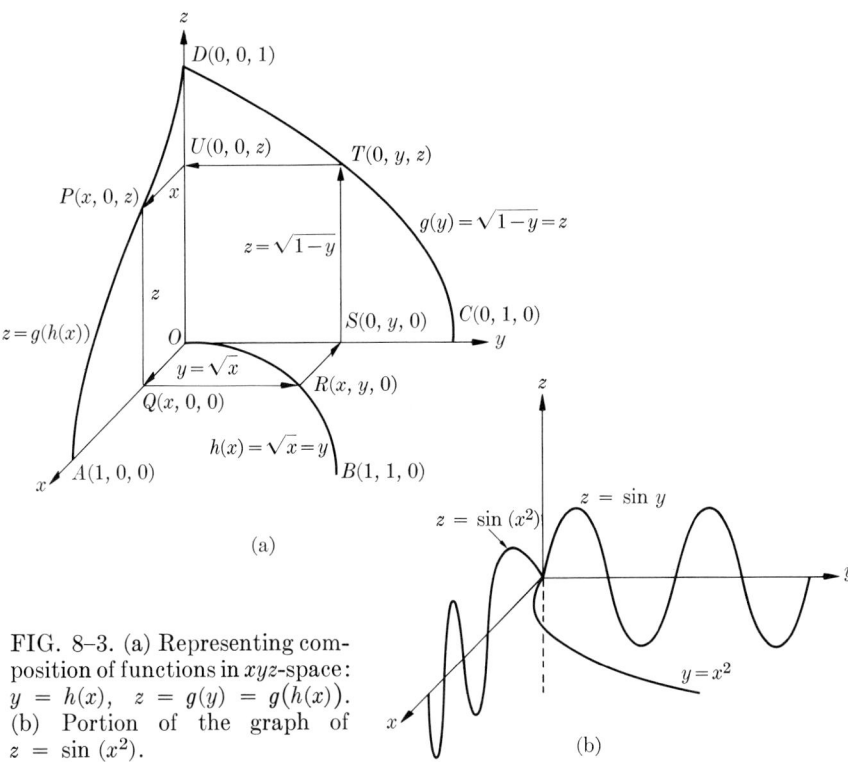

FIG. 8–3. (a) Representing composition of functions in xyz-space: $y = h(x)$, $z = g(y) = g(h(x))$. (b) Portion of the graph of $z = \sin(x^2)$.

Figure 8–3(a) shows such a representation for the example
$$h(x) = \sqrt{x}, \qquad g(y) = \sqrt{1-y},$$
$$f(x) = \sqrt{1-\sqrt{x}}.$$
To find the value of $f(x)$ for a particular value of x, $0 \leq x \leq 1$, let Q be the point with abscissa x on the x-axis, and follow the diagram $Q \to R$, $R \to S$, $S \to T$, $T \to U$, $U \to P$. Here $R(x, y, 0)$ is on the graph of $y = h(x)$, $S(0, y, 0)$ has the same y-coordinate as R, and $T(0, y, z)$ is on the graph of $z = g(y)$. Because $U(0, 0, z)$ and $P(x, 0, z)$ have the same z-coordinates as T, the x- and z-coordinates of P satisfy the equations
$$z = g(y) = g(h(x)),$$
so that P is on the graph of $z = f(x)$ in the xz-coordinate plane (that is, $y = 0$). For the example shown in Fig. 8–3(a), the curve $A \ldots P \ldots D$ is the graph of
$$z = \sqrt{1-\sqrt{x}}, \qquad 0 \leq x \leq 1.$$
See also Fig. 8–3(b), illustrating the function $z = \sin(x^2)$.

EXERCISES 8–1

Find the compositions $f_1 = g \circ h$ and $f_2 = h \circ g$ for the following pairs of functions g and h. In each instance, indicate the domains of f_1 and f_2.

1. $g(x) = 2x + 1$, x any real number
   $h(x) = \sqrt{x}$, $x \geq 0$

2. $g(x) = \dfrac{x}{x+1}$, $x \neq -1$
   $h(x) = \dfrac{1}{x}$, $x \neq 0$

3. $g(x) = \sqrt{x-1}$, $x \geq 1$
   $h(x) = \dfrac{1}{x}$, $x \neq 0$

8–2 CONTINUITY OF COMPOSITE FUNCTIONS

The question we plan to investigate in this section is this: is the composition of a continuous function with a continuous function also continuous? The question needs to be put more precisely, because a function may be continuous at some places in its domain and discontinuous at others. Once we formulate the problem precisely, it is quite easy to supply the answer, which is, "yes."

Theorem 8–1. Let h be a function that is continuous at c. Let $b = h(c)$ and suppose g is a function that is continuous at b. Let $f = g \circ h$, that is, $f(x) = g(h(x))$. Then f is continuous at c.

Proof. Let $\epsilon > 0$. We need to show that there exists a positive number δ such that

$$|f(x) - f(c)| < \epsilon \quad \text{whenever} \quad |x - c| < \delta \quad \text{and} \quad x \in D_f.$$

Since the function g is continuous at b, and $\epsilon > 0$, there is a number $\delta_1 > 0$ such that

$$|g(y) - g(b)| < \epsilon \quad \text{whenever} \quad |y - b| < \delta_1 \quad \text{and} \quad y \in D_g.$$

Now, for the function h at c let δ_1 play the role usually played by epsilon. Since h is continuous at c and $\delta_1 > 0$, there exists $\delta_2 > 0$ such that

$$|h(x) - h(c)| < \delta_1 \quad \text{whenever} \quad |x - c| < \delta_2 \quad \text{and} \quad x \in D_h.$$

Now combine: if $x \in D_f$ and $|x - c| < \delta_2$ and $y = h(x)$, then $x \in D_h$ and

$$|f(x) - f(c)| = |g(h(x)) - g(h(c))|$$
$$= |g(y) - g(b)|$$
$$< \epsilon,$$

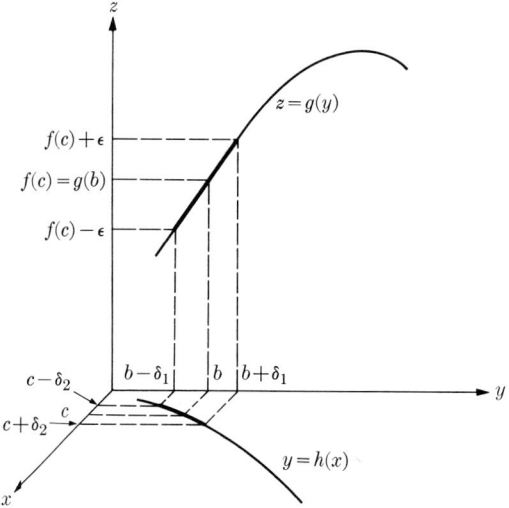

FIG. 8-4. Continuity of composite function.

because $y \in D_g$ and
$$|y - b| = |h(x) - h(c)| < \delta_1.$$
Thus given $\epsilon > 0$, follow the steps described above and let $\delta = \delta_2$. Q.E.D.

Figure 8-4 illustrates graphs of the functions $y = h(x)$, $z = g(y)$ in the xy-plane and the yz-plane, respectively. Given c, on the x-axis, and $\epsilon > 0$, we locate $f(c)$, $f(c) - \epsilon$, and $f(c) + \epsilon$ on the z-axis. We then use the graph of g, and the fact that g is continuous at $b = h(c)$, to find an interval
$$b - \delta_1 < y < b + \delta_1$$
on the y-axis such that when y is in this interval, then $g(y)$ is between $f(c) - \epsilon$ and $f(c) + \epsilon$. Then we turn to the graph of h and use the continuity of h at c to find an interval
$$c - \delta_2 < x < c + \delta_2$$
on the x-axis such that when x is in this interval, then $y = h(x)$ is between $b - \delta_1$ and $b + \delta_1$.

Example 1. Let $h(x) = x^2$, and $g(y) = 3y - 2$. Suppose c is a real number and $\epsilon > 0$. Then
$$b = h(c) = c^2$$
and
$$f(c) = g(h(c)) = 3h(c) - 2 = 3c^2 - 2.$$

COMPOSITE FUNCTIONS AND INVERSE FUNCTIONS

We follow the steps of the proof for Theorem 8–1.

1. To find $\delta_1 > 0$ such that
$$|g(y) - g(b)| = |(3y - 2) - (3b - 2)|$$
$$= 3|y - b| < \epsilon$$
when $|y - b| < \delta_1$, we take
$$\delta_1 = \frac{\epsilon}{3}.$$

2. To find δ_2 such that
$$|h(x) - h(c)| = |x^2 - c^2|$$
$$= |x - c| \cdot |x + c| < \delta_1 = \frac{\epsilon}{3}$$
when $|x - c| < \delta_2$, we take
$$\delta_2 = \min\left(1, \frac{\epsilon}{6|c| + 3}\right).$$

[The $6|c| + 3$ gets into the act this way: first, we pin down x to no more than 1 unit away from c. Then
$$|x + c| \leq |x| + |c| \leq |c| + 1 + |c| = 2|c| + 1,$$
and hence,
$$|x^2 - c^2| = |x - c| \cdot |x + c| \leq |x - c|(2|c| + 1)$$
and the rightmost member of this is
$$< \epsilon/3 \quad \text{if} \quad |x - c| < \epsilon/3(2|c| + 1).]$$

3. Let $\delta = \delta_2$. (Forget about y and δ_1 now if you like.) With
$$\delta = \min\left(1, \frac{\epsilon}{6|c| + 3}\right)$$
we have a positive number δ that fulfills the requirement:
$$|f(x) - f(c)| = |(3x^2 - 2) - (3c^2 - 2)|$$
$$= 3|x^2 - c^2|$$
$$= 3|x - c| |x + c| < \epsilon,$$
provided
$$|x - c| < \delta.$$

EXERCISES 8-2

In each of the following, sketch the graphs of $y = h(x)$, $z = g(y)$, and $z = f(x)$ with $f(x) = g(h(x))$, as illustrated in the text. Then, for each given real number c and $\epsilon > 0$, find a $\delta > 0$ such that $|f(x) - f(c)| < \epsilon$ when $|x - c| < \delta$.

1. $h(x) = 2x - 3$, $g(y) = 5 - 4y$, c arbitrary
2. $h(x) = \sin x$, $g(y) = 3y$, c arbitrary
3. $h(x) = \dfrac{1}{x}$, $g(y) = \dfrac{1}{y}$, $x \neq 0$, $y \neq 0$, $c > 0$
4. $h(x) = \sin x$, $g(y) = y^2$, c arbitrary
5. $h(x) = x^2$, $g(y) = \sin y$, c arbitrary

8-3 INVERSE FUNCTIONS

Under some conditions, the effect of one mapping can be undone by a second mapping. That is, if the first mapping is g and the second is f, the composition $f \circ g$ applied to any number may yield just that number.

Example 1. The functions f and g for which

$$f(x) = x^3, \qquad g(x) = x^{1/3}, \tag{1}$$

have the set of all real numbers both as domain and as range. The composite functions $h_1 = f \circ g$ and $h_2 = g \circ f$ satisfy

$$h_1(x) = f(g(x)) = (g(x))^3 = (x^{1/3})^3 = x \tag{2a}$$

and

$$h_2(x) = g(f(x)) = (f(x))^{1/3} = (x^3)^{1/3} = x. \tag{2b}$$

Since $h_1(x) = x$ and $h_2(x) = x$ for all real x, we may write $h_1 = h_2 = I$, where I is the *identity* function

$$I(x) = x. \tag{3}$$

In other words,

$$f \circ g = I \quad \text{and} \quad g \circ f = I. \tag{4}$$

In terms of mappings, Eqs. (2a, b) and (4) tell us that whatever the mapping g (or f) does to a particular value of x, the other mapping, f (or g), undoes. This is certainly obvious for the functions of Eq. (1), because cubing a cube root, or extracting the real cube root of a cube, will get us back where we started.

Whenever two functions f and g satisfy Eq. (4), we say that g is the *inverse* of f, and f is the *inverse* of g.

COMPOSITE FUNCTIONS AND INVERSE FUNCTIONS 8–3

Example 2. It is evident that, if we take the reciprocal of the reciprocal of any number $x \neq 0$, we get back the original number. This suggests (and it is easy to verify) that if f is the function

$$f(x) = 1/x, \quad x \neq 0, \tag{5}$$

then

$$f \circ f = I_{(x \neq 0)}, \tag{6}$$

where the restriction $x \neq 0$ is written as a subscript to remind us that whereas the identity function I has for domain *all* real numbers, both f and $f \circ f$ have zero excluded from their domains. That is, whereas the unrestricted identity function I is the set of *all* ordered pairs (x, x), and has as its graph the entire line $y = x$, the function of Eq. (6) does not include $(0, 0)$, and its graph is the line $y = x$ *excluding* the origin.

Definition 2. Left Inverse. Suppose that f is a function with domain D_f and range R_f, and g is a function with domain $D_g = R_f$ and range $R_g = D_f$. If, for every $x \in D_f$, $g(f(x)) = x$, then the composition $g \circ f$ is the identity function on D_f, and g is called a left inverse of f. (See Fig. 8–5.)

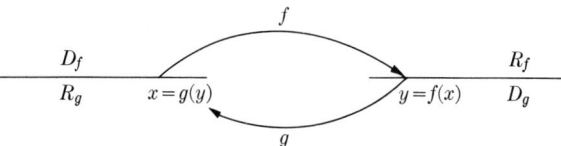

FIG. 8–5. If $g(f(x)) = x$ for each x in D_f, g is a left inverse of $f : g \circ f = I$ on D_f.

Remark 1. We can also interpret these conditions on inverse functions in terms of sets and graphs. The function g is the set of ordered pairs $(y, g(y))$ with $y \in D_g = R_f$. But for each such $y \in R_f$ there is an $x \in D_f$ such that $f(x) = y$ and $g(y) = g(f(x)) = x$. Moreover, this x is unique, because g is a function so that $g(y)$ has a single value for any given y. Thus, if $f(x_1) = f(x_2) = y$, then

$$g(y) = g(f(x_1)) = x_1$$

and

$$g(y) = g(f(x_2)) = x_2,$$

so that $x_1 = x_2$. Thus f gives a one-to-one mapping from D_f to R_f. As x runs over the domain D_f, the image $f(x)$ runs just once over the range R_f, and conversely. Thus we can write

$$g = \{(y, g(y)) : y \in D_g\} = \{(f(x), x) : x \in D_f\}. \tag{7}$$

Equation (7) simply says that to get all the ordered pairs that make the function g, we just change the order in each of the pairs $(x, f(x))$ belonging to f, thus obtaining the pairs $(f(x), x)$. This means that the graph of g is simply the curve we get by reflecting the graph of f in the line $y = x$ as mirror. Figure 8–6 illustrates this principle for the functions $f(x) = x^3$, $g(x) = x^{1/3}$.

Since the graph of $y = 1/x$ is just repeated when we reflect it across the line $y = x$, the function f such that $f(x) = 1/x$ is its own inverse, as we observed before.

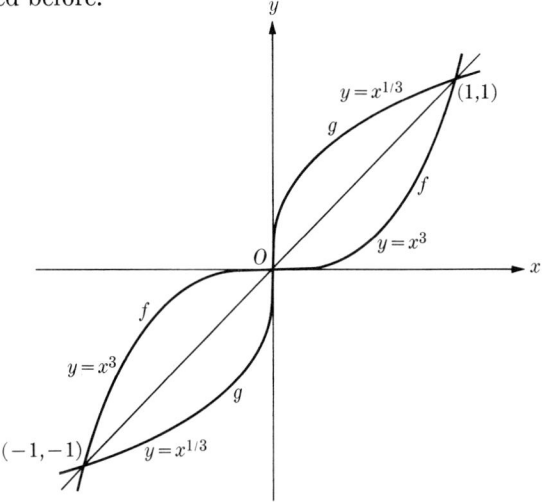

FIG. 8–6. Graphs of $f(x) = x^3$ and $g(x) = x^{1/3}$ are mirror images of each other with respect to the line $y = x$.

Remark 2. When g is a left inverse of f, as in Fig. 8–5, we also say that f is a right inverse of g. This is merely another way of saying the same thing, that $g \circ f = I$ is the identity function on the domain of f, or on the range of g, meaning that $x = g(y)$ if and only if $y = f(x)$. Therefore, for each $y \in D_g$ there is a unique $x \in D_f$ such that

$$y = f(x),$$

$$g(y) = g(f(x)) = x,$$

and

$$f(x) = f(g(y)) = y.$$

In other words,

$$g \circ f = I \quad \text{on the domain of } f,$$

and

$$f \circ g = I \quad \text{on the domain of } g.$$

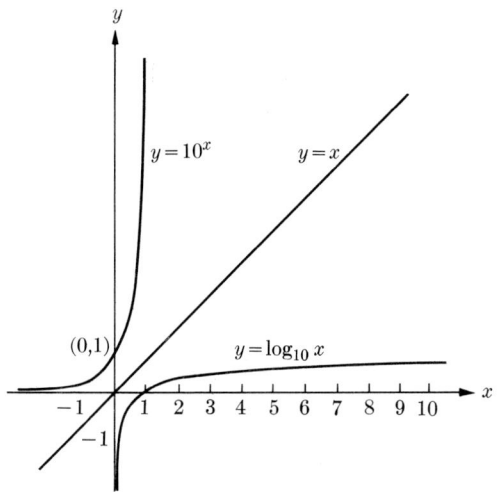

FIG. 8–7. Graphs of inverse functions $f(x) = \log_{10} x$ and $g(x) = 10^x$.

Thus g is both a right inverse and a left inverse of f, while f is also an inverse of g.

Example 3. Suppose

$$f(x) = \log_{10} x, \quad x > 0,$$

and

$$g(y) = 10^y, \quad y \text{ any real number.}$$

For each $x > 0$ there exists a unique $y = \log_{10} x$ such that $10^y = 10^{\log_{10} x} = x$. That is,

$$g(f(x)) = x \quad \text{for all} \quad x > 0. \tag{8}$$

Similarly,

$$f(g(y)) = f(10^y) = \log_{10}(10^y) = y \quad \text{for any } y. \tag{9}$$

Therefore,

$$g \circ f = I \quad \text{on the domain} \quad x > 0,$$

$$f \circ g = I \quad \text{on the domain of all real numbers } y.$$

More generally, if $a > 0$ and $a \neq 1$, the exponential function with base a,

$$f(x) = a^x, \quad x \text{ any real number,}$$

and the logarithmic function to the same base a,

$$g(x) = \log_a x, \quad x > 0,$$

are inverse functions. Figure 8–7, for $a = 10$, is representative of the graphs of such inverse functions for $a > 1$. (If $0 < a < 1$, the graph of $y = a^x$ has negative slope.)

Remark 3. We have seen above that if the function

$$f = \{(x, y) : y = f(x), x \in D_f\} \tag{10}$$

has an inverse, then this inverse, g, is

$$g = \{(y, x) : y = f(x), x \in D_f\}. \tag{11}$$

Equation (11) says that

$$g(y) = x,$$

if

$$x \in D_f \quad \text{and} \quad f(x) = y. \tag{12}$$

Equation (12) points the way to the following algebraic technique for finding the inverse of a function:

Solve the equation $y = f(x)$ for x in terms of y. The result is $x = g(y)$. If we want the inverse function g in the standard form $y = g(x)$, we simply interchange the letters x and y in the result.

Example 4. Let f be the linear function

$$f(x) = 3x + 2, \quad -\infty < x < \infty. \tag{13a}$$

If we solve

$$y = 3x + 2$$

for x, we get

$$x = \frac{y - 2}{3} = g(y). \tag{13b}$$

If we now interchange the letters x and y, we get for $y = g(x)$ the equation

$$y = g(x) = \frac{x - 2}{3}, \quad -\infty < x < \infty. \tag{13c}$$

(Note that Eq. (13a) says, "to find $f(x)$, take any value of x, multiply it by 3 and add 2," and the inverse operation (13b) or (13c) says, "subtract 2 and divide by 3." Obviously, each of these two functions is the inverse of the other.)

COMPOSITE FUNCTIONS AND INVERSE FUNCTIONS   8-3

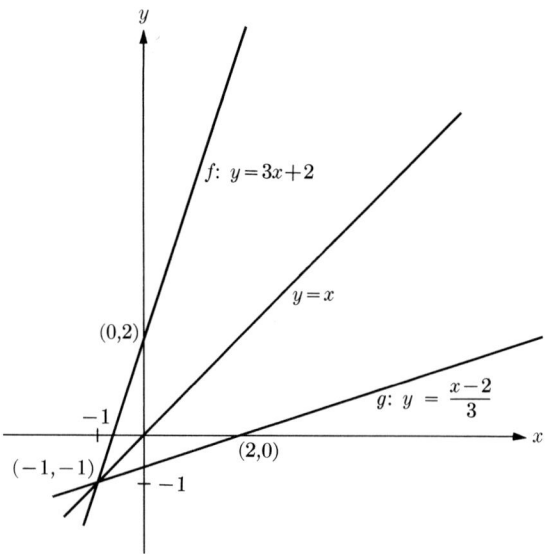

FIG. 8-8. Graphs of inverse functions $f(x) = 3x + 2$, $g(x) = (x - 2)/3$ are symmetric about the line $y = x$.

Figure 8-8 shows the graphs of the inverse functions, Eqs. (13a, c).

Remark 4. Nonexistence of an inverse function. If f is a function,

$$f = \{(x, y) : y = f(x), x \in D_f\}, \tag{14}$$

the *inverse relation*,

$$f^{-1} = \{(y, x) : y = f(x), x \in D_f\}, \tag{15}$$

may *not* be a function, because there may be more than one value of x in D_f mapping into the same value of $y = f(x)$, but no two ordered pairs, (y, x_1) and (y, x_2), with the same first element y and $x_1 \neq x_2$ can belong to a *function* f^{-1}.

Example 5. Suppose

$$f(x) = x^2, \qquad -\infty < x < \infty.$$

Then the ordered pairs $(-2, 4)$ and $(2, 4)$ both belong to f. Hence the reversed pairs

$$(4, -2) \quad \text{and} \quad (4, 2)$$

both belong to the inverse relation f^{-1}. Therefore, f^{-1} is not a *function*

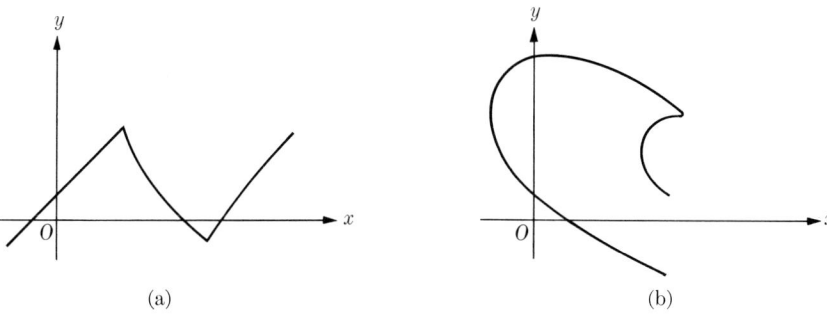

FIG. 8–9. (a) Graph of a function. (b) Graph of a relation that is not a function.

g, because if it were, we would have $g(4) = -2$ and $g(4) = +2$, which violates the requirement that a function be single valued.

There is a simple graphical way to test whether a given curve in the xy-plane represents a function. If every line drawn perpendicular to the x-axis either intersects the curve in just one point, or not at all, the graph represents a function (single valued). But if some line perpendicular to the x-axis meets the curve in more than one point (Fig. 8–9(b)), the relation represented by the graph is not a function.

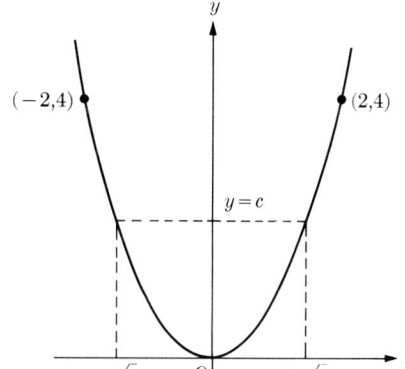

FIG. 8–10. If $f(x) = x^2$, $-2 \leq x \leq 2$, then f^{-1} is a relation that is not a function.

Now if we interchange the roles of x and y in the foregoing graphical test, we get a way to determine from the graph of f whether f^{-1} is also a function. If any line perpendicular to the y-axis meets the graph of f in more than one point, then f^{-1} is not a function. This is just another way of saying that a necessary and sufficient condition for f^{-1} to be a function is that the mapping from the domain of f onto the range of f be one-to-one. For example (Fig. 8–10), the line $y = c$, $0 < c \leq 4$, meets the graph of $f(x) = x^2$ at the points $(-\sqrt{c}, c)$ and $(+\sqrt{c}, c)$. For f such that

$$f(x) = x^2, \quad -2 \leq x \leq 2,$$

the inverse relation is the set

$$f^{-1} = \{(y, x) : y = x^2,\ -2 \le x \le 2\}.$$

This is the union of the two sets

$$g_1 = \{(y, x) : y = x^2,\ -2 \le x \le 0\}$$

and

$$g_2 = \{(y, x) : y = x^2,\ 0 \le x \le 2\},$$

each of which is a function. The functions g_1 and g_2 can also be expressed as

$$g_1 = \{(y, -\sqrt{y}) : 0 \le y \le 4\} \quad \text{and} \quad g_2 = \{(y, \sqrt{y}) : 0 \le y \le 4\}.$$

Alternatively, using x instead of y to represent any number in the domain of g_1 or g_2, we have

$$g_1 = \{(x, -\sqrt{x}) : 0 \le x \le 4\} \quad \text{and} \quad g_2 = \{(x, \sqrt{x}) : 0 \le x \le 4\}.$$

Remark 5. If a function f is continuous over the domain $[a, b]$ and is also one-to-one, then $f(a) \ne f(b)$. This means that either

i) $\quad f(a) < f(b)$

or

ii) $\quad f(a) > f(b).$

In the first case it is fairly easy to see that it must also be true that $f(a) < f(x) < f(b)$ for every $x \in (a, b)$. For, if this were not so, then either $f(x) \le f(a)$ for some x, or $f(x) \ge f(b)$ for some x. Applying the intermediate-value theorem to the first of these possibilities, on $[x, b]$, leads to the conclusion that f takes on the value $f(a)$ for some c between x and b, which is false if f is one-to-one. The possibility $f(x) \ge f(b)$ leads, in the same way, to the false conclusion that f takes on the value $f(b)$ for some c' between a and x. In other words, we see that, if f is one-to-one on $[a, b]$ and $f(a) < f(b)$, then

$$a < x < b \Rightarrow f(a) < f(x) < f(b).$$

The same argument, applied on $[a, x]$, shows that

$$a < x' < x < b \Rightarrow f(a) < f(x') < f(x),$$

which means that f is strictly increasing on $[a, b]$.

If the second case, namely $f(a) > f(b)$, holds, and f is continuous and one-to-one on $[a, b]$, one concludes in like manner that f is strictly decreasing on $[a, b]$.

EXERCISES 8-3

Find the inverses of the following functions f, and specify the domain of each inverse function. Sketch graphs of each function and its inverse.

1. $f(x) = 2x - 3$, $-\infty < x < \infty$
2. $f(x) = 2x - 3$, $-1 \leq x \leq 1$
3. $f(x) = x^2$, $0 \leq x \leq 2$
4. $f(x) = \dfrac{1}{x+1}$, $x \neq -1$
5. $f(x) = \dfrac{1}{\sqrt{x}}$, $x > 0$
6. $f(x) = 2^{1/x}$, $x > 0$
7. $f(x) = x^2 + 2x - 3$, $x \leq -1$
8. Does the function f,
$$f(x) = x^2 + 2x - 3, \quad -3 \leq x \leq 1,$$
have an inverse function? Explain.

For each of the following functions f, determine whether the inverse relation f^{-1} is a function.

9. $f(x) = \dfrac{x}{x+1}$, $x \neq -1$
10. $f(x) = x + \dfrac{1}{x}$, $x \neq 0$
11. $f(x) = \sin x$, $0 \leq x \leq \pi$
12. $f(x) = \sin x$, $-\pi/2 \leq x \leq \pi/2$
13. $f(x) = \tan x$, $0 < |x - \pi/2| < \pi/2$
14. $f(x) = \tan x$, $|x| < \pi/2$

8-4 CONTINUITY OF INVERSE FUNCTIONS

Let f be a function and let c be an inner point of the domain of f at which f is continuous. Suppose that g is an inverse function of f. Is it necessarily true that g is continuous at $f(c)$?

Example 1. Let $f(x) = x^3$, $c = 2$. Then $g(x) = x^{1/3}$ and $f(c) = 8$. The question is whether g is continuous at 8. It is easy to supply the answer for this example by going directly to the ϵ, δ-condition for continuity. Suppose that $\epsilon > 0$. We look for $\delta > 0$ such that

$$|g(x) - g(8)| < \epsilon \quad \text{whenever} \quad |x - 8| < \delta. \tag{1}$$

Now g is a strictly increasing function, and there exist numbers x_1 and x_2 such that

$$g(x_1) = (x_1)^{1/3} = 2 - \epsilon, \tag{2a}$$
$$g(x_2) = (x_2)^{1/3} = 2 + \epsilon. \tag{2b}$$

Indeed, we need only take

$$x_1 = (2 - \epsilon)^3, \quad x_2 = (2 + \epsilon)^3. \tag{3}$$

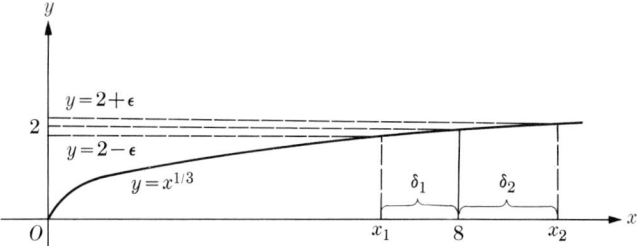

FIG. 8-11. The function f, $f(x) = x^3$, is continuous at $x = 2$. Its inverse function g, $g(x) = x^{1/3}$, is continuous at 8.

Since $\epsilon > 0$, $x_1 < 8 < x_2$, and the numbers

$$\delta_1 = 8 - x_1, \qquad \delta_2 = x_2 - 8 \qquad (4)$$

are both positive. Let $\delta = \min(\delta_1, \delta_2)$. Then $\delta > 0$, and

$$\text{if } |x - 8| < \delta, \quad \text{then} \quad x_1 < x < x_2,$$

and

$$g(x_1) = 2 - \epsilon < g(x) < 2 + \epsilon = g(x_2),$$

so that

$$|g(x) - g(8)| = |g(x) - 2| < \epsilon.$$

Figure 8-11 illustrates the various steps of the foregoing argument.

Does the example provide a clue for the general situation? Are there special features [other than the explicit formulas $f(x) = x^3$, $g(x) = x^{1/3}$] that might be helpful? We exploited the fact that g was strictly increasing, which is a consequence of the fact that the original function f is strictly increasing. The argument would have needed only slight changes if g had been strictly decreasing. Can we prove a general continuity theorem for the inverse of any strictly increasing (or decreasing) continuous function? The following theorem and proof show that the answer is affirmative.

Theorem 8-2. Let f be a strictly increasing function, continuous on the domain $a \leq x \leq b$. Then
1. f has an inverse function g, and
2. g is continuous on the domain $f(a) \leq x \leq f(b)$.

Proof. Because f is strictly increasing, $f(a) < f(x) < f(b)$ if $a < x < b$. Also, $f(x_1) < f(x_2)$ if $a \leq x_1 < x_2 \leq b$, so that an equation of the form

$f(x) = d$ has at most one solution if $f(a) < d < f(b)$. By the intermediate-value theorem, on the other hand, such an equation has at least one solution because f is continuous on $[a, b]$. Combining these facts, we see that if

$$f(a) \leq y \leq f(b),$$

the equation $f(x) = y$ has a unique solution

$$x = g(y), \quad a \leq x \leq b.$$

Thus we define g to be the function

$$g = \{(y, x) : y = f(x), f(a) \leq y \leq f(b)\},$$

or, interchanging the roles of x and y, we have

$$g = \{(x, y) : x = f(y), f(a) \leq x \leq f(b)\}.$$

Clearly g is a strictly increasing function, because if $y_1 < y_2$ and $f(x_1) = y_1$, $f(x_2) = y_2$, then x_1 cannot be equal to, or greater than x_2, and therefore $x_1 < x_2$; that is, $g(y_1) < g(y_2)$. By construction, g is an inverse function for f, and it remains to be shown that g is continuous on its domain.

First, suppose that d is an interior point of the domain of g,

$$f(a) < d < f(b).$$

Then there exists a unique c, $a < c < b$, such that $f(c) = d$ and $g(d) = c$. Let $\epsilon > 0$. If we were to follow Example 1, we would look for x_1 and x_2 such that $g(x_1) = g(d) - \epsilon$ and $g(x_2) = g(d) + \epsilon$. We need to modify this slightly because $g(d) - \epsilon$ might be less than a, or $g(d) + \epsilon$ greater than b. So we take

$$\epsilon_1 = c - a, \quad \epsilon_2 = b - c,$$

and use, in place of the original ϵ, the possibly smaller positive number

$$\epsilon_0 = \min(\epsilon, \epsilon_1, \epsilon_2). \tag{5}$$

Then

$$a \leq c - \epsilon_0 < c < c + \epsilon_0 \leq b, \tag{6}$$

and there exist unique values of x_1 and x_2 such that

$$x_1 = f(c - \epsilon_0) < f(c) = d < f(c + \epsilon_0) = x_2 \tag{7}$$

and

$$f(a) \leq x_1 < x_2 \leq f(b).$$

The equations and inequalities (7) are equivalent to

$$g(x_1) = c - \epsilon_0 < c = g(d) < c + \epsilon_0 = g(x_2). \tag{8}$$

The steps of the proof expressed in (7) and (8) are illustrated by Fig. 8–12. We start at d on the vertical axis, follow the arrows over to the curve and down to c on the horizontal axis. Then, using (5), we locate $c - \epsilon_0$ and $c + \epsilon_0$ on this axis. From these points, follow the arrows up to the curve and over to x_1 and x_2 on the vertical axis.

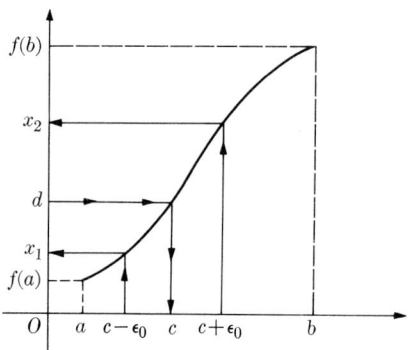

FIG. 8–12. The domain of g is shown on the vertical axis, the domain of f on the horizontal axis. The curve shown is the graph of f. It becomes the graph of g if the vertical axis is labeled x and the horizontal axis is labeled y.

We are now ready for the final step of the proof. The numbers $d - x_1$ and $x_2 - d$ are both positive, by (7), and we let δ be their minimum,

$$\delta = \min(d - x_1, x_2 - d). \tag{9}$$

Then $\delta > 0$, and whenever x is within δ-distance of d it must be between x_1 and x_2 so that $g(x)$ is between $c - \epsilon_0$ and $c + \epsilon_0$. More formally, if

$$|x - d| < \delta,$$

then

$$x_1 < x < x_2,$$

$$c - \epsilon_0 = g(x_1) < g(x) < g(x_2) = c + \epsilon_0,$$

and

$$|g(x) - c| = |g(x) - g(d)| < \epsilon_0 \leq \epsilon.$$

Therefore g is continuous at d.

We have thus proved that g is continuous at each interior point of its domain. Some modifications are necessary if either $d = f(a)$ or $d = f(b)$. Essentially, if $d = f(b)$, then $c = b$, and we omit ϵ_2 from the definition of ϵ_0 and also omit all reference to $c + \epsilon_0$ or x_2. Similar modifications, which become quite clear by reference to Fig. 8–12 with appropriate changes, apply if $d = f(a)$. (These special endpoint parts of the proof are given as exercises. They will help the reader to be sure that he understands the proof as given above.)

Remark. Although Theorem 8–2 is stated for an increasing function f, the conclusion is also valid if f is strictly decreasing.

Example 2. Let $f(x) = 1/\sqrt{x}$, $x > 0$. Then f is a decreasing function:

$$x_2 > x_1 > 0 \quad \text{implies} \quad f(x_2) < f(x_1).$$

To find the inverse function, solve

$$y = 1/\sqrt{x}$$

for

$$x = 1/y^2, \quad x > 0, y > 0,$$

then permute x and y and get

$$y = 1/x^2, \quad y > 0, x > 0.$$

Thus g is the function such that

$$g(x) = 1/x^2, \quad x > 0.$$

Note that g is also a strictly decreasing function, and it is continuous wherever the denominator is not zero. Therefore, g is continuous everywhere on its domain.

EXERCISES 8–4

1. Supply the rest of the proof of Theorem 8–2 for continuity of g at d if
   a) $d = f(a)$,
   b) $d = f(b)$.
2. In what ways would the equations and inequalities (7) and (8) be changed if f were a strictly decreasing function (instead of increasing)? How would Eq. (9) be changed? (Use a graph if you find it helpful.)
3. The Remark at the end of Section 8–3 contains the following theorem. Write out the proof in detail.

   Theorem. If f is continuous on $[a, b]$ and is one-to-one from $[a, b]$ to $[f(a), f(b)]$, then f is strictly increasing [if $f(a) < f(b)$] or strictly decreasing [if $f(a) > f(b)$].
4. Following the notation in the proof of Theorem 8–2, find explicit expressions for c, ϵ_0, x_1, x_2, and δ for the following data:
   a) $f(x) = x^2$, $a = 0$, $b = 3$, $d = 4$, $\epsilon = 0.2$
   b) $f(x) = 1/x$, $a = 1$, $b = 2$, $d = \frac{3}{4}$, $\epsilon = \frac{1}{2}$

9
DIFFERENTIABLE FUNCTIONS

9-1 DIFFERENTIABILITY AND CONTINUITY

In previous chapters we studied the properties of continuity and differentiability of functions. Is there a connection between these two properties? In order to investigate this question, we recall the formal definitions of continuity and differentiability, and consider some specific examples.

Let f be defined on a domain $a \leq x \leq b$ and c be an inner point of that domain, then we say that

1. f is *continuous* at c if

$$\lim_{x \to c} f(x) = f(c) \tag{1}$$

(Cf. with Section 6–1, Definition 1), and

2. f is *differentiable* at c if the difference quotient

$$\frac{f(x) - f(c)}{x - c} \tag{2}$$

has a *limit* as $x \to c$. In that case we write

$$f'(c) = \lim_{x \to c} \frac{f(x) - f(c)}{x - c}, \tag{3}$$

and call $f'(c)$ the derivative of f at c. (Cf. Section 1–3, Eq. (2).)

Example 1. The function

$$f(x) = |x|, \quad -\infty < x < \infty, \tag{4}$$

is everywhere continuous. It is differentiable for $x \neq 0$. But it is not differentiable at 0, because the difference quotient (2) for the absolute-value function and for $c = 0$ is

$$\frac{f(x) - f(c)}{x - c} = \frac{|x|}{x} = \begin{cases} +1 & \text{if } x > 0, \\ -1 & \text{if } x < 0, \end{cases}$$

and this has no limit as $x \to 0$ (Fig. 9–1).

Example 2. Let

$$f(x) = \begin{cases} x^2 & \text{for } x \text{ rational}, \\ 0 & \text{for } x \text{ irrational}, \end{cases} \quad -\infty < x < \infty. \tag{5}$$

The graph of f consists of points on the parabola $y = x^2$ for x rational, and points on the x-axis for x irrational. (See Fig. 9–2.) For any $c \neq 0$, if we take $\epsilon = \frac{1}{2}c^2$, then $\epsilon > 0$. Now, no matter how small a positive

DIFFERENTIABLE FUNCTIONS

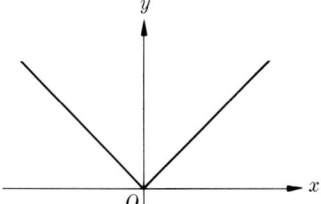

FIG. 9–1. The function $f(x) = |x|$ is not differentiable at $x = 0$.

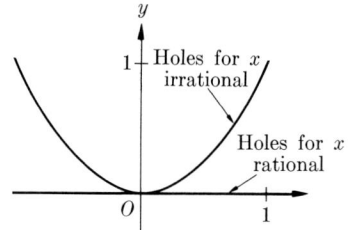

FIG. 9–2. $f(x) = x^2$ for x rational, $f(x) = 0$ for x irrational. Continuous and differentiable at $x = 0$ only.

number we might try for δ, the interval $c - \delta < x < c + \delta$ contains both rational and irrational members close to c. Hence, there exists x in every such neighborhood such that

$$|f(x) - f(c)| = \begin{cases} x^2 > \epsilon & \text{if } x \text{ is rational, } c \text{ irrational,} \\ c^2 > \epsilon & \text{if } x \text{ is irrational, } c \text{ rational.} \end{cases}$$

Therefore f is discontinuous at every $c \neq 0$. But f is continuous at $x = 0$, because

$$\lim_{x \to 0} f(x) = \lim_{x \to 0} \begin{Bmatrix} x^2 \\ \text{or} \\ 0 \end{Bmatrix} = 0 = f(0).$$

Also, f is *differentiable* at $c = 0$, because the difference quotient is

$$\frac{f(x) - f(0)}{x - 0} = \begin{cases} x & \text{if } x \text{ is rational,} \\ 0 & \text{if } x \text{ is irrational,} \end{cases}$$

and

$$\lim_{x \to 0} \frac{f(x) - f(0)}{x - 0} = \lim_{x \to 0} \begin{Bmatrix} x \\ \text{or} \\ 0 \end{Bmatrix} = 0.$$

Even though the graph in Fig. 9–2 is erratic, jumping back and forth between the parabola and the x axis, there is a unique line ($y = 0$) tangent to the graph at the origin.

Example 3. Let

$$f(x) = \begin{cases} x \sin (1/x) & \text{if } x \neq 0, \\ 0 & \text{if } x = 0. \end{cases} \qquad (6)$$

The graph (Fig. 9–3) has infinitely many oscillations back and forth between the lines $y = x$ and $y = -x$ in any neighborhood containing

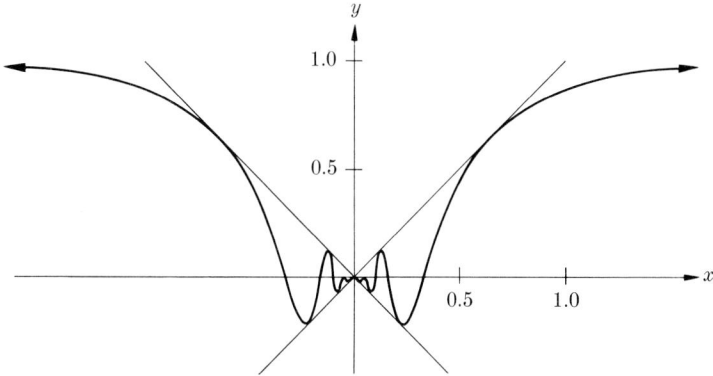

FIG. 9-3. Portion of the graph of $y = \begin{cases} x \sin(1/x), & x \neq 0, \\ 0, & x = 0. \end{cases}$

$x = 0$. The graph is continuous for $x \neq 0$, because

a) $1/x$ is continuous for $x \neq 0$,
b) $\sin(1/x)$ is continuous where $1/x$ is,
c) $x \sin(1/x)$ is a product of functions both continuous for $x \neq 0$.

The function is also continuous at $x = 0$, because

$$|f(x)| = |x \sin(1/x)| \leq |x|, \quad \text{for all } x \neq 0,$$

so that, if $\epsilon > 0$ and $\delta = \epsilon$, then

$$|f(x) - f(0)| \leq |x| < \epsilon \quad \text{whenever} \quad |x - 0| < \delta.$$

Therefore,

$$\lim_{x \to 0} f(x) = 0 = f(0).$$

However, this function is not differentiable at $x = 0$, because

$$\frac{f(x) - f(0)}{x - 0} = \sin(1/x), \quad x \neq 0,$$

has no limit as $x \to 0$, since the sine oscillates back and forth between $+1$ and -1 in every interval $1/(2n\pi) \geq x \geq 1/(2n\pi + 2\pi)$ for n any positive integer.

Examples 1 and 3 show that a function may be continuous at a point without being differentiable there. But we have no examples of the opposite behavior. The following theorem shows that such examples are nonexistent.

DIFFERENTIABLE FUNCTIONS

Theorem 9–1. Let f be defined on a domain having c as an inner point. If f is *differentiable* at c, then f is continuous at c.

Proof. By hypothesis, the limit

$$f'(c) = \lim_{x \to c} \frac{f(x) - f(c)}{x - c}$$

exists. Hence, the following limit also exists:

$$\lim_{x \to c} [f(x) - f(c)] = \lim_{x \to c} \frac{f(x) - f(c)}{x - c} (x - c)$$

$$= \lim_{x \to c} \frac{f(x) - f(c)}{x - c} \lim_{x \to c} (x - c)$$

$$= f'(c) \cdot 0 = 0.$$

Therefore,

$$\lim_{x \to c} f(x) = \lim_{x \to c} [f(c) + f(x) - f(c)]$$

$$= \lim_{x \to c} f(c) + \lim_{x \to c} [f(x) - f(c)]$$

$$= f(c). \qquad \text{Q.E.D.}$$

Remark 1. The above proof is really too formal. The idea is this: if the quotient

$$\frac{f(x) - f(c)}{x - c}$$

has a limit as $x \to c$, then the numerator must approach zero when the denominator does; therefore

$$f(x) \to f(c) \text{ as } x \to c.$$

EXERCISES 9–1

1. Which of the following functions are continuous, and which are discontinuous on the given domain.

   a) $f(x) = \begin{cases} x, & x \neq 0 \\ 1, & x = 0 \end{cases}$ b) $f(x) = |x - 1|, \quad x \in R$

   c) $f(x) = [x], \quad |x| \leq 5$ d) $f(x) = \sqrt{a^2 - x^2}, \quad |x| \leq a$

   e) $f(x) = |\cos x|, \quad |x| \leq 2\pi$ f) $f(x) = |\sec x|, \quad |x| \leq 2\pi$

2. For what values of x, if any, in each of the above, does $f'(x)$ fail to exist?
3. Given $f(x) = x|x|$, where $x \in R$. Does $f'(0)$ exist? Give a reason.

4. Given
$$f(x) = \begin{cases} \dfrac{|x|}{x}, & x \neq 0, \\ 0, & x = 0, \end{cases}$$
does $f'(0)$ exist? Give a reason.

5. Sketch graphs of the functions given in Exercises 3 and 4. On the same axes, sketch graphs of the derived functions.

9–2 ROLLE'S THEOREM

We now consider a function f, defined and continuous at each point of the closed interval $a \leq x \leq b$, and differentiable at each point of the open interval $a < x < b$. Rolle's theorem asserts that, if $f(a) = f(b) = 0$, then there exists at least one c, $a < c < b$, such that $f'(c) = 0$. (See Fig. 9–4.) The proof is simple.

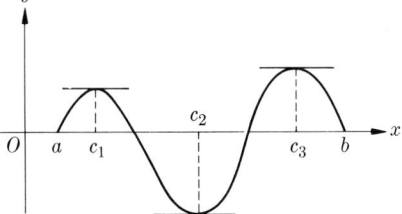

FIG. 9–4. Rolle's theorem illustrated: here there are three solutions c_1, c_2, and c_3 of $f'(c) = 0$, $a < c < b$.

CASE 1. If $f(x) = 0$ for all x, $a \leq x \leq b$, then also $f'(c) = 0$ for all c in the open interval $a < c < b$, and c can be any number between a and b.

CASE 2. Suppose $f(x)$ is different from zero for some values of x between a and b. By Theorem C-I of Section 6–1, f has a maximum and a minimum, and not both of these are zero, so one of them occurs at a point c strictly between a and b. Since $a < c < b$, and $f'(c)$ exists, and $f(c)$ is a maximum (or minimum) value of f, $f'(c) = 0$ (Theorem 5–1, Section 5–3). We have therefore proved the following theorem.

Theorem 9–2. Rolle's Theorem. Let f be a real-valued function, continuous on $a \leq x \leq b$, and differentiable on $a < x < b$. Then, if $f(a) = f(b) = 0$, there exists a number c, $a < c < b$, such that

$$f'(c) = 0. \qquad (1)$$

The chord through the points $(a, 0)$ and $(b, 0)$ in Fig. 9–4 lies on the x-axis, and the c referred to in Eq. (1) gives a point $(c, f(c))$ where the tangent to the graph is parallel to that chord. Can this result be generalized? Yes, as the following theorem asserts.

DIFFERENTIABLE FUNCTIONS

Theorem 9–3. Mean-Value Theorem. Let f be a real-valued function continuous on $a \leq x \leq b$ and differentiable on $a < x < b$. Then

$$f'(c) = \frac{f(b) - f(a)}{b - a} \quad \text{for some } c, \quad a < c < b. \tag{2}$$

Proof.* Consider the chord AB (Fig. 9–5) through the points $A(a, f(a))$ and $B(b, f(b))$. Let $P(x, f(x))$ be an arbitrary point on the graph of f, with $a \leq x \leq b$. Let $Q(x, 0)$ be the foot of the perpendicular from P to the x-axis, and let $R(x, y)$ be the point of intersection of AB and PQ. The slope of AB is

$$m = \frac{f(b) - f(a)}{b - a}, \tag{3}$$

and the ordinate, y, of R satisfies the equation

$$y - f(a) = m(x - a),$$

or

$$y = f(a) + m(x - a). \tag{4}$$

The signed vertical distance PR between the curve and the chord at abscissa x is

$$F(x) = y_R - f(x)$$

or, from (3) and (4),

$$F(x) = f(a) + \frac{f(b) - f(a)}{b - a}(x - a) - f(x). \tag{5}$$

Since the vertical distance PR is zero when $x = a$ or $x = b$,

$$F(a) = 0 \quad \text{and} \quad F(b) = 0. \tag{6}$$

[Check that these results also follow directly from Eq. (5) if we substitute $x = a$ or $x = b$.] The function F is the difference between the linear function (4) and the function f. Hence F satisfies all the hypotheses of Rolle's theorem, so

$$F'(c) = 0 \quad \text{for some } c, \quad a < c < b. \tag{7}$$

But from (5),

$$F'(x) = \frac{f(b) - f(a)}{b - a} - f'(x),$$

---

* See Exercise 10 for an outline of an alternative proof.

so (7) gives

$$0 = \frac{f(b) - f(a)}{b - a} - f'(c) \quad \text{for some } c, \quad a < c < b,$$

which is the conclusion, (2), we wished to establish. Q.E.D.

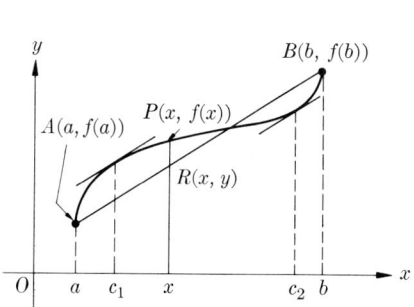

FIG. 9–5. Illustrating the mean-value theorem:

$$f'(c) = \frac{f(b) - f(a)}{b - a}$$

for $c = c_1$ and $c = c_2$.

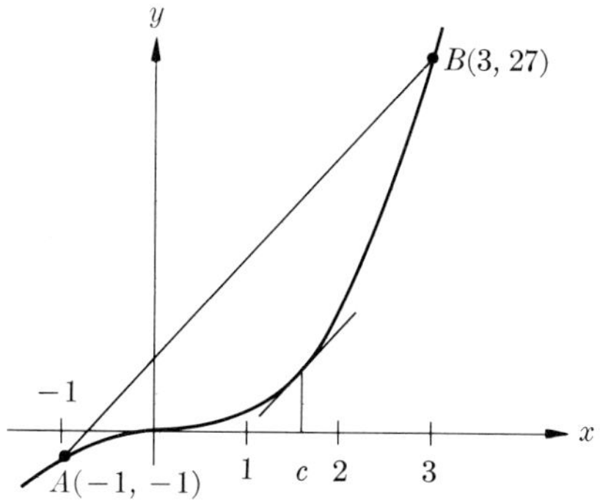

FIG. 9–6. Illustrating the mean-value theorem:

$f(x) = x^3, a = -1, b = 3, c = \sqrt{\frac{7}{3}}.$

Example 1. Let $a = -1$, $b = 3$, $f(x) = x^3$. Then $f(a) = (-1)^3 = -1$, $f(b) = 27$, $f'(c) = 3c^2$,

$$\frac{f(b) - f(a)}{b - a} = \frac{28}{4} = 7,$$

and Eq. (2) becomes

$$3c^2 = 7 \quad \text{for some } c, \quad -1 < c < 3.$$

Since

$$-\sqrt{\tfrac{7}{3}} \approx -1.527, \quad \sqrt{\tfrac{7}{3}} \approx +1.527,$$

the appropriate choice is $c = \sqrt{\frac{7}{3}}$. (See Fig. 9–6.)

Example 2. If $f(x) = |x|$, $a = -1$, $b = 3$, then $f(a) = 1$, $f(b) = 3$, and

$$\frac{f(b) - f(a)}{b - a} = \frac{2}{4} = \frac{1}{2}.$$

But

$$f'(c) = \begin{cases} -1 & \text{if } c < 0, \\ \text{does not exist} & \text{if } c = 0, \\ +1 & \text{if } c > 0, \end{cases}$$

and the mean-value theorem does not apply. Why not?

DIFFERENTIABLE FUNCTIONS

Michel Rolle (1652–1716) wrote on algebra and geometry. His publications appear in the form of memoirs and two books.

The theorem that $f'(x) = 0$ has at least one real root lying between two successive roots of $f(x) = 0$ is due to him. It was first published in "Demonstration d'une Methode pour resoudre les Egalitez de tous degrez," Paris, 1691.

EXERCISES 9–2

1. Draw the graph of $f: x \to x^3 - x^2 - 6x$. Does Rolle's theorem apply? Discuss in detail.

2. Given $f: x \to x^3 - 3x + 1$:
   a) Find the slope of the line L through points of the graph of f with abscissas -2 and 2. Sketch the graphs of f and L.
   b) Find those points of the curve where the slope is the same as that of L.
   c) Are the abscissas of these points between -2 and 2?
   d) For $|x| \leq 2$, can $f(x)$ assume the values $-\frac{1}{2}$, 0, 2, -2, 4, 2.9, 3.2? Give a reason for your answer.

In each of the following, state whether or not the mean-value theorem holds for the given interval, and give a reason for your answer.

3. $y = x^{2/3}$ for $[-2, 1]$; for $[0, 3]$

4. $y = \dfrac{x}{1 + x^2}$ for $[-2, 2]$; for $[0, 3]$

5. $y = \dfrac{1}{1 - x^2}$ for $[-2, 2]$; for $[-2, -1]$; for $[1, 2]$

6. $y = \tan x$ for $[-\pi/2, \pi/2]$; for $[-\pi/4, \pi/4]$

7. $y = a_n x^n + a_{n-1} x^{n-1} + \cdots + a_1 x + a_0$, $a_n \neq 0$, for $(-\infty, \infty)$

8. Use the mean-value theorem (MVT) to show that $\sin x < x$ for $x > 0$.

9. Use the MVT to show that

$$\frac{h}{1+h} < \ln(1+h) < h, \quad \text{where } h > -1 \text{ and } h \neq 0.$$

10. Deduce the MVT from Rolle's theorem, following the steps indicated: Let f be a real-valued function, continuous on the closed interval $a \leq x \leq b$ and differentiable on the open interval $a < x < b$.

   i) Introduce the linear function g which agrees with f at a and b. What is the slope of g?
   ii) Let h be the function defined by $h(x) = f(x) - g(x)$. Sketch the graphs of f, g, and h.
   iii) Does h satisfy the hypotheses of Rolle's theorem? State precisely and give reasons.
   iv) Complete the proof.

11. Use the MVT to prove
$$e^x > 1 + x, \qquad x \neq 0$$
12. Derive the following inequalities by using the MVT:
   a) $|\sin x_1 - \sin x_2| \leq |x_1 - x_2|$
   b) $\tan x > x, \quad 0 < x < \pi/2$
13. Given
$$y = \begin{cases} x \sin(1/x), & x \neq 0, \\ 0, & x = 0 \end{cases}$$
(see Fig. 9–3). At what values of x other than zero is $y = 0$?

9–3 EXTENSIONS OF THE MEAN-VALUE THEOREM

We turn our attention now to a theorem that shows how the value of a function f at any x "near" a can be approximated if we know the value of f and its first n derivatives at a. The idea is to use a polynomial p that matches f "very well" at $x = a$. More specifically, we try to determine the coefficients $a_0, a_1, \ldots, a_n$ so that the polynomial

$$p(x) = a_0 + a_1 x + a_2 x^2 + \cdots + a_n x^n \tag{1}$$

and its first n derivatives have the same values as $f(x)$ and its first n derivatives, respectively, at $x = a$:

$$\begin{aligned} p(a) &= f(a), \\ p'(a) &= f'(a), \\ &\vdots \\ p^{(n)}(a) &= f^{(n)}(a). \end{aligned} \tag{2}$$

The $n + 1$ equations (2) can be used to evaluate $a_0, a_1, \ldots, a_n$ in terms of $f(a), f'(a), \ldots, f^{(n)}(a)$. The algebra is simpler, though, if we write the polynomial (1) in powers of $x - a$, which we can surely do by replacing x by $(x - a) + a$ and expanding terms. There is no need to carry out the substitution. The result is an equation of the form

$$p(x) = c_0 + c_1(x - a) + c_2(x - a)^2 + \cdots + c_n(x - a)^n, \tag{3}$$

with $n + 1$ new coefficients. It is easy to evaluate these new coefficients using Eqs. (2). First, we compute a few derivatives from Eq. (3):

$$p'(x) = c_1 + 2c_2(x - a) + \cdots + nc_n(x - a)^{n-1},$$
$$p''(x) = 2c_2 + 6c_3(x - a) + \cdots + n(n - 1)c_n(x - a)^{n-2}.$$

DIFFERENTIABLE FUNCTIONS

Thus,
$$p(a) = c_0, \quad p'(a) = c_1, \quad p''(a) = 2c_2, \qquad (4)$$

because $x - a$, $(x - a)^2$, and so on, are all zero when $x = a$. In fact, if we differentiate $p(x)$ k times, the terms of degree less than k, like

$$c_0, c_1(x - a), \ldots, c_{k-1}(x - a)^{k-1}$$

have kth derivatives equal to zero. The kth derivative of $c_k(x - a)^k$ is

$$k!c_k.$$

The kth derivative of terms of degree higher than k, like

$$c_{k+1}(x - a)^{k+1}, \ldots, c_n(x - a)^n$$

have positive powers of $x - a$, like

$$(x - a)^1, \ldots, (x - a)^{n-k},$$

and all these are zero when $x = a$. Therefore, the kth derivative of $p(x)$, at $x = a$, is just $k!c_k$:

$$p^{(k)}(a) = k!c_k, \qquad k = 1, 2, \ldots, n. \qquad (5)$$

If we interpret the zeroth derivative $p^{(0)}(a)$ as the value of the function itself, $p(a)$, and define $0! = 1$, then Eqs. (5) include Eqs. (4), when $k = 0$, 1, and 2. Now we are ready for the matching of values of p and its derivatives with f and its derivatives at $x = a$, as in Eqs. (2):

$$\begin{aligned}
p(a) &= c_0 = f(a), \\
p'(a) &= c_1 = f'(a), \\
p''(a) &= 2c_2 = f''(a), \\
&\vdots \\
p^{(k)}(a) &= k!c_k = f^{(k)}(a), \quad 0 \le k \le n.
\end{aligned} \qquad (6)$$

From Eqs. (6) we get

$$c_k = \frac{f^{(k)}(a)}{k!}, \qquad k = 0, 1, 2, \ldots, n. \qquad (7)$$

If we substitute (7) into (3), we get the following theorem.

Theorem 9-4. Let f be n times differentiable at $x = a$. Then, the unique polynomial p, of degree n, satisfying

$$p(a) = f(a), \quad p'(a) = f'(a), \quad \ldots, \quad p^{(n)}(a) = f^{(n)}(a),$$

is

$$p(x) = f(a) + f'(a)(x - a) + \frac{f''(a)}{2!}(x - a)^2 + \cdots$$
$$+ \frac{f^{(n)}(a)}{n!}(x - a)^n, \tag{8a}$$

or

$$p(x) = \sum_{k=0}^{n} \frac{f^{(k)}(a)}{k!}(x - a)^k. \tag{8b}$$

If n is large, Eqs. (8a, b) provide a polynomial whose graph has a high degree of contact with the graph of f at $x = a$, where "contact" is measured by "matching of derivatives." Two questions arise: (1) If we hold x fixed and let $n \to \infty$, is it true that

$$\lim_{n \to \infty} p(x) = f(x)?$$

(2) If we hold n fixed, and vary x, what can we say about the size of the difference $f(x) - p(x)$ in terms of the size of $x - a$?

An answer to the second question will also help us answer the first question. The result, known as *Taylor's theorem with remainder*, after Brook Taylor (1685–1731), who rediscovered it in 1715*, is a general theorem found earlier (c. 1670) by James Gregory (1638–1675) but not published at that time. As we state it, the theorem expresses $f(b)$ in terms of the polynomial $p(b)$, Eqs. (8a, b) with $x = b$, and a remainder term.

Theorem 9–5. Taylor's Theorem with Remainder. Let a and b be real numbers. Let f and its first n derivatives be continuous on the closed interval $[a, b]$, and suppose $f^{(n+1)}$ exists on the open interval (a, b). Then there exists a number c in (a, b) such that

$$f(b) = f(a) + f'(a)(b - a) + \frac{f''(a)}{2!}(b - a)^2 + \cdots$$
$$+ \frac{f^{(n)}(a)}{n!}(b - a)^n + \frac{f^{(n+1)}(c)}{(n + 1)!}(b - a)^{n+1}. \tag{9}$$

Proof. Our proof follows that published in the *American Mathematical Monthly*, **60** (1953), p. 415, by James Wolfe. The idea of the proof is this. Start with the polynomial

$$p(x) = \sum_{k=0}^{n} \frac{f^{(k)}(a)}{k!}(x - a)^k \tag{10}$$

---

* "Methodus Incrementorum directa et inversa," London, 1715.

described in Theorem 9–4. This polynomial matches f very well at $x = a$, but may be off by quite a bit at $x = b$. We can add an additional term, of the form

$$r(x) = C(x - a)^{n+1}, \tag{11}$$

and choose C so that

$$p(b) + r(b) = f(b). \tag{12}$$

This is easy; just take

$$C = (b - a)^{-(n+1)}(f(b) - p(b)), \quad \text{if} \quad b \neq a, \tag{13}$$

or

$$C = 0, \quad \text{if} \quad b = a.$$

Actually, if $b = a$, all terms after the first on the right-hand side of Eq. (9) are zero, and the theorem reduces to saying $f(a) = f(a)$, which is true, but not exciting. So we henceforth assume $b \neq a$.

Now observe that no matter what constant is chosen for C, the polynomial in Eq. (11) is continuous, together with all of its derivatives. Moreover, r and its first n derivatives have the value 0 at $x = a$:

$$r(a) = r'(a) = r''(a) = \cdots = r^{(n)}(a) = 0. \tag{14}$$

This means that the new polynomial

$$q(x) = p(x) + r(x), \tag{15}$$

of degree $(n + 1)$, still has high contact with f at $x = a$, for any choice of C; and, for the particular choice of C given by Eq. (13), we also have

$$q(b) = f(b). \tag{16}$$

Now consider the function F defined by

$$F(x) = f(x) - q(x), \quad \text{for } x \text{ in } [a, b]. \tag{17}$$

Because

$$q(a) = p(a) + r(a) = f(a) + 0 = f(a),$$

and

$$q(b) = f(b),$$

we have

$$F(a) = F(b) = 0. \tag{18}$$

Also, F is continuous (together with its first n derivatives) on $[a, b]$, so Rolle's theorem ensures the existence of c_1 in (a, b) such that

$$F'(c_1) = 0. \tag{19a}$$

But also

$$F'(a) = 0, \tag{19b}$$

because

$$q'(a) = p'(a) + r'(a) = f'(a) + 0 = f'(a).$$

Now apply Rolle's theorem to the derived function, F', on $[a, c_1]$, to conclude that there exists c_2 in (a, c_1) such that

$$F''(c_2) = 0. \tag{20a}$$

Again

$$F''(a) = 0, \tag{20b}$$

because

$$q''(a) = p''(a) + r''(a) = f''(a) + 0 = f''(a).$$

So Rolle's theorem, applied to F'' on $[a, c_2]$, yields c_3 in (a, c_2) such that

$$F'''(c_3) = 0.$$

This method, continued, produces $c_1, c_2, \ldots, c_n$ such that

$$F'(c_1) = F''(c_2) = F'''(c_3) = \cdots = F^{(n)}(c_n) = 0.$$

And now comes the last step. We have

$$F^{(n)}(a) = 0, \qquad F^{(n)}(c_n) = 0,$$

and $F^{(n)}$ is continuous on $[a, c_n]$ and differentiable on (a, c_n), so Rolle's theorem applied to $F^{(n)}$ on $[a, c_n]$ yields c_{n+1}, between a and c_n, such that

$$F^{(n+1)}(c_{n+1}) = 0. \tag{21}$$

But

$$F^{(n+1)}(x) = f^{(n+1)}(x) - q^{(n+1)}(x) \tag{22a}$$

and

$$\begin{aligned} q^{(n+1)}(x) &= p^{(n+1)}(x) + r^{(n+1)}(x) \\ &= 0 + (n+1)!C. \end{aligned} \tag{22b}$$

Substituting (22a, b) in (21), with $x = c_{n+1}$, we get

$$f^{(n+1)}(c_{n+1}) - (n+1)!C = 0$$

or

$$C = \frac{f^{(n+1)}(c_{n+1})}{(n+1)!}. \tag{23}$$

We know that c_{n+1} is in (a, b) because c_1 is in (a, b), c_2 in (a, c_1), ..., c_{n+1} in (a, c_n). Thus, if we take $c = c_{n+1}$ and substitute (23) into (11) and (12), we get

$$f(b) = p(b) + r(b)$$
$$= p(b) + \frac{f^{(n+1)}(c)}{(n+1)!}(b-a)^{n+1},$$

which is equivalent to Eq. (9). Q.E.D.

Example 1. Let $f(x) = x^3$, $a = 1$, $b = 1.2$, and take $n = 2$. By Taylor's theorem,

$$f(b) = f(a) + f'(a)(b-a) + \frac{f''(a)}{2!}(b-a)^2 + \frac{f'''(c)}{3!}(b-a)^3$$

for some c, $1 < c < 1.2$. But

$$f(x) = x^3, \quad f'(x) = 3x^2, \quad f''(x) = 6x, \quad f'''(x) = 6,$$

so

$$f(a) = f(1) = 1, \quad f'(a) = 3, \quad f''(a) = 6, \quad f'''(c) = 6,$$

and $f(b) = (1.2)^3$, so

$$(1.2)^3 = 1 + 3(0.2) + \frac{6}{2!}(0.2)^2 + \frac{6}{3!}(0.2)^3,$$

which is just another form of the binomial expansion

$$(1+h)^3 = 1 + 3h + 3h^2 + h^3, \quad \text{with} \quad h = 0.2.$$

Example 2. In the study of the exponential function $f(x) = e^x$, it is shown that the derivative is $f'(x) = e^x = f(x)$. Also, $f(0) = e^0 = 1$. Thus, if we take $a = 0$, and $b > 0$, Taylor's theorem becomes

$$f(b) = f(0) + f'(0)b + \frac{f''(0)}{2!}b^2 + \cdots + \frac{f^{(n)}(0)}{n!}b^n + \frac{f^{(n+1)}(c)}{(n+1)!}b^{n+1}$$

or

$$e^b = 1 + b + \frac{b^2}{2!} + \cdots + \frac{b^n}{n!} + \frac{e^c}{(n+1)!}b^{n+1}. \tag{24}$$

Since $e = 2.718\ldots < 3$ and $0 < c < b$, the last term in (24) can be estimated:

$$\frac{e^c}{(n+1)!} b^{n+1} < \frac{3}{(n+1)!}, \quad \text{if} \quad 0 < b \leq 1.$$

Since

$$\lim_{n \to \infty} \frac{3}{(n+1)!} = 0,$$

Eq. (24) gives, for $0 < b \leq 1$,

$$e^b = \lim_{n \to \infty} \left(1 + b + \frac{b^2}{2!} + \cdots + \frac{b^n}{n!}\right). \tag{25a}$$

In particular, with $b = 1$, we get

$$e = \sum_{n=0}^{\infty} \frac{1}{n!} = \frac{1}{0!} + \frac{1}{1!} + \frac{1}{2!} + \frac{1}{3!} + \cdots \tag{25b}$$

Remark 1. The c in Eq. (9) of Taylor's theorem depends upon a, b, n, and the function f. If the hypotheses of the theorem are satisfied over the interval $[a, b]$ for a particular n, they are also satisfied for the interval $[a, x]$ for that n (or any smaller positive integer) provided x is in $[a, b]$. Therefore, under those hypotheses, we can replace b by x and conclude that

$$f(x) = f(a) + f'(a)(x - a) + \frac{f''(a)}{2!}(x - a)^2 + \cdots$$

$$+ \frac{f^{(n)}(a)}{n!}(x - a)^n + R_n(x, a), \tag{26a}$$

where $R_n(x, a)$, the remainder after the term containing $(x - a)^n$, is given by

$$R_n(x, a) = \frac{f^{(n+1)}(c)}{(n+1)!}(x - a)^{n+1} \tag{26b}$$

for some c in (a, x). If, in (26a), we omit the remainder term, then we no longer have an equality (in general), but an approximation of the form

$$f(x) \approx f(a) + f'(a)(x - a) + \frac{f''(a)}{2!}(x - a)^2 + \cdots$$

$$+ \frac{f^{(n)}(a)}{n!}(x - a)^n. \tag{27}$$

The right-hand side of (27) is just the polynomial $p(x)$ of Theorem 9–4, Eqs. (8a, b). There is a polynomial of this type for each positive integer

DIFFERENTIABLE FUNCTIONS

n for which the derivatives of f, to order n, exist at $x = a$. Thus, for

$$f(x) = e^x, \quad a = 0,$$

some approximations to the graph of $y = e^x$ are shown in Fig. 9-7.

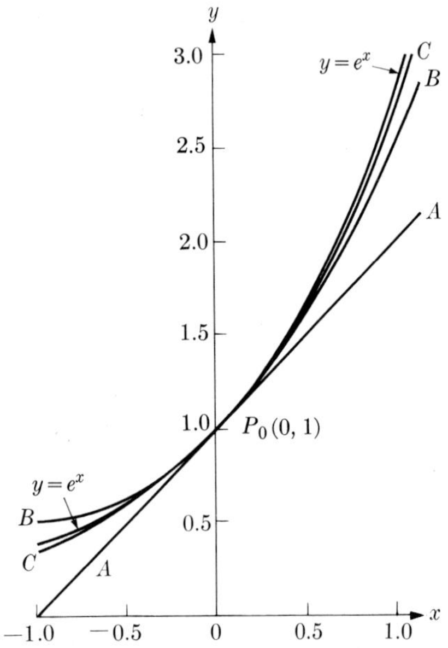

Remark 2. Since

$$R_n(x, a) = f(x) - p(x)$$

is the difference between $f(x)$ and the approximating polynomial $p(x)$, Eq. (26b) gives us an answer to our question about the size of $f(x) - p(x)$ in terms of $(x - a)$. For fixed n, the difference between $f(x)$ and $p(x)$ is (approximately) proportional to $(x - a)^{n+1}$. We must say "(approximately) proportional" instead of "proportional," because in most cases $f^{(n+1)}(c)$ varies when x varies. Often there is a bound, say

FIG. 9-7. Approximations to $y = e^x$:
(A) $y = 1 + x$,
(B) $y = 1 + x + (x^2/2!)$,
(C) $y = 1 + x + (x^2/2!) + (x^3/3!)$.

$$|f^{(n+1)}(x)| \leq K_n \quad \text{for all } x \text{ in } [a, b].$$

Then, of course, we have

$$|R_n(x, a)| \leq \frac{K_n}{(n+1)!} |x - a|^{n+1}. \tag{28}$$

Some examples of bounded derivatives:
1. $f(x) = \sin x$ or $\cos x$, $|f^{(n+1)}(x)| \leq 1$;
2. $f(x) = e^x$, $a = 0$, $|f^{(n+1)}(x)| = e^x \leq e^b$, if $x \leq b$;
3. $f(x) = \ln(1 + x)$, $a = 0$, $|f^{(n+1)}(x)| \leq n!$, if $x > 0$.

We have partially answered the first question following Eq. (8b). For a fixed x,

$$\lim_{n \to \infty} p(x) = f(x),$$

provided $R_n(x, a) \to 0$ as $n \to \infty$. This condition is satisfied, for example, whenever $|f^{(n+1)}(x)|$ is bounded on the interval $[a, b]$. In the logarithmic example,

$$|R_n(x, 0)| \leq \frac{n!}{(n+1)!} |x|^{n+1} = \frac{|x|^{n+1}}{n+1},$$

and $|R_n|$ approaches 0 as $n \to \infty$, provided $0 \leq x < 1$.

EXERCISES 9-3

1. Give an example of a function that is continuous everywhere, has a first derivative everywhere, but has no second derivative at $x = 0$.

2. If
$$f(x) = \begin{cases} 1 + 3x & \text{for } x < 1, \\ x^2 + Ax + B & \text{for } x \geq 1, \end{cases}$$
what values of A and B will make:

   a) f continuous at $x = 1$? b) f differentiable at $x = 1$?

   Will these same values of A and B make $f''(1)$ exist? Explain.

3. If $f(0) = 0$ and $f(x) = x^{3/2} \sin(1/x)$ for $x \neq 0$, show that:

   a) $f'(x)$ exists for all x; b) f' is not continuous at $x = 0$.

   [*Hint*: For $x \neq 0$, $f'(x) = -x^{-1/2} \cos(1/x) + \frac{3}{2} x^{1/2} \sin(1/x)$, but
$$f'(0) = \lim_{x \to 0} \frac{f(x) - f(0)}{x - 0}.]$$

4. For the mean-value theorem,
$$f(b) = f(a) + f'(c)(b - a), \quad c \text{ in } (a, b),$$
find c given:

   a) $f(x) = \sqrt{x}$, $a = 0$, $b = 4$;
   b) $f(x) = x^2$, $a = 0$, $b = 2$;
   c) $f(x) = \sin x$, $a = \pi/6$, $b = \pi/2$;
   d) $f(x) = x^2 + hx + k$, $a = x_1$, $b = x_2$.

5. In the notation of Theorem 9-4, find the polynomial $p(x)$, of degree n, given $f(x) = \sqrt{x}$ and $a = 1$: (a) for $n = 1$; (b) for $n = 2$; (c) for $n = 3$. Sketch graphs of $f(x)$ and the three approximating polynomials for $0 \leq x \leq 3$. Comment on the goodness of fit.

6. Use the mean-value theorem to prove:

   Theorem 9-6. If f is continuous on $[a, b]$ and differentiable on (a, b), and $f'(x) = 0$ for all x in (a, b), then $f(x) = f(a)$ for all x in $[a, b]$.

   Theorem 9-7. If F_1 and F_2 are continuous on $[a, b]$ and differentiable on (a, b) and $F_1'(x) = F_2'(x)$ for all x in (a, b), then for some constant C, $F_1(x) = F_2(x) + C$ for all x in $[a, b]$.

   Theorem 9-8. If f is continuous on $[a, b]$ and $f'(x) > 0$ for all x in (a, b), then $f(x_2) > f(x_1)$ if $a \leq x_1 < x_2 \leq b$. (That is, "if f' is positive, then f is an *increasing* function.")

10
ADDITIONAL DIFFERENTIATION FORMULAS

10-1 INCREMENT OF A FUNCTION

If f is the function such that
$$f(x) = x^2,$$
then for any two numbers x and a,
$$\begin{aligned} f(x) - f(a) &= (x+a)(x-a) \\ &= (2a + (x-a))(x-a) \\ &= (f'(a) + \alpha)(x-a), \end{aligned} \quad (1)$$
where
$$f'(a) = 2a$$
is the derivative of f, at a, and
$$\alpha = x - a$$
approaches 0 as x approaches a. This way of writing the *increment*
$$\Delta f = f(x) - f(a)$$
of the function f in terms of the derivative $f'(a)$ and the increment
$$\Delta x = x - a$$
of the domain variable holds for *any* differentiable function.

Theorem 10–1. Let f be a function that is differentiable at a. Then, for any x in the domain of f, there exists α such that
$$f(x) - f(a) = (f'(a) + \alpha)(x - a), \quad (2)$$
and
$$\lim_{x \to a} \alpha = 0. \quad (3)$$

Proof. By definition of "differentiable at a" we have
$$f'(a) = \lim_{x \to a} \frac{f(x) - f(a)}{x - a}. \quad (4)$$
That is, given a and the function f, to any $\epsilon > 0$ there corresponds $\delta > 0$ such that
$$\left| \frac{f(x) - f(a)}{x - a} - f'(a) \right| < \epsilon \quad (5a)$$
whenever
$$0 < |x - a| < \delta, \quad \text{and} \quad x \in D_f. \quad (5b)$$

177

ADDITIONAL DIFFERENTIATION FORMULAS 10–1

If we define α to be the quantity inside the absolute-value bars in (5a), when $x \neq a$, Eq. (2) is satisfied. The left-hand side of (5a) is meaningless when $x = a$, because it contains 0/0, but Eq. (2) can still be satisfied, in the form $0 = 0$, if we take $\alpha = 0$ when $x = a$. To summarize: we define α so that

$$\alpha = \begin{cases} \dfrac{f(x) - f(a)}{x - a} - f'(a), & \text{if } x \neq a, \ x \in D_f, \\ 0, & \text{if } x = a. \end{cases} \tag{6}$$

Then the inequalities (5) and this definition of α imply Eq. (3), and Eq. (2) also holds. Q.E.D.

Example 1. $f(x) = x^2$. See Eq. (1).

Example 2. $f(x) = 1/x$, $a \neq 0$. Then

$$\frac{f(x) - f(a)}{x - a} = \frac{1/x - 1/a}{x - a} = \frac{(a - x)}{xa(x - a)} = \frac{-1}{xa}$$

and

$$f'(a) = \frac{-1}{a^2}.$$

Thus, for $x \neq a$, and $x \neq 0$,

$$\alpha = \frac{f(x) - f(a)}{x - a} - f'(a) = -\frac{1}{xa} + \frac{1}{a^2} = \frac{-a + x}{xa^2}.$$

Therefore,

$$\alpha = \begin{cases} \dfrac{x - a}{a^2 x}, & \text{if } x \neq a, \ x \neq 0, \\ 0, & \text{if } x = a, \end{cases}$$

and

$$\lim_{x \to a} \alpha = \lim_{x \to a} \frac{x - a}{a^2 x} = \frac{0}{a^3} = 0.$$

Finally, since

$$\alpha + f'(a) = \frac{f(x) - f(a)}{x - a}, \quad \text{when } x \neq a, \text{ and } x \neq 0,$$

we also have

$$f(x) - f(a) = (f'(a) + \alpha)(x - a),$$

as required by Eq. (2).

178

10–1 INCREMENT OF A FUNCTION

Remark 1. Equation (2) is sometimes written in the alternative form

$$\Delta y = \left(\frac{dy}{dx} + \alpha\right) \Delta x, \quad (7)$$

with

$$\Delta y = f(x) - f(a), \quad (8a)$$
$$\Delta x = x - a, \quad (8b)$$

and

$$\frac{dy}{dx} = f'(a). \quad (8c)$$

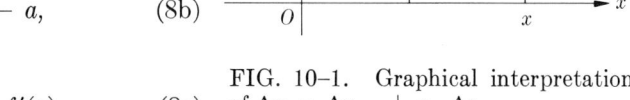

FIG. 10–1. Graphical interpretation of $\Delta y = \Delta y_{\tan} + \alpha \cdot \Delta x$.

The right-hand side of Eq. (7) can be multiplied out to give

$$\Delta y = \frac{dy}{dx} \Delta x + \alpha \, \Delta x. \quad (9)$$

The first term on the right-hand side of (9) is easily identified. It is the rise $\Delta y_{\tan}$ along the line tangent to the graph of f at $(a, f(a))$ corresponding to a run $\Delta x = x - a$. Thus, the total increment Δy, Eq. (9), can be written as a sum

$$\Delta y = \Delta y_{\tan} + \alpha \cdot \Delta x, \quad (10a)$$

where

$$\Delta y_{\tan} = f'(a) \cdot \Delta x, \quad (10b)$$

and

$$\alpha \to 0 \quad \text{when} \quad \Delta x \to 0. \quad (10c)$$

Figure 10–1 illustrates Eq. (10a). The key feature is this: the total increment Δy consists of a major part* $\Delta y_{\tan}$ which is proportional to the first power of Δx, and a minor part $\alpha \cdot \Delta x$ which is the product of Δx and a number α that approaches zero when Δx does. For Examples 1 and 2 we find:

1. $\quad f(x) = x^2, \quad \Delta y_{\tan} = 2a \cdot \Delta x, \quad \alpha \cdot \Delta x = (\Delta x)^2,$

2. $\quad f(x) = 1/x, \quad \Delta y_{\tan} = \dfrac{-1}{a^2} \cdot \Delta x, \quad \alpha \cdot \Delta x = \dfrac{(\Delta x)^2}{a^2 x}.$

---

* An exception occurs when $f'(a) = 0$. Then, of course, $\Delta y_{\tan}$ is zero and is no longer the "major part" of Δy. Even if $f'(a) \neq 0$, $\Delta y_{\tan}$ may be smaller than $\alpha \cdot \Delta x$ when Δx is large. The terms "major part" and "minor part" describe the relative importance of $\Delta y_{\tan}$ and $\alpha \cdot \Delta x$ when $f'(a) \neq 0$ and $|\Delta x|$ is small.

EXERCISES 10-1

For each of the following functions express $\Delta y = f(a + \Delta x) - f(a)$ as a sum of the form $\Delta y_{\tan} + \alpha \cdot \Delta x$, where $\Delta y_{\tan} = f'(a)\,\Delta x$.

1. $f(x) = 2x^2 + 1$
2. $f(x) = x^3$
3. $f(x) = \sqrt{x},\ a > 0,\ a + \Delta x > 0$
4. $f(x) = 1/x^2,\ a \neq 0$
5. $f(x) = \sin x$. [Use Eq. (9), Section 9-3, with $n = 1$.]

10-2 DERIVATIVES OF COMPOSITE FUNCTIONS: THE CHAIN RULE

A derivative gives a rate of change. Thus, if $dy/dx = 5$ and $dx/dt = 3$, we might say "y changes five times as fast as x and x changes three times as fast as t," and conclude that "y changes fifteen times as fast as t." This suggests that

$$\frac{dy}{dt} = \frac{dy}{dx} \cdot \frac{dx}{dt}. \tag{1}$$

In this section, we shall see that Eq. (1), which is sometimes called the *chain rule* for derivatives, is indeed valid when properly interpreted, and under appropriate conditions.

Example 1. A particle moves along the line $y = 5x - 2$ in such a way that, at time t, its x-coordinate satisfies $x = 3t$. Show that Eq. (1) holds.

Solution. From the given data, we get

$$\frac{dy}{dx} = 5, \qquad \frac{dx}{dt} = 3. \tag{2a}$$

On the other hand, we also have

$$y = 5x - 2 = 15t - 2,$$

so that

$$\frac{dy}{dt} = 15. \tag{2b}$$

The numbers in Eqs. (2a, b) satisfy Eq. (1).

Example 2. A particle moves along the curve $y = x^2$. Its x-coordinate satisfies $x = 2t^2 + t + 1$. Show that Eq. (1) is satisfied when $t = 2$.

Solution. When $t = 2$, we have $x = 11$, $y = 121$, $dy/dx = 2x = 22$, and $dx/dt = 4t + 1 = 9$. Therefore,

$$\left(\frac{dy}{dx}\right)\left(\frac{dx}{dt}\right) = (22)(9) = 198.$$

On the other hand, by substitution we get

$$y = x^2 = (2t^2 + t + 1)^2$$
$$= 4t^4 + 4t^3 + 5t^2 + 2t + 1,$$

so that

$$\frac{dy}{dt} = 16t^3 + 12t^2 + 10t + 2.$$

At $t = 2$, we therefore have

$$\frac{dy}{dt} = 128 + 48 + 20 + 2 = 198 = \frac{dy}{dx} \cdot \frac{dx}{dt}.$$

Remark 1. In Example 2, we verified the chain rule at time $t = 2$. For the same example, Eq. (1) holds for any value of t, because

$$\frac{dy}{dt} = 16t^3 + 12t^2 + 10t + 2$$
$$= 2(2t^2 + t + 1)(4t + 1)$$
$$= 2x(4t + 1)$$
$$= \left(\frac{dy}{dx}\right)\left(\frac{dx}{dt}\right).$$

To prepare for a careful statement of the chain rule and its proof, we introduce some notation.

Let f be a function

$$f = \{(x, y) : y = f(x),\ x \in D_f\}, \tag{3}$$

which is differentiable at an interior point x_0 of its domain D_f. Let g be a function

$$g = \{(t, x) : x = g(t),\ t \in D_g\}, \tag{4a}$$

which is differentiable at an interior point t_0 of its domain such that

$$g(t_0) = x_0. \tag{4b}$$

Let $h = f \circ g$ be the composition of f with g:

$$h = \{(t, y) : y = f(g(t)),\ t \in D_h\}. \tag{5}$$

Theorem 10–2. Under the hypotheses described in the preceding paragraph and Eqs. (3), (4), (5), if t_0 is an inner point of the domain of h, then

$$h'(t_0) = f'(x_0) \cdot g'(t_0). \tag{6}$$

Proof. Equations (3), (4), (5) say that, for any t in the domain of h, we have
$$y = h(t) = f(x),$$
provided
$$x = g(t).$$
By definition,
$$h'(t_0) = \lim_{t \to t_0} \frac{h(t) - h(t_0)}{t - t_0},$$
provided this limit exists. Now
$$h(t) - h(t_0) = f(g(t)) - f(g(t_0))$$
$$= f(x) - f(x_0),$$
so that
$$\frac{h(t) - h(t_0)}{t - t_0} = \frac{f(x) - f(x_0)}{t - t_0}. \tag{7}$$

We would like to write the last term in Eq. (7) thus:
$$\frac{f(x) - f(x_0)}{t - t_0} = \frac{f(x) - f(x_0)}{x - x_0} \cdot \frac{x - x_0}{t - t_0}$$
$$= \frac{f(x) - f(x_0)}{x - x_0} \cdot \frac{g(t) - g(t_0)}{t - t_0}, \tag{8}$$
but Eq. (8) is valid only if $x \neq x_0$ and $t \neq t_0$. We cannot guarantee this. However, we do know, from Theorem 10–1, with $a = x_0$, that
$$f(x) - f(x_0) = (f'(x_0) + \alpha)(x - x_0) \tag{9a}$$
and
$$\lim_{x \to x_0} \alpha = 0. \tag{9b}$$

Substitute from (9a) into the right-hand side of Eq. (7) and get
$$\frac{h(t) - h(t_0)}{t - t_0} = (f'(x_0) + \alpha) \frac{x - x_0}{t - t_0}$$
$$= (f'(x_0) + \alpha) \frac{g(t) - g(t_0)}{t - t_0}. \tag{10}$$

Equation (10) brings us quickly to our goal, because $x = g(t)$ approaches $x_0 = g(t_0)$ when $t \to t_0$. Therefore,
$$\lim_{t \to t_0} \alpha = \lim_{x \to x_0} \alpha = 0 \quad \text{and} \quad \lim_{t \to t_0} \frac{g(t) - g(t_0)}{t - t_0} = g'(t_0),$$

10-2 DERIVATIVES OF COMPOSITE FUNCTIONS: THE CHAIN RULE

so that, from Eq. (10),

$$\lim_{t \to t_0} \frac{h(t) - h(t_0)}{t - t_0} = \lim_{t \to t_0} [f'(x_0) + \alpha] \cdot \lim_{t \to t_0} \frac{g(t) - g(t_0)}{t - t_0}$$
$$= f'(x_0)g'(t_0). \qquad (11)$$

Equation (11) says that h has a derivative at t_0 and

$$h'(t_0) = f'(x_0)g'(t_0). \qquad \text{Q.E.D.}$$

Example 3. Find dy/dx if $y = \sin(x^2)$.

Solution. Let $g(x) = \sin x$, $h(x) = x^2$. Then $y = f(x)$, where

$$f(x) = g(h(x)).$$

Therefore,

$$\frac{dy}{dx} = f'(x) = g'(h(x)) \cdot h'(x).$$

But $g'(x) = \cos x$, so

$$g'(h(x)) = \cos(x^2)$$

and

$$h'(x) = 2x,$$

so that

$$\frac{dy}{dx} = 2x \cos(x^2).$$

Remark 2. The chain rule is frequently written in the form

$$\frac{dy}{dx} = \frac{dy}{du} \frac{du}{dx}. \qquad (12)$$

Equation (12) applies when y can be expressed as a differentiable function of u, and u as a differentiable function of x:

$$y = g(u), \qquad u = h(x).$$

In this notation, Example 3 becomes

$$y = \sin u, \qquad u = x^2,$$
$$\frac{dy}{du} = \cos u, \qquad \frac{du}{dx} = 2x,$$

so that

$$\frac{dy}{dx} = \frac{dy}{du} \frac{du}{dx} = \cos(x^2) \cdot 2x.$$

183

EXERCISES 10-2

Find dy/dx for each of the following composite functions $y = f(x)$.

1. $y = (2x + 3)^{10}$
2. $y = \sin^2 x$
3. $y = \cos(x^3)$
4. $y = (\sin x + \cos x)^2$
5. $y = (5x^2 + 7x - 3)^3$
6. $y = (x - \sin x)^2$
7. $y = \sin^2(2x)$
8. $y = \cos(2 \cos x)$
9. $y = \sin^2(\frac{1}{2}x)$
10. $y = \tan \dfrac{3x+5}{4}$
11. $y = \sin(\sin x)$
12. $y = \sec\left(\dfrac{2x}{5}\right)$
13. $y = \cos^3(x^2 + 1)$
14. $y = (\sin x^2)\cot(x+2)$, $x \neq -2$
15. $y = \left(\sin \dfrac{2}{x^2}\right)^3$, $x \neq 0$

10-3 DERIVATIVES OF INVERSE FUNCTIONS

If f and g are inverse functions, their graphs are mirror images of each other with respect to the 45° line $y = x$. Thus, if L_1 is the line tangent to the graph of f at (c, d) and L_2 is the mirror image of L_1 with respect to the 45° line, it is reasonable to expect L_2 to be tangent to the graph of g at (d, c). [Figure 10–2 illustrates this idea.] Since (rise)/(run) on L_2 corresponds to (run)/(rise) on L_1, we see that the slope of L_2 is just the

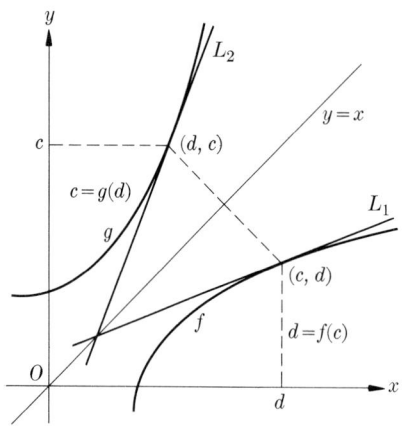

FIG. 10–2. Graphs of inverse functions f and g and tangents L_1 and L_2 at corresponding points (c, d) and (d, c); $d = f(c)$, $c = g(d)$.

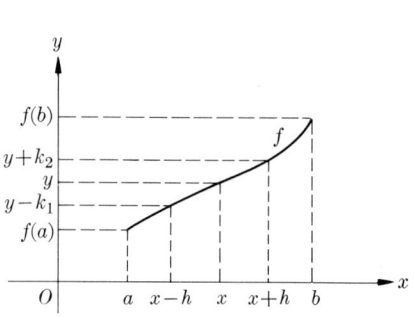

FIGURE 10–3

reciprocal of the slope of L_1: $m' = 1/m$. Since L_1 is tangent to f at $(c, f(c))$,

$$m = f'(c).$$

And, if L_2 is tangent to the graph of g at $(d, g(d))$, then

$$m' = g'(d).$$

Hence the geometric evidence suggests that

$$g'(f(c)) = \frac{1}{f'(c)}. \tag{1}$$

Naturally, Eq. (1) requires the existence of the derivative of f at c, and also requires that $f'(c) \neq 0$. In Fig. 10–2, f' is positive and f is a strictly increasing function. The following theorem establishes Eq. (1) for such functions.

Theorem 10–3. Let f be a function, with domain $a < x < b$, whose derivative exists and is positive on that domain. Then f has an inverse, g, and if $y = f(x)$, then

$$g'(y) = \frac{1}{f'(x)}, \quad \text{for} \quad a < x < b. \tag{2}$$

Or, in another notation, if $y = f(x)$, then $x = g(y)$ and

$$\frac{dx}{dy} = \frac{1}{dy/dx}.$$

Proof. Because f' is positive on D_f, f is one-to-one from its domain to its range R_f. Thus the rule

$$g(y) = x \quad \text{if and only if} \quad y = f(x), \quad x \in D_f$$

defines a function g whose domain D_g is the range of f:

$$D_g = R_f.$$

Figure 10–3 will help us to follow the remaining steps in the proof. Fix $y \in D_g$ and $x = g(y) \in D_f$. Since the domain of f is (by hypothesis) the open interval $a < x < b$, there exists a positive number h such that the closed interval $[x - h, x + h]$ is in the domain of f. Let

$$y - k_1 = f(x - h), \quad y + k_2 = f(x + h).$$

Then k_1 and k_2 are positive numbers, so y is an inner point of the domain of g, and that domain contains the closed interval $[y - k, y + k]$, where

$k = \min(k_1, k_2)$. For each $\Delta y \neq 0$, such that $|\Delta y| < k$, the intermediate-value theorem applied to f shows that there exists $\Delta x \neq 0$, such that $|\Delta x| < h$, and

$$f(x + \Delta x) = y + \Delta y, \qquad g(y + \Delta y) = x + \Delta x.$$

To prove that $g'(y)$ exists, we must show that the difference quotient

$$\frac{g(y + \Delta y) - g(y)}{\Delta y}$$

has a limit as $\Delta y \to 0$. But this is easy, because

$$g(y) = x, \qquad g(y + \Delta y) = x + \Delta x,$$
$$y = f(x), \qquad y + \Delta y = f(x + \Delta x),$$

so that

$$\frac{g(y + \Delta y) - g(y)}{\Delta y} = \frac{(x + \Delta x) - x}{f(x + \Delta x) - f(x)} = \frac{\Delta x}{f(x + \Delta x) - f(x)} \qquad (3)$$

and

$$\lim_{\Delta x \to 0} \frac{f(x + \Delta x) - f(x)}{\Delta x} = f'(x). \qquad (4\text{a})$$

By hypothesis, $f'(x) \neq 0$. Therefore, taking reciprocals in Eq. (4a) we get

$$\lim_{\Delta x \to 0} \frac{\Delta x}{f(x + \Delta x) - f(x)} = \frac{1}{f'(x)}. \qquad (4\text{b})$$

Since f is continuous on D_f and g is continuous on D_g, $\Delta x \to 0$ when $\Delta y \to 0$, and conversely. Therefore, from Eqs. (3) and (4b) we get

$$\lim_{\Delta y \to 0} \frac{g(y + \Delta y) - g(y)}{\Delta y} = \lim_{\Delta x \to 0} \frac{\Delta x}{f(x + \Delta x) - f(x)},$$

or

$$g'(y) = \frac{1}{f'(x)}. \qquad \text{Q.E.D.}$$

Example 1. Let $a = 0$, $b > 0$, $f(x) = x^2$. Then $f'(x) = 2x > 0$ for $a < x < b$, and

$$\text{if} \quad y = x^2 = f(x), \quad \text{then} \quad x = \sqrt{y} = g(y)$$

for $0 < x < b$ and $0 < y < b^2$. By Theorem 10–3,

$$g'(y) = \frac{1}{f'(x)} = \frac{1}{2x} = \frac{1}{2\sqrt{y}}.$$

DERIVATIVES OF INVERSE FUNCTIONS

Since g is f^{-1}, Eq. (2) of Theorem 10–3, can be written as

$$g' = (f^{-1})' = \frac{1}{f' \circ f^{-1}} \tag{5}$$

and reads: the derivative of the inverse of f is the reciprocal of the composition of the derivative of f with f^{-1}. This holds for all functions for which the right-hand side of (5) is meaningful.

Applied to Example 1 above:

$$f(x) = x^2, \quad f^{-1}(x) = \sqrt{x},$$

$$f'(x) = 2x, \quad (f^{-1})'(x) = \frac{1}{2\sqrt{x}}.$$

Example 2

$$f(x) = \sqrt{x}, \ x > 0, \quad f^{-1}(x) = x^2,$$

$$f'(x) = \frac{1}{2\sqrt{x}}, \quad (f^{-1})'(x) = 2x.$$

What we have just shown is that if $g(y) = y^{1/n}$, and $n = 2$, then $g'(y) = (1/n)y^{(1/n)-1}$. This is a special case of Corollary 10–3.1.

Corollary 10–3.1. Let n be a positive integer, $y > 0$, and

$$g(y) = y^{1/n}. \tag{6a}$$

Then

$$g'(y) = \frac{1}{n} y^{(1/n)-1}. \tag{6b}$$

Proof. Let the inverse of g be f, defined for $x > 0$ such that $f(x) = x^n$ or $y = x^n$. Then by Theorem 10–3,

$$g'(y) = \frac{1}{f'(x)} \quad \text{or} \quad g'(y) = \frac{1}{nx^{n-1}}.$$

Since $x^n = y$, $x = y^{1/n}$, and $x^{n-1} = y^{(n-1)/n} = y^{1-(1/n)}$. Therefore,

$$g'(y) = \frac{1}{ny^{1-(1/n)}} \quad \text{or} \quad g'(y) = \frac{1}{n} y^{(1/n)-1}. \quad \text{Q.E.D.}$$

Remark 1. In Theorem 10–3, we assumed that $f'(x) > 0$ for $a < x < b$. We could equally well have proved the theorem under the assumption $f'(x) < 0$ on the domain D_f. Under the first hypothesis f is a strictly increasing function, while the second hypothesis fits a strictly decreasing function. Only minor changes in the proof are needed in this alternative form. Indeed, one can achieve the desired result by applying Theorem 10–3 to the functions $f_1 = -f$ and $g_1 = -g$:

$$g_1'(y) = \frac{1}{f_1'(x)} \Rightarrow g'(y) = \frac{1}{f'(x)}.$$

Henceforth, when we invoke Theorem 10–3, we may do so under either of the assumptions

$$f'(x) > 0 \quad \text{for all} \quad a < x < b,$$

or

$$f'(x) < 0 \quad \text{for all} \quad a < x < b.$$

Corollary 10–3.2. Let n be a negative integer and $g(y) = y^{1/n}$, $y > 0$. Then

$$g'(y) = \frac{1}{n} y^{(1/n)-1}.$$

Proof. For $x > 0$, let $f(x) = x^n$. Then

$$f'(x) = nx^{n-1} < 0 \quad \text{for} \quad x > 0.$$

The functions f and g are inverses on $x > 0$, $y > 0$, because

$$y = x^n = f(x) \quad \text{if and only if} \quad x = y^{1/n} = g(y).$$

Hence

$$g'(y) = \frac{1}{f'(x)} = \frac{1}{nx^{n-1}} = \frac{1}{n} y^{(1/n)-1}. \qquad \text{Q.E.D.}$$

Remark 2. These applications of Theorem 10–3 have been made easier by using the notation $g(y) = y^{1/n}$. However, it is also true that, if $g(x) = x^{1/n}$, then $g'(x) = (1/n)x^{(1/n)-1}$, because there is nothing sacred about which letter of the alphabet we use for the variable in the definition of g. Moreover, instead of writing the exponent as $1/n$, with n a positive or negative integer, we could write the exponent as n and say that n is the reciprocal of a positive or negative integer. In other words, Corollaries 10–3.1 and 10–3.2 simply extend the familiar formula

$$\frac{d}{dx}(x^n) = nx^{n-1} \qquad (7)$$

from its earlier applications, where n was a positive integer (and all values of x), or a negative integer (and $x \neq 0$), to the situation where n is the reciprocal of a positive or negative integer (and $x > 0$). We are now in a position to extend Eq. (7) to any *rational* exponent n, for $x > 0$.

Theorem 10–4. Let n be a rational number. Let f be the function
$$f(x) = x^n, \quad x > 0.$$
Then
$$f'(x) = nx^{n-1}. \tag{8}$$

Proof. There exist integers p and q, with $q > 0$, such that $n = p/q$. Therefore,
$$f(x) = x^{p/q} = (x^{1/q})^p, \tag{9}$$
so that f is the composition of functions g and h:
$$g(x) = x^p, \quad h(x) = x^{1/q} \tag{10a}$$
and
$$f(x) = g(h(x)) = (h(x))^p = (x^{1/q})^p. \tag{10b}$$

Therefore, by Theorem 10–2 (the chain rule),
$$\begin{aligned}
f'(x) &= g'(h(x)) \cdot h'(x) \\
&= p(h(x))^{p-1} \cdot \frac{1}{q} x^{(1/q)-1} \\
&= p(x^{1/q})^{p-1} \cdot \frac{1}{q} x^{(1/q)-1} \\
&= \frac{p}{q} x^{(p/q)-1} = nx^{n-1}. \quad \text{Q.E.D.}
\end{aligned}$$

Example 3. If $y = x^{2/3}$, and $x > 0$, then $dy/dx = \frac{2}{3} x^{-1/3}$.

Corollary 10–4.1. If n is a rational number and u is a differentiable function of x, then
$$\frac{d}{dx}(u^n) = nu^{n-1} \frac{du}{dx},$$
wherever $u > 0$.

Proof. Combine the chain rule and Theorem 10–4:
$$\frac{dy}{dx} = \frac{dy}{du} \frac{du}{dx} = nu^{n-1} \frac{du}{dx}. \quad \text{Q.E.D.}$$

Remark 3. The restriction $x > 0$ is not necessary in Example 3, because either positive or negative values of x have real cube roots and Theorem 10–4 could be extended to cover negative as well as positive values of x for all rational values of n for which x^n is real valued. The corresponding change in Corollary 10–3.2 would be to allow y to be negative (or positive) when n is an odd integer. The key feature there is that the equation $y = x^n$, for n an odd integer, establishes a one-to-one correspondence between the set of all non-zero real numbers x and the set of all non-zero real numbers y. (This is true for $n = 1, 3, 5, \ldots$ or $n = -1, -3, -5, \ldots$)

Remark 4. Our procedure for establishing Eq. (8) has been more sophisticated than that found in Thomas, *Elements of Calculus and Analytic Geometry*, pp. 68–69. The method used there was to apply the technique of implicit differentiation. That technique is illustrated in the next example.

Example 4 (*Implicit differentiation*). Observe that the equation

$$x^3 + 3xy + y^5 = 5 \tag{11}$$

is satisfied by $x = 1$, $y = 1$. Suppose that, for some open interval of values of x containing $x = 1$, there exists a corresponding set of values of y, which we represent as $y = f(x)$, such that the points (x, y) lie on the graph of Eq. (11). Assuming that the function f so defined implicitly by Eq. (11) is differentiable at $x = 1$, then each of the terms

$$x^3, \qquad 3xy = 3xf(x), \qquad y^5 = (f(x))^5,$$

is also differentiable there, and

$$\frac{d}{dx}(x^3 + 3xy + y^5) = \frac{d}{dx}(5) = 0 \tag{12}$$

for $y = f(x)$, because of Eq. (11). Therefore, we should have

$$3x^2 + 3\left(x\frac{dy}{dx} + y\right) + 5y^4 \frac{dy}{dx} = 0$$

or

$$(3x + 5y^4)\frac{dy}{dx} = -(3x^2 + 3y)$$

or

$$\frac{dy}{dx} = -\frac{3x^2 + 3y}{3x + 5y^4}$$

$$= -\tfrac{6}{8} \quad \text{at} \quad x = 1, \ y = 1.$$

Example 5. Given $x^2 + y^2 = 25$, find dy/dx at $x = 3$, $y = 4$.

Solution. If x and y satisfy the given equation, then

$$\frac{d}{dx}(x^2 + y^2) = \frac{d}{dx}(25) = 0,$$

so
$$2x + 2y \frac{dy}{dx} = 0$$
or
$$\frac{dy}{dx} = \frac{-x}{y} = \frac{-3}{4} \quad \text{at} \quad (3, 4).$$

Observe that this is the negative reciprocal of the slope of the line from the origin to $P(3, 4)$.

Figure 10–4 illustrates the corresponding general result for the circle $x^2 + y^2 = a^2$.

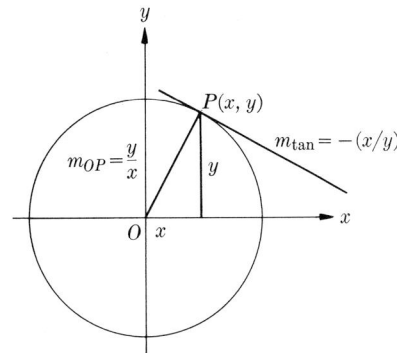

FIG. 10–4. For the circle $x^2 + y^2 = a^2$, the slope of the tangent at $P(x, y)$ is the negative reciprocal of the slope of OP: $dy/dx = -(x/y)$, $m_{OP} = y/x$.

Example 6. If $y = x^{p/q}$, then
$$y^q = x^p,$$
and implicit differentiation with respect to x yields
$$qy^{q-1} \frac{dy}{dx} = px^{p-1},$$
or
$$\frac{dy}{dx} = \frac{p}{q} y^{1-q} x^{p-1} = \frac{p}{q} (x^{p/q})^{1-q} x^{p-1} = \frac{p}{q} x^{(p/q)-1}.$$

Remark 5. It is not easy to establish the validity of implicit differentiation without using partial derivatives. (See, for example: Olmsted, *Advanced Calculus*, Appleton-Century-Crofts (1961), p. 289; or Fine, *Calculus*, Macmillan (1937), p. 248.) Sufficient conditions for validity are these: Write the defining equation in the form $f(x, y) = 0$, and suppose that $x = x_0$, $y = y_0$ satisfy this equation. Let $f_x(x_0, y_0)$ and $f_y(x_0, y_0)$ be the values of the partial derivatives of f with respect to x and y, respectively, at (x_0, y_0). If f, f_x, and f_y are continuous in a neighborhood containing (x_0, y_0) as an inner point and $f_y(x_0, y_0) \neq 0$, then there exists a function $y = \phi(x)$ such that
$$f(x, \phi(x)) \equiv 0, \quad \phi(x_0) = y_0,$$
and
$$\phi'(x_0) = -\frac{f_x(x_0, y_0)}{f_y(x_0, y_0)}.$$

ADDITIONAL DIFFERENTIATION FORMULAS

This answer is what one obtains by differentiating the equation $f(x, y) = 0$ implicitly, with respect to x, then solving for dy/dx and substituting $x = x_0$ and $y = y_0$.

Example 7. In Example 4, let

$$f(x, y) = x^3 + 3xy + y^5 - 5$$

$$x_0 = 1, \quad y_0 = 1.$$

To find the partial derivative of f with respect to x, just treat y as a constant and differentiate f with respect to x:

$$f_x(x, y) = 3x^2 + 3y.$$

Similarly, holding x constant and differentiating f with respect to y, we get

$$f_y(x, y) = 3x + 5y^4.$$

If $y = \phi(x)$ satisfies the equation $f(x, y) = 0$ and $\phi(1) = 1$, then

$$\phi'(1) = \left(\frac{dy}{dx}\right)_{(1,1)} = -\frac{f_x(1, 1)}{f_y(1, 1)}$$

$$= -\frac{3x^2 + 3y}{3x + 5y^4}\bigg]_{(1,1)} = -\frac{6}{8}.$$

Here, the functions f, f_x, and f_y are simply polynomials in x and y which are continuous everywhere, and $f_y(1, 1) = 8 \neq 0$.

As an operational rule: go ahead and differentiate implicitly, but don't ever try to divide by zero.

EXERCISES 10–3

In each of the following functions indicate on what interval the function is increasing and on what interval the function is decreasing.

1. $f(x) = 3x + 1$
2. $f(x) = -2x + 1$
3. $f(x) = x^5$
4. $f(x) = x^2 + 5$
5. $f(x) = 2x^3 + 3$
6. $f(x) = \dfrac{1}{1 + x^2}$
7. $f(x) = \dfrac{x}{1 + x^2}$
8. $f(x) = x^{2/3}$

Each of Exercises 9 and 10 describes a relation which defines y as a function of x. On what interval(s) are these functions increasing? decreasing?

9. $x^2y - x + y = 0$
10. $x^2 - 6x + y^2 - 2y = 15$ and $y > 1$

11–18. In each of the functions of Exercises 1 through 8 above, restrict, if necessary, the domain so that an inverse exists and find the derivative of the inverse function.

19. The equation $x^2y - x + y = 0$ defines y as a function, $y = f(x)$. Find $f'(x)$. For a suitably restricted domain, f has an inverse. In the restricted domain, find the derivative of the inverse function. Draw the graphs of f, f^{-1}, and the derived function f'.

20. Find dy/dx for $x^2 - 6x + y^2 - 2y = 15$ at $(-1, 4)$ and $(6, 5)$.

10–4 ALGEBRAIC FUNCTIONS AND THEIR DERIVATIVES

Let f be a function such that if (x, y) is a point on the graph of f, then x and y satisfy an irreducible algebraic equation of the form

$$P_0(x)y^n + P_1(x)y^{n-1} + \cdots + P_{n-1}(x)y + P_n(x) = 0, \qquad (1)$$

where $P_0(x), P_1(x), \ldots, P_n(x)$ are polynomials with real coefficients and n is a positive integer. Then f is called an *algebraic* function.

Example 1. The function f such that

$$f(x) = \sqrt{x}, \qquad x \geq 0,$$

is an algebraic function, because if

$$y = \sqrt{x},$$

then

$$y^2 - x = 0.$$

This is an irreducible equation of the form of Eq. (1) with $n = 2$, $P_0(x) = 1$, $P_1(x) = 0$, $P_2(x) = -x$.

Example 2. The absolute-value function

$$y = |x|$$

satisfies the equation

$$y^2 - x^2 = 0. \qquad (2)$$

However, the left-hand side of Eq. (2) is the product of linear factors $(y - x)$ and $(y + x)$, so Eq. (2) is not irreducible. Nor does $y = |x|$ satisfy (everywhere) either of the irreducible equations

$$y - x = 0 \quad \text{or} \quad y + x = 0.$$

Thus, the absolute-value function is not an algebraic function.

ADDITIONAL DIFFERENTIATION FORMULAS

Remark 1. From our earlier discussion of the absolute-value function, we know that, if $y = |x|$, then

$$\frac{dy}{dx} = \begin{cases} +1 & \text{if } x > 0, \\ -1 & \text{if } x < 0, \end{cases}$$

and dy/dx does not exist if $x = 0$. One easily verifies that this is equivalent to

$$\frac{dy}{dx} = \frac{x}{|x|} = \frac{x}{y} \quad \text{if} \quad y = |x| \neq 0. \tag{3}$$

Implicit differentiation of Eq. (2), with respect to x, also leads to Eq. (3):

$$2y\frac{dy}{dx} - 2x = 0; \quad \frac{dy}{dx} = \frac{x}{y} \quad \text{if } y \neq 0.$$

Remark 2. Equation (2) has for its graph all points on the two lines

$$L_1: y = x \quad \text{and} \quad L_2: y = -x,$$

shown in Fig. 10–5. At all points on L_1 we have

$$y = x, \quad \frac{dy}{dx} = 1 \quad \left(= \frac{x}{y} \quad \text{if } y \neq 0\right),$$

and on L_2

$$y = -x, \quad \frac{dy}{dx} = -1 \quad \left(= \frac{x}{y} \quad \text{if } y \neq 0\right).$$

Thus, three different functions,

i) $y = |x|$,
ii) $y = x$,
iii) $y = -x$,

have elements (x, y) that satisfy

$$y^2 - x^2 = 0. \tag{4a}$$

In each case, it is also true that

$$\frac{dy}{dx} = \frac{x}{y} \quad \text{if} \quad y \neq 0. \tag{4b}$$

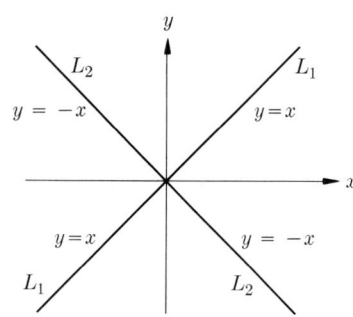

FIG. 10–5. The graph of the equation $y^2 - x^2 = 0$ is the pair of lines L_1, L_2.

In other words, implicit differentiation of Eq. (4a) leads to a correct result, Eq. (4b). If we let $f(x, y) = y^2 - x^2$ so that Eq. (4a) fits the discussion of implicit functions satisfying $f(x, y) = 0$, then the condition $\partial f/\partial y \neq 0$ is $2y \neq 0$ or $y \neq 0$, as in Eq. (4b).

10-4 ALGEBRAIC FUNCTIONS AND THEIR DERIVATIVES

Remark 3. Formal application of implicit differentiation can be misleading. In particular, before we can differentiate the y^2 on the left-hand side of Eq. (4a), with respect to x, we must assume that y is a differentiable function of x. We have seen three examples above where the procedure works. An example where implicit differentiation is meaningless is provided by the function f such that

$$y = f(x) = \begin{cases} x, & \text{if } x \text{ is rational,} \\ -x, & \text{if } x \text{ is irrational.} \end{cases}$$

The graph of this function consists of the lines L_1 and L_2 of Fig. 10–5 with points having rational coordinates deleted from L_2 and those with irrational coordinates deleted from L_1. The resulting function, f, is continuous only at $x = 0$, and is nowhere differentiable. Hence, although Eq. (4a) still applies, Eq. (4b) does not.

Example 3. If $y = f(x) = \sqrt{x^2 + 1}$, then f is the composition of $g(x) = \sqrt{x}$, $x > 0$, and $h(x) = x^2 + 1$:

$$f(x) = g(h(x)) = \sqrt{h(x)} = \sqrt{x^2 + 1}.$$

Since $x^2 + 1$ is positive, for all real x, and both g and h are differentiable, the chain rule guarantees the existence of $f'(x)$ and provides the result

$$f'(x) = g'(h(x)) \cdot h'(x) = \frac{1}{2\sqrt{h(x)}} \cdot 2x = \frac{x}{f(x)}.$$

If we apply implicit differentiation to the equation $y^2 = x^2 + 1$, relying on the fact that $y = f(x)$ is differentiable, we get

$$2y \frac{dy}{dx} = 2x$$

or

$$\frac{dy}{dx} = \frac{x}{y} = \frac{x}{f(x)}.$$

Summary. For certain explicitly given algebraic functions, like $y = \sqrt{x}$ or $y = \sqrt{x^2 + 1}$, we can apply results on composite functions to establish their differentiability. To calculate their derivatives, we can use the chain rule, or implicitly differentiate the corresponding algebraic equation (like $y^2 = x$ or $y^2 = x^2 + 1$). For more complicated algebraic functions we use implicit differentiation to calculate what the derivative is, *if it exists*. Sufficient conditions for the existence of the derivative have been given for implicit functions in general, and these apply to algebraic functions as a special case.

EXERCISES 10-4

Use implicit differentiation to find dy/dx for each of the following. Show restrictions that, if met in addition to the given equation, would guarantee the existence of the derivative.

1. $x^2 + xy + y^2 = 3$
2. $x^3y + x^2 + y^2 = 10$
3. $y \sin x + x \sin y = \pi$
4. $(x^2 + 1)y + (y^2 - 1)x = 2$
5. $y^3 \sin^2 3x + x = 5y^2$
6. $x \sin 2x - y \sin 2y = 0$
7. $\sqrt{y/x} + \sqrt{x/y} = 6$, $xy \neq 0$
8. $x^{2/3} + y^{2/3} = a^{2/3}$, $|x| \leq a$, $|y| \leq a$
9. $x^2 + 2xy + y^2 = 22$
10. $x^3 - y^3 + 4xy = 0$
11. $\sqrt{2x} + \sqrt{3y} - 5 = 0$
12. $x^3 - 2xy + y^3 = 0$
13. $y^2 = \dfrac{x-y}{x+y}$

For each of the following find the slope of the curve at the given point:

14. $x^2 + 3xy + 2y^2 - 2x + y + 1 = 0$ at $(1, -2)$
15. $x \sin y = 3x^2 - 5$ at $(\sqrt{2}, \pi/4)$

11
DIFFERENTIATION OF TRIGONOMETRIC FUNCTIONS AND THEIR INVERSES

11-1 DIFFERENTIATION OF THE TRIGONOMETRIC FUNCTIONS*

We have already seen that

$$\frac{d}{dx}\sin x = \cos x. \tag{1}$$

Equation (1) was derived directly from the definition of derivative:

$$\frac{d}{dx}\sin x = \lim_{\Delta x \to 0} \frac{\sin(x + \Delta x) - \sin x}{\Delta x}.$$

If u is a differentiable function of x, we can combine Eq. (1) and the chain rule to get the more general formula:

$$\frac{d}{dx}\sin u = \cos u \cdot \frac{du}{dx}. \tag{2}$$

Example 1. $\dfrac{d}{dx}\sin(x^2 + 1) = 2x\cos(x^2 + 1).$

The derivatives of the other trigonometric functions could also be deduced by applying the definition of derivative. However, it is easier to use various trigonometric identities and the general formulas for derivatives of quotients. The list of formulas for derivatives of the six trigonometric functions follows:

$$\frac{d}{dx}\sin u = \cos u \frac{du}{dx}, \tag{3a}$$

$$\frac{d}{dx}\cos u = -\sin u \frac{du}{dx}, \tag{3b}$$

$$\frac{d}{dx}\tan u = \sec^2 u \frac{du}{dx}, \tag{3c}$$

$$\frac{d}{dx}\cot u = -\csc^2 u \frac{du}{dx}, \tag{3d}$$

$$\frac{d}{dx}\sec u = \sec u \tan u \frac{du}{dx}, \tag{3e}$$

$$\frac{d}{dx}\csc u = -\csc u \cot u \frac{du}{dx}. \tag{3f}$$

The proofs are easy, but before we take them up, we observe a pattern in Eqs. (3a) through (3f). If we know the derivatives of the sine, tangent, and secant, then we easily get the derivatives of the corresponding co-

---

* Our use of the term "trigonometric functions" is synonymous with the term "circular functions." Radian measure is assumed throughout.

11-1 DIFFERENTIATION OF THE TRIGONOMETRIC FUNCTIONS

functions if we do two things:

i) change the sign, and
ii) replace every function by its cofunction. (The cofunction of the cosine is the sine.)

This pattern results from applying the chain rule to the identities

$$\cos u = \sin(\pi/2 - u), \quad \cot u = \tan(\pi/2 - u),$$
$$\csc u = \sec(\pi/2 - u).$$

These identities express the relation between a trigonometric function and its cofunction.

Proofs. For any differentiable function u,

$$\frac{d}{dx}\cos u = \frac{d}{dx}\sin\left(\frac{\pi}{2} - u\right) = \cos\left(\frac{\pi}{2} - u\right)\frac{d}{dx}\left(\frac{\pi}{2} - u\right)$$
$$= \sin u \left(-\frac{du}{dx}\right)$$
$$= -\sin u \frac{du}{dx},$$

which establishes Eq. (3b).

To prove Eq. (3c), write $\tan u$ as $(\sin u)/(\cos u)$, and apply the formula for the derivative of a quotient. Thus,

$$\frac{d}{dx}\left(\frac{\sin u}{\cos u}\right) = \frac{\cos u \frac{d}{dx}(\sin u) - \sin u \cdot \frac{d}{dx}(\cos u)}{\cos^2 u}$$
$$= \frac{\cos u \cdot \cos u \frac{du}{dx} - \sin u \cdot \left(-\sin u \frac{du}{dx}\right)}{\cos^2 u}$$
$$= \frac{\cos^2 u + \sin^2 u}{\cos^2 u}\frac{du}{dx} = \sec^2 u \frac{du}{dx}.$$

Equation (3e) follows in like manner from the identity

$$\sec u = \frac{1}{\cos u}.$$

In Eqs. (3c) through (3f) it is not only necessary that u be a differentiable function of x but also that $\cos u \neq 0$ [in (3c) and (3e)], or that $\sin u \neq 0$ [in (3d) and (3f)].

Another method of deriving (3c) is based on the definition of derivative and the trigonometric identity

$$\tan(a+b) = \frac{\tan a + \tan b}{1 - \tan a \tan b}.$$

We have $f(x) = \tan x$, $f(x + \Delta x) = \tan(x + \Delta x)$, and therefore

$$f(x + \Delta x) - f(x) = \tan(x + \Delta x) - \tan x$$
$$= \frac{\tan x + \tan \Delta x}{1 - \tan x \tan \Delta x} - \tan x$$
$$= \frac{(1 + \tan^2 x)\tan \Delta x}{1 - \tan x \tan \Delta x}.$$

Thus

$$\frac{f(x + \Delta x) - f(x)}{\Delta x} = \frac{(1 + \tan^2 x)}{1 - \tan x \tan \Delta x} \cdot \frac{\tan \Delta x}{\Delta x}. \quad (4)$$

Taking the limit of both sides of (4) as $\Delta x \to 0$, and noting that

$$\lim_{\Delta x \to 0} \tan \Delta x = 0, \quad \lim_{\Delta x \to 0} \frac{\tan \Delta x}{\Delta x} = 1 \quad \text{(why?)},$$

we have

$$\frac{dy}{dx} = 1 + \tan^2 x = \sec^2 x.$$

Applying the chain rule, we have (3c).

Example 2. $\dfrac{d}{dx} \tan 3x = \sec^2 3x \cdot \dfrac{d}{dx}(3x) = 3 \sec^2 3x.$

Example 3. $\dfrac{d}{dx} \sec^3 5x = 3 \sec^2 5x \cdot \dfrac{d}{dx} \sec 5x$
$$= 3 \sec^2 5x \cdot (\sec 5x \tan 5x) \frac{d}{dx}(5x)$$
$$= 15 \sec^3 5x \tan 5x.$$

EXERCISES 11-1

Find dy/dx when:

1. $y = \sin 4x$
2. $y = \cos^2 3x$
3. $y = \tan^3 x$
4. $y = \cot(3x)$
5. $y = \cot^2(3x)$
6. $y = \sec 3x \tan 3x$
7. $y = 5 \tan 2x + \cos 3x$
8. $y = \dfrac{1 + \sin 2x}{1 - \sin 2x}$

9. $y = \sin x \cos 3x$

10. $y = \tan 2x \sec 3x$

11. $y = \cot 3x \sin 2x$

12. $y = \csc^2 5x$

13. Show that the graph of
$$y = x + \cos x$$
has no maxima and no minima. Find points on the graph where the slope is: (a) zero, (b) a maximum, (c) a minimum. Sketch the portion of the graph for $-2\pi \leq x \leq 2\pi$.

14. Differentiate both sides of the identity $\sin 2x = 2 \sin x \cos x$. Does this produce another equation that is true for every value of x?

15. Prove that $\tan x > x$ for $0 < x < \pi/2$.

16. Using the result of Exercise 15, prove that $(\sin x)/x$ is a strictly decreasing function for $0 < x < \pi/2$.

17. Prove that $x > \sin x > (2/\pi)x$ for $0 < x < \pi/2$. Sketch the graphs of $y = x$, $y = \sin x$, $y = 2x/\pi$ for $0 \leq x \leq \pi/2$.

18. Derive Eq. (3d) from Eq. (3c) and the identity
$$\cot u = \tan (\pi/2 - u).$$

19. Derive Eq. (3d) from Eqs. (3a, b) and the identity
$$\cot u = \frac{\cos u}{\sin u}.$$

20. Derive Eq. (3f) from Eq. (3e) and the identity
$$\csc u = \sec (\pi/2 - u).$$

21. Derive Eq. (3f) from Eq. (3a) and the identity
$$\csc u = \frac{1}{\sin u}.$$

11–2 INVERSE TRIGONOMETRIC FUNCTIONS

The range of the function $y = \sin x$ is the interval $-1 \leq y \leq 1$, if the domain includes an x-interval of length 2π. Figure 11–1 shows the graph for $-\pi \leq x \leq \pi$. Because $\sin (\pi - x) = \sin x$, every number c between 0 and 1 in the range is the image of 2 distinct values of x between 0 and π. Therefore, the mapping from $[-\pi, \pi]$ onto $[-1, 1]$ is not one-to-one, and the inverse relation of $y = \sin x$, $-\pi \leq x \leq \pi$, is not a function. However, if we suitably restrict the domain, say to the interval $-\pi/2 \leq x \leq \pi/2$, then apply the mapping $y = \sin x$ to this restricted domain, the result is a one-to-one mapping from $[-\pi/2, \pi/2]$ onto $[-1, 1]$ as shown in Fig. 11–2.

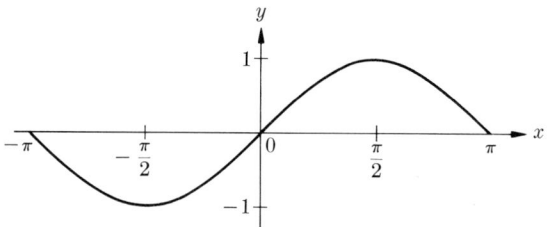

FIG. 11-1. Portion of the graph of $y = \sin x$, $-\pi \leq x \leq \pi$.

Suppose, therefore, that f is the function

$$f(x) = \sin x, \qquad -\pi/2 \leq x \leq \pi/2. \tag{1}$$

Then f is strictly increasing, because

$$f'(x) = \cos x$$

is positive for $-\pi/2 < x < \pi/2$. The inverse function g is what we get if we solve $y = \sin x$ for x in terms of y [i.e., $x = g(y)$] and then interchange x and y [i.e., $y = g(x)$]; or, what amounts to the same thing, if we first interchange x and y and then solve for y in terms of x:

$$x = \sin y, \qquad y = \sin^{-1} x, \tag{2a}$$

with

$$-\pi/2 \leq y \leq \pi/2 \quad \text{and} \quad -1 \leq x \leq 1. \tag{2b}$$

In Eq. (2a), $\sin^{-1} x$ is to be read "the inverse sine of x," and the superscript -1 is not to be treated as an exponent. [This means that although we

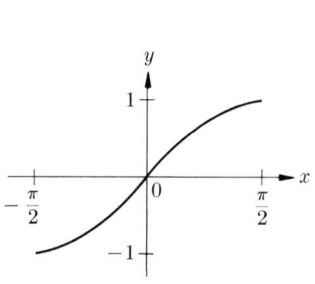

FIG. 11-2. The mapping of the equation $y = \sin x$ is one-to-one from the domain $-\pi/2 \leq x \leq \pi/2$ onto the range $-1 \leq y \leq 1$.

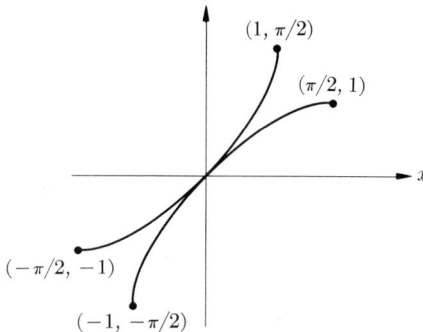

FIG. 11-3. Graphs of $y = \sin x$ and $y = \sin^{-1} x$.

write $(\sin x)^2$ as $\sin^2 x$, we do *not* write $(\sin x)^{-1} = \csc x$ as $\sin^{-1} x$.] The notation arc sin x is also used for the inverse sine function.

As we saw previously, the graphs of f and f^{-1} are symmetric with respect to the line $y = x$. Figure 11–3 shows this relationship for the graphs of $y = \sin x$, $|x| \leq \pi/2$, and $y = \sin^{-1} x$. Such a graph conveys much information about each function, its domain, range, and extrema, for example. Note that since the slope of f at $(0, 0)$ is 1, the slope of f^{-1} must also be 1 there. The horizontal tangent to f at $(\pi/2, 1)$ would lead one to conjecture a vertical tangent to f^{-1} at $(1, \pi/2)$.

Derivative of $\sin^{-1} x$.

Theorem 10–3 and Eq. (5) of Section 10–3 can be applied directly to get the derivative of $\sin^{-1} x$. To this end, let

$$f(x) = \sin x, \qquad -\pi/2 < x < \pi/2,$$

so that f is strictly increasing on its domain and

$$f^{-1}(x) = \sin^{-1} x, \qquad -1 < x < 1.$$

By Eq. (5) of Section 10–3,

$$(f^{-1})'(x) = \frac{1}{f'(f^{-1}(x))}.$$

For the present application, $f(x) = \sin x$, $f'(x) = \cos x$, and

$$(f^{-1})'(x) = \frac{1}{\cos(\sin^{-1} x)}. \qquad (3)$$

The right-hand side of Eq. (3) can readily be simplified as we now show. For, if

$$y = \sin^{-1} x, \qquad -1 \leq x \leq 1, \quad -\pi/2 \leq y \leq \pi/2,$$

then

$$\sin y = x \quad \text{and} \quad \cos y = \pm\sqrt{1 - x^2}.$$

However, the cosine is positive for $-\pi/2 < y < \pi/2$, so we have

$$\cos y = \cos(\sin^{-1} x) = \sqrt{1 - x^2}.$$

Therefore, Eq. (3) leads to the following equation:

$$\frac{d(\sin^{-1} x)}{dx} = \frac{1}{\sqrt{1 - x^2}}. \qquad (4)$$

TRIGONOMETRIC FUNCTIONS AND THEIR INVERSES

Remark 1. Note that the endpoints $x = -1$ and $x = +1$ are excluded in (4) to avoid division by zero. Thus our conjecture that the tangent is vertical at the point $(1, \pi/2)$ is substantiated.

Remark 2. If u is a differentiable function of x, the chain rule combined with Eq. (4) leads to the more general formula

$$\frac{d}{dx}\sin^{-1} u = \frac{du/dx}{\sqrt{1-u^2}}, \qquad -1 < u < 1. \tag{5}$$

Example 1. Find dy/dx when $y = \sin^{-1}(2x)$.

Solution

$$\frac{dy}{dx} = \frac{2}{\sqrt{1-4x^2}}.$$

Example 2. Find a function $y = f(x)$ such that

$$\frac{dy}{dx} = \frac{1}{\sqrt{4-x^2}}.$$

Solution. Write

$$\frac{dy}{dx} = \frac{1}{2}\frac{1}{\sqrt{1-(x/2)^2}},$$

and let $u = x/2$. Then we have $du/dx = \frac{1}{2}$ and

$$\frac{dy}{dx} = \frac{du/dx}{\sqrt{1-u^2}} = \frac{d}{dx}(\sin^{-1} u).$$

Therefore,

$$y = \sin^{-1}(x/2)$$

is a solution of the problem. (As a matter of fact, every solution is of the form $y = \sin^{-1}(x/2) + C$, where C is an arbitrary constant.)

Remark 3. We get Eq. (5) more simply if we differentiate implicitly:

$$y = \sin^{-1} u \qquad \text{or} \qquad \sin y = u.$$

Then

$$\cos y \frac{dy}{dx} = \frac{du}{dx},$$

so that

$$\frac{dy}{dx} = \frac{du/dx}{\cos y}.$$

Since $y = \sin^{-1} u$ implies $-\pi/2 \leq y \leq \pi/2$, we have $\cos y \geq 0$, so

$$\cos y = \sqrt{1 - \sin^2 y} = \sqrt{1 - u^2},$$

and

$$\frac{dy}{dx} = \frac{du/dx}{\sqrt{1 - u^2}}, \quad |u| < 1.$$

The Inverse Cosine and Its Derivative

The cosine of an acute angle is the sine of the complementary angle. If we state this relation in terms of radian measure,

$$\cos y = \sin(\pi/2 - y), \qquad (6)$$

we are led to a relation between the inverse cosine and inverse sine, as follows. If

$$x = \cos y = \sin(\pi/2 - y) \qquad (7a)$$

and

$$-\pi/2 \leq (\pi/2 - y) \leq \pi/2, \qquad (7b)$$

then

$$\pi/2 - y = \sin^{-1} x$$

or

$$y = \pi/2 - \sin^{-1} x. \qquad (7c)$$

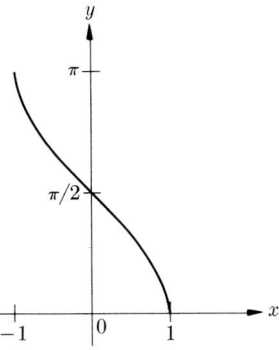

FIG. 11-4. The graph of $y = \cos^{-1} x$, $-1 \leq x \leq 1$, $0 \leq y \leq \pi$.

We let Eqs. (7a) and (7c) guide us to the definition

$$y = \cos^{-1} x = \pi/2 - \sin^{-1} x \qquad (7d)$$

with restrictions on y obtained from (7b):

$$0 \leq y \leq \pi. \qquad (7e)$$

(To get (7e), subtract $\pi/2$ throughout (7b), then multiply by -1.)

The graph of the inverse cosine function

$$y = \cos^{-1} x = \pi/2 - \sin^{-1} x, \quad -1 \leq x \leq 1, \; 0 \leq y \leq \pi, \qquad (8)$$

is shown in Fig. 11-4.

The derivative may be computed from Eqs. (8) and (4):

$$\frac{d}{dx} \cos^{-1} x = \frac{d}{dx}\left(\frac{\pi}{2} - \sin^{-1} x\right) = -\frac{1}{\sqrt{1 - x^2}}.$$

Checking our result:

$$f(x) = \cos x, \quad f'(x) = -\sin x,$$

$$f^{-1}(x) = \cos^{-1} x, \quad (f^{-1})'(x) = \frac{1}{-\sin(\cos^{-1} x)} = -\frac{1}{\sqrt{1-x^2}}.$$

The restrictions indicated above must be observed.

More generally, if u is a differentiable function of x, then

$$\frac{d}{dx}(\cos^{-1} u) = \frac{-du/dx}{\sqrt{1-u^2}}, \quad -1 < u < 1. \tag{9}$$

The Inverse Tangent and Its Derivative

The function

$$y = \tan x, \quad -\pi/2 < x < \pi/2, \tag{10}$$

is continuous, strictly increasing, and maps the interval $-\pi/2 < x < \pi/2$ one-to-one onto the set of all real numbers $-\infty < y < \infty$. Therefore, its inverse exists and is also continuous and strictly increasing. The inverse function is obtained by solving

$$x = \tan y, \quad -\pi/2 < y < \pi/2, \tag{11a}$$

for y in terms of x, which we denote by

$$y = \tan^{-1} x, \quad -\infty < x < \infty. \tag{11b}$$

Figures 11–5(a) and (b) show graphs of Eqs. (10) and (11).

It is interesting to note that the use of the function $\tan^{-1} x$, together with a Taylor Series, has made it possible to compute the value of π to 100,000 decimal places. This was done in 1962 at the David Taylor Model Basin, Washington, D.C., by Daniel Shanks and John W. Wrench, Jr., using an IBM 7090 system. The computation required 8 hours 43 minutes of machine time!

If we differentiate Eq. (11a) implicitly with respect to x, we get

$$1 = \sec^2 y \, \frac{dy}{dx}$$

or

$$\frac{dy}{dx} = \frac{1}{\sec^2 y} = \frac{1}{1 + \tan^2 y} = \frac{1}{1 + x^2}.$$

Therefore, since Eq. (11a) is just a different way of stating (11b),

$$\frac{d}{dx} \tan^{-1} x = \frac{1}{1 + x^2}. \tag{12}$$

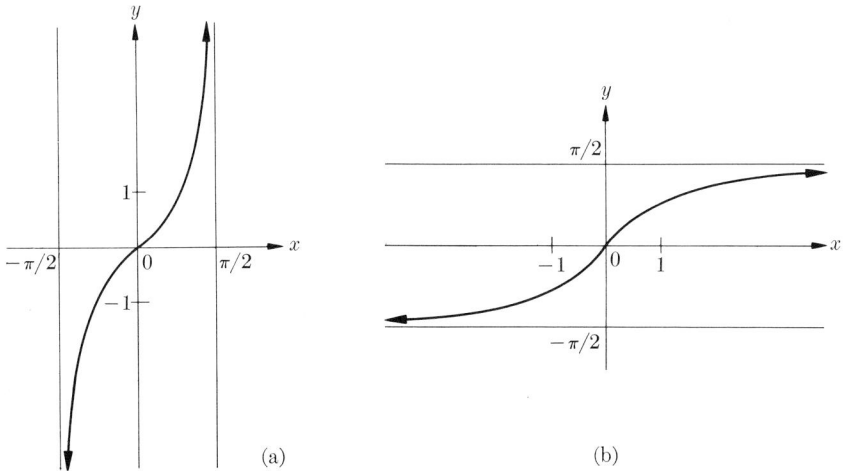

FIG. 11-5. Graphs of (a) $y = \tan x$ and (b) $y = \tan^{-1} x$ or $x = \tan y$.

Combining this with the chain rule, we also have

$$\frac{d}{dx} \tan^{-1} u = \frac{du/dx}{1 + u^2}, \qquad (13)$$

provided u is a differentiable function of x.

The reader is probably familiar with the relation between differentiation and integration. The process of reversing differentiation is the key to the solution of many problems. (See, for example, Section 12-3.) We will examine in the next two examples the relation between the derivative and the antiderivative of two functions.

Example 3. $\dfrac{d}{dx} \tan^{-1}(3x) = \dfrac{3}{1 + 9x^2}.$

Example 4. Find a function $y = f(x)$ that satisfies

$$\frac{dy}{dx} = \frac{1}{9 + x^2}.$$

Solution. Rewrite the given equation in the form

$$\frac{dy}{dx} = \frac{1}{9} \frac{1}{1 + (x/3)^2}$$

and let $u = x/3$. Then $du/dx = \frac{1}{3}$, so that

$$\frac{dy}{dx} = \frac{1}{3} \frac{du/dx}{1 + u^2} = \frac{d}{dx} \left(\frac{1}{3} \tan^{-1} u \right).$$

Hence,
$$y = \frac{1}{3}\tan^{-1}\left(\frac{x}{3}\right)$$
is a solution of the given equation.

Inverse Cotangent

Because of the identity
$$\cot y = \tan(\pi/2 - y), \qquad (14)$$
we define
$$y = \cot^{-1} x \qquad (15a)$$
to be equivalent to
$$\cot y = x = \tan(\pi/2 - y) \qquad (15b)$$
and
$$\pi/2 - y = \tan^{-1} x. \qquad (15c)$$

The range of $\tan^{-1} x$, Eq. (15c), is the open interval
$$-\pi/2 < \tan^{-1} x < \pi/2. \qquad (15d)$$

Substituting from (15c) into (15d), we get $-\pi/2 < \pi/2 - y < \pi/2$, or
$$0 < y < \pi. \qquad (15e)$$

Equations (15a), (15c), and the restriction (15e) yield:
$$y = \cot^{-1} x = \pi/2 - \tan^{-1} x, \qquad 0 < y < \pi. \qquad (16)$$

From Eqs. (16) and (13) we get
$$\frac{d}{dx}\cot^{-1} u = -\frac{du/dx}{1 + u^2}. \qquad (17)$$

The inverse of a trigonometric function, f^{-1}, and the inverse of its cofunction, co-f^{-1}, are related in the following way: $f^{-1} + \text{co-}f^{-1} = \pi/2$. For example,
$$\sin^{-1} x + \cos^{-1} x = \pi/2, \qquad -1 \leq x \leq 1,$$
as we noted previously. The two accompanying graphs, Fig. 11–6(a) and (b), give a clear picture of this situation for $\sin^{-1} x$ and $\tan^{-1} x$. In Fig. 11–6(a), for example, let line BC be perpendicular to the x-axis at

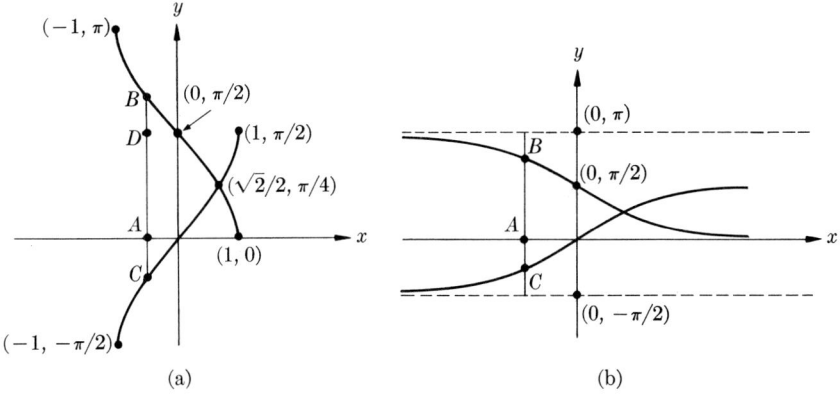

FIG. 11-6. (a) Graphs of $y = \sin^{-1} x$ and $y = \cos^{-1} x$. (b) Portion of the graph of $y = \tan^{-1} x$ and $y = \cot^{-1} x$; $BC \perp x$-axis; $AB + AC = \pi/2$, or $\tan^{-1} x + \cot^{-1} x = \pi/2$.

point A, and intersect the graph of $\sin^{-1} x$ at C and that of $\cos^{-1} x$ at B. Then $|DB| = |AC|$, and $BC = AD + |DB| + |AC| = AD + DB - AC = AD = \pi/2$, or $\sin^{-1} x + \cos^{-1} x = \pi/2$. This is true, of course, for any position of BC such that point A lies in the interval $[-1, 1]$. A similar observation can be made in Fig. 11-6(b), except that point A may be any point on the x-axis. The same relation applies to $\csc^{-1} x + \sec^{-1} x$, $|x| \geq 1$, and the reader is invited to verify this from his own graph.

Inverse Secant

The secant and cosine are reciprocals:

$$\cos y = \frac{1}{\sec y}. \tag{18a}$$

If, therefore,

$$x = \sec y, \quad y = \sec^{-1} x, \tag{18b}$$

we should also have, from Eq. (18a),

$$\cos y = 1/x, \quad y = \cos^{-1}(1/x). \tag{18c}$$

We therefore define

$$\sec^{-1} x = \cos^{-1}(1/x). \tag{19a}$$

Because $\cos^{-1}(1/x)$ is defined only for $-1 \leq 1/x \leq 1$, and $1/x$ cannot be zero, the domain for (19a) consists of the semi-infinite intervals

$$-\infty < x \leq -1, \quad 1 \leq x < \infty. \tag{19b}$$

The *range* for (19a) is the set of all possible values of

$$y = \cos^{-1}(1/x)$$

when $|x| \geq 1$. These are all values of the $\cos^{-1}$ function *except* $\cos^{-1} 0$, which is $\pi/2$. [In Eq. (19a), $1/x$ cannot be zero for any real value of x.] Therefore, the range for $y = \sec^{-1} x$ is

$$0 \leq y \leq \pi, \qquad y \neq \pi/2. \tag{19c}$$

Figure 11-7 shows the graph of

$$y = \sec^{-1} x,$$

where

$$|x| \geq 1, \quad 0 \leq y \leq \pi, \quad y \neq \pi/2.$$

It is just the same as the graph of

$$x = \sec y,$$

where

$$0 \leq y \leq \pi, \quad y \neq \pi/2.$$

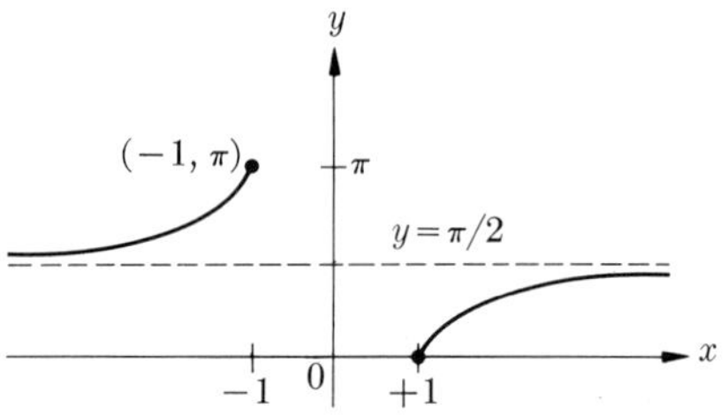

FIG. 11-7. Graph of the equation $y = \sec^{-1} x = \cos^{-1}(1/x)$, $|x| \geq 1$.

We get the derivative of the inverse secant from Eq. (19a):

$$\frac{d}{dx}\sec^{-1} x = \frac{d}{dx}\cos^{-1}\frac{1}{x} = \frac{-\frac{d}{dx}\left(\frac{1}{x}\right)}{\sqrt{1-(1/x)^2}}$$

$$= \frac{(1/x^2)}{\sqrt{(x^2-1)/x^2}} = \frac{1}{|x|\sqrt{x^2-1}}.$$

Combining this and the chain rule, we have

$$\frac{d}{dx}\sec^{-1} u = \frac{du/dx}{|u|\sqrt{u^2-1}}, \qquad |u| > 1. \tag{20}$$

Equation (20) is used mostly for finding a function whose derivative can be put in the form of the right-hand side of this equation.

Example 5. Find a function $y = f(x)$ such that

$$\frac{dy}{dx} = \frac{1}{x\sqrt{4x^2-1}}, \qquad x > \frac{1}{2}.$$

Solution. Let $u = 2x$. Then $du/dx = 2$ and

$$\frac{dy}{dx} = \frac{2}{2x\sqrt{(2x)^2 - 1}} = \frac{du/dx}{u\sqrt{u^2 - 1}}.$$

Since the condition $x > \frac{1}{2}$ means that $2x = u > 1$, $|u| = u$, and

$$\frac{dy}{dx} = \frac{du/dx}{|u|\sqrt{u^2 - 1}} = \frac{d}{dx} \sec^{-1} u,$$

so

$$y = \sec^{-1}(2x)$$

is a solution.

Inverse Cosecant

To complete the discussion, we define

$$\csc^{-1} x = \sin^{-1}(1/x), \qquad |x| \geq 1. \tag{21}$$

This leads to

$$\frac{d}{dx} \csc^{-1} x = \frac{-1}{|x|\sqrt{x^2 - 1}}, \qquad |x| > 1,$$

and

$$\frac{d}{dx} \csc^{-1} u = -\frac{du/dx}{|u|\sqrt{u^2 - 1}}, \qquad |u| > 1. \tag{22}$$

Because

$$\sin^{-1}(1/x) = \pi/2 - \cos^{-1}(1/x),$$

we also have

$$\csc^{-1} x = \pi/2 - \sec^{-1} x. \tag{23}$$

The range of $\csc^{-1} x$ is

$$-\pi/2 \leq \csc^{-1} x \leq \pi/2, \qquad \csc^{-1} x \neq 0. \tag{24}$$

To conclude our discussion of inverse trigonometric functions, we show two graphs, Figs. 11–8 and 11–9. The former shows the graphs of the sine function and the cosecant function on the same set of axes. Note that $(\sin x)(\csc x) = 1$ for all values of x except $x = n\pi$. Thus, if the line AC is drawn perpendicular to the x-axis, $AB \cdot AC = 1$, except for A at the points $x = n\pi$.

TRIGONOMETRIC FUNCTIONS AND THEIR INVERSES   11-2

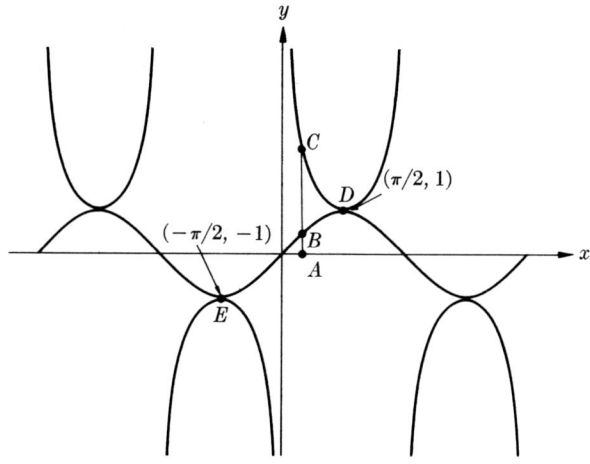

FIG. 11-8. Portion of the graphs of $y = \sin x$ and $y = \csc x$; $CA \perp x$-axis; $CA \cdot BA = 1$, or $\sin x \cdot \csc x = 1$.

For the inverse functions, the domain in Fig. 11–8 must be restricted, $|x| \leq \pi/2$. Restricting the domain this way, and reflecting across the line $y = x$ those portions of the graphs left in Fig. 11–8, we obtain the graphs of the inverse functions in Fig. 11–9. The curve between E' and D' is $y = \sin^{-1} x$, the rest of the graph is $y = \csc^{-1} x$. Note that the vertical line AC is reflected into the horizontal line $A'C'$, which still has the property that $A'B' \cdot A'C' = 1$, or $\sin^{-1}(x) = \csc^{-1}(1/x)$ for a suitably chosen x.

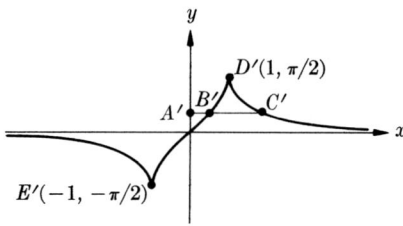

FIG. 11-9. Portion of the graphs of $y = \sin^{-1} x$ and $y = \csc^{-1} x$; $C'A' \perp y$-axis, $C'A' \cdot B'A' = 1$, or $\sin^{-1}(x) = \csc^{-1}(1/x)$.

EXERCISES 11-2

Find dy/dx given:

1. $y = \sin^{-1} 2x$
2. $y = \sin^{-1}(x^2)$
3. $y = \cos^{-1}(3x)$
4. $y = 2 \tan^{-1}(3x)$
5. $y = 3 \cot^{-1}(2x)$
6. $y = \frac{1}{a} \tan^{-1}\left(\frac{x}{a}\right)$, $a > 0$
7. $y = \sin^{-1}\left(\frac{x}{a}\right)$, $a > 0$
8. $y = \cos^{-1}(\sin x)$ (Sketch the graph for $0 \leq x \leq \pi$.)

9. $y = \sin^{-1}(\cos x)$. (Sketch the graph for $-\pi/2 \leq x \leq \pi/2$.)

10. $y = \tan^{-1}(\cot x)$

11. $y = \cot^{-1}(\tan x)$

12. $y = \sec^{-1}(\tan x)$

13. $y = \csc^{-1}(\sec x)$

14. Assuming that $x = \sec y$ defines y as a differentiable function of x, and $0 < y < \pi$, $y \neq \pi/2$, find dy/dx, in terms of x, by implicit differentiation.

15. Show (geometrically) that

$$\tan^{-1} x = \sin^{-1}\left(\frac{x}{\sqrt{1+x^2}}\right), \quad -\infty < x < \infty.$$

Use this result and Eq. (5) to derive Eq. (13).

16. Given $f(x) = |\sin x| - |\cos x|$.
   a) Sketch the graph for $0 \leq x \leq 2\pi$.
   b) What are domain, range, period?
   c) Is $f'(x)$ continuous for all x in D_f?

17. Sketch graphs of $y = \csc^{-1} x$ and $y = \sec^{-1} x$ on the same axes. Verify the fact that

$$\csc^{-1} x + \sec^{-1} x = \pi/2, \quad |x| \geq 1.$$

12
AREA AND INTEGRATION

12-1 INTRODUCTION

In this chapter we shall study the question of area: how can we define the area of a region bounded by a finite number of curves? In particular, we shall consider regions of the form

$$R = \{(x, y) : a \leq x \leq b,\ 0 \leq y \leq f(x)\}, \tag{1}$$

where f is a function with domain $[a, b]$ whose values are nonnegative real numbers lying in some interval $m \leq f(x) \leq M$. We shall also assume that f is piecewise continuous. That is, we assume that there exists a finite set of numbers $X_1, X_2, \ldots, X_t$ such that

$$a = X_1 < X_2 < \cdots < X_t = b \tag{2}$$

and such that f is continuous on each of the subintervals

$$(X_1, X_2),\ \ldots,\ (X_{t-1}, X_t). \tag{3}$$

Figure 12-1 illustrates the graph of such a piecewise continuous function.

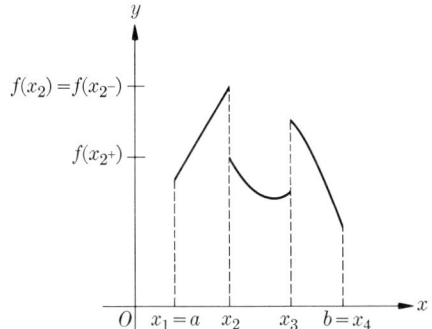

FIG. 12-1. Graph of a function that is piecewise continuous on an interval $a \leq x \leq b$.

Remark 1. If $f(x)$ tends to a limit as x approaches X_2 from the left (that is, over values of x that are less than X_2), we designate that number by $f(X_2-)$:

$$f(X_2-) = \lim_{x \to X_2-} f(x). \tag{4a}$$

The value of f at X_2 may or may not be equal to this number. If $f(X_2) = f(X_2-)$, we say that f is continuous *from the left* at X_2. Similarly, we define limit from the right,

$$f(X_2+) = \lim_{x \to X_2+} f(x), \tag{4b}$$

and say that f is continuous *from the right* at X_2 if $f(X_2) = f(X_2+)$. Of course, f may fail to be continuous both from the left and from the right, because $f(X_2)$ may be different from both $f(X_2-)$ and $f(X_2+)$.

Remark 2. Removable Discontinuity. If $f(x)$ approaches the same limit L as x approaches a number c (interior to the domain of f) from the left or from the right, but for some reason the value of the function at c is not equal to L, we say that f has a *removable discontinuity* at c. Such functions are sometimes found as textbook examples, but in most serious mathematics a removable discontinuity is *removed* by simply redefining the value of f at such a point c to make $f(c) = L$. For example, if

$$f(x) = \frac{\sin x}{x} \quad \text{for} \quad x \neq 0,$$

then

$$\lim_{x \to 0^-} f(x) = \lim_{x \to 0^+} f(x) = 1.$$

If the value of f at zero is any number other than 1, the function would have a removable discontinuity at $x = 0$. Most mathematicians would just redefine the value of the function to be $f(0) = 1$, thereby making the newly defined f continuous. We did something like this earlier [see Eq. (6), Section 10–1] when we were considering the increment of a differentiable function and comparing it with the increment along a tangent line. In that instance we didn't even allow a removable discontinuity to occur: we just defined the value of the function α to be the appropriate thing (namely 0) at $x = a$.

Resuming the discussion of area, we ask how one might go about assigning a number to represent the area of a region like that shown in Fig. 12–1? We reduce this to a sequence of simpler questions; namely, how can we assign areas to the separate portions that lie over the individual subintervals $(X_1, X_2), \ldots, (X_{t-1}, X_t)$? Once we have succeeded in doing this, we would probably agree that we should just add the results to get the area of the entire region. Accordingly, we now turn our attention to the question of assigning an area to a region R of the form described in Eq. (1), when the upper boundary is the curve

$$C : y = f(x), \quad a \leq x \leq b, \tag{5}$$

and f is now assumed to be *continuous on the entire domain* $a \leq x \leq b$. From Theorem C-I, Section 6–3, we know that such a function has both a minimum m and a maximum M:

$$m = \min_{a \leq x \leq b} f(x), \quad M = \max_{a \leq x \leq b} f(x). \tag{6}$$

Therefore, the region R contains all points inside the inscribed rectangle of altitude m, base $b - a$, and is contained in the circumscribed rectangle having the same base, and altitude M (Fig. 12–2). Thus, if $A(R)$ denotes

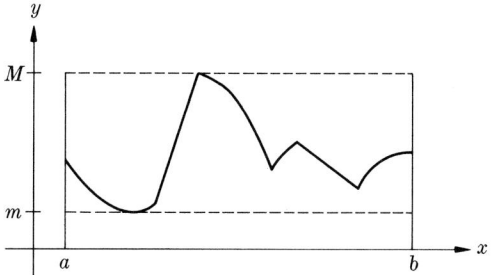

FIG. 12–2. Region R with inscribed and circumscribed rectangles: $m(b - a) \leq A(R) \leq M(b - a)$.

a number measuring the area of R, we want it to be true that

$$m(b - a) \leq A(R) \leq M(b - a). \tag{7}$$

Additivity

Another requirement that we would like a definition of area to satisfy is called *additivity*. This means that if R is split into two subregions R_1 and R_2, the area of R should be the sum of the areas of these subregions. In particular, if c is a number between a and b, and

$$R_1 = \{(x, y) : a \leq x \leq c,\ 0 \leq y \leq f(x)\},$$
$$R_2 = \{(x, y) : c \leq x \leq b,\ 0 \leq y \leq f(x)\},$$

then it should be true that

$$A(R) = A(R_1) + A(R_2). \tag{8}$$

Figure 12–3 shows inscribed and circumscribed rectangles for regions R_1 and R_2. The minimum of $f(x)$ on $a \leq x \leq c$ is m_1, the maximum is M_1. On the other subinterval the minimum is m_2, and the maximum is

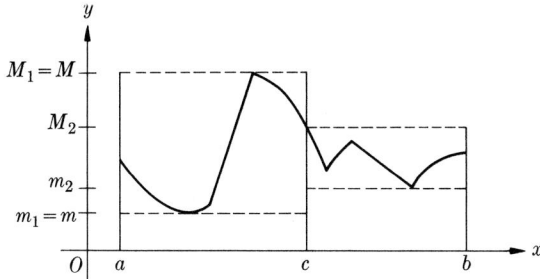

FIG. 12–3. Regions R_1 and R_2 with inscribed and circumscribed rectangles.

AREA AND INTEGRATION

M_2. The inequalities (7) should apply to both R_1 and R_2. In other words, it should be true that

$$m_1(c - a) \leq A(R_1) \leq M_1(c - a) \tag{9a}$$

and

$$m_2(b - c) \leq A(R_2) \leq M_2(b - c). \tag{9b}$$

Both geometrically and analytically it is evident that the sum of the areas of the two inscribed rectangles in Fig. 12–3 is greater than or equal to the area of the one inscribed rectangle in Fig. 12–2. Here is the analytic reasoning: for every x in the entire interval (a, b) we have $f(x) \geq m$. Since m_1 and m_2 are values of $f(x)$ in subintervals of (a, b), it is necessarily true that

$$m_1 \leq m \quad \text{and} \quad m_2 \leq m. \tag{10}$$

Multiply the first of these inequalities by the positive number $c - a$ and the second by $b - c$ and add. The result is

$$m_1(c - a) + m_2(b - c) \geq m(c - a) + m(b - c) = m(b - a). \tag{11}$$

The left-hand side of formula (11) is the sum of the areas of the two inscribed rectangles in R_1 and R_2, while the right-hand side is the area of the one rectangle inscribed in R. Thus, when we introduced the additional subdivision point c, we either left the inscribed area unchanged or else we increased it. In a similar way, it is easy to show that the circumscribed area either remained the same or *decreased*.

Suppose, now, that the process of introducing additional subdivision points is continued. At each stage, the inscribed areas either remain the same or *increase* and the circumscribed areas remain the same or *decrease*. And all the time, the area of the region R is to be between the corresponding inscribed and circumscribed areas. We might therefore represent the situation by writing

increasing inscribed areas $\leq A(R) \leq$ *decreasing* circumscribed areas.

Hence, if we could ever bring the inscribed and circumscribed areas arbitrarily close together, we would have a very close approximation to $A(R)$. The foregoing discussion is intended to motivate the more formal definition we now present.

Formal Definition of Area.* Let R be the region described by Eq. (1). Let $P = \{a, x_1, x_2, \ldots, x_{n-1}, b\}$ be an arbitrary *partition* of $[a, b]$

---

* For a more general treatment, see *Advanced Calculus* by R. C. Buck and E. F. Buck, McGraw Hill (1965).

with
$$a = x_0 < x_1 < x_2 < \cdots < x_n = b. \tag{12}$$

In the ith subinterval $[x_{i-1}, x_i]$ let the minimum value of $f(x)$ be m_i and the maximum M_i. Define the *lower* sum $L_f(P)$, read "L sub f of P," by the equation

$$L_f(P) = \sum_{i=1}^{n} m_i(x_i - x_{i-1}), \tag{13a}$$

and the upper sum $U_f(P)$, read "U sub f of P," by the equation

$$U_f(P) = \sum_{i=1}^{n} M_i(x_i - x_{i-1}). \tag{13b}$$

The *supremum*, over all partitions P, of the lower sums $L_f(P)$ is called the *inner area* of A and is denoted by $\underline{A}(R)$, read "A lower bar of R." Similarly, the *outer* area of R, denoted by $\overline{A}(R)$ is the *infimum* of all upper sums $U_f(P)$:

$$\underline{A}(R) = \sup L_f(P), \qquad \overline{A}(R) = \inf U_f(P). \tag{14}$$

Finally, if the inner area of R and the outer area of R are equal, we call their common value the area of R:

$$A(R) = \underline{A}(R) = \overline{A}(R) \quad \text{if} \quad \underline{A}(R) = \overline{A}(R). \tag{15}$$

Example 1. If $f(x) = k$ is constant for $a \leq x \leq b$, then all lower sums and all upper sums have the constant value $k(b - a)$, so

$$\underline{A}(R) = \overline{A}(R) = A(R) = k(b - a).$$

Example 2. If $a = 0$ and $b = 1$, and $f(x) = 1$ when x is rational, and $f(x) = 0$ when x is irrational, then all lower sums $L_f(P)$ are equal to 0 [because the minimum value of $f(x)$ in each subinterval is zero], so the inner area is 0:

$$\underline{A}(R) = \text{lub } L_f(P) = 0.$$

On the other hand, the maximum value of $f(x)$ in each subinterval in any partition P is 1, so for each P, $U_f(P) = 1$, and the glb over all P of such sums is 1. Therefore,

$$\overline{A}(R) = 1.$$

Thus, for this example, the inner area of the region $0 \leq x \leq 1$, $0 \leq y \leq f(x)$ is zero, while the outer area is 1. Therefore, since the two are not equal, we do not assign an area to the region.

The formal definition of area given above makes use of the idea of an *arbitrary partition* of $[a, b]$. It is sometimes convenient to partition a closed interval so that all the subintervals are of equal length. Such a partition is called a *regular* partition:

Definition. A regular partition, P_n, of $[a, b]$ is one in which

$$a_i - a_{i-1} = (b - a)/n, \qquad i = 1, 2, \ldots, n.$$

EXERCISES 12-1

1. By analogy with Eq. (11), show that, in the notation of the text,

$$M_1(c - a) + M_2(b - c) \leq M(b - a).$$

If $n = 5$, and P_5 is a regular partition of $[1, 2]$ into 5 subintervals of equal length, find $L_f(P_5)$ and $U_f(P_5)$ for each of Exercises 2 through 7.

2. $f(x) = 3, \quad 1 \leq x \leq 2$

3. $f(x) = \begin{cases} 1, & \text{if } 1 \leq x \leq 1.5 \\ 2, & \text{if } 1.5 < x \leq 2 \end{cases}$

4. $f(x) = x, \quad 1 \leq x \leq 2$

5. $f(x) = 1/x, \quad 1 \leq x \leq 2$

6. $f(x) = [10x + 5], \quad 1 \leq x < 2$, where the brackets denote "greatest integer in."

7. $f(x) = |x - 1.5|, \quad 1 \leq x \leq 2$

8. Using the definition of *regular* partition given above, let n be an arbitrary positive integer, $a = 0$, $b > 0$, and P_n a regular partition of $[0, b]$ into y subintervals of equal lengths. Show that, if $f(x) = x$ for $0 \leq x \leq b$, then

$$L_f(P_n) = \frac{b^2}{2}\left(1 - \frac{1}{n}\right)$$

and

$$U_f(P_n) = \frac{b^2}{2}\left(1 + \frac{1}{n}\right).$$

What can you conclude with respect to:
a) $U_f(P_n) - L_f(P_n)$?
b) $\lim_{n \to \infty} L_f(P_n)$?
c) The order relations among the four upper and lower sums corresponding to two regular partitions P_n and P_m if $m < n$.

12-2 INTEGRATION OF CONTINUOUS FUNCTIONS

In Section 12–1 we defined the area of a region R that is bounded below by the x-axis, above by the graph of a continuous function, and on the left and right by straight line segments parallel to the y-axis. Thus, if R is defined by

$$R = \{(x, y): a \leq x \leq b, 0 \leq y \leq f(x)\} \tag{1}$$

and f is positive and continuous on $[a, b]$, the area of R is the lub of sums of areas of the inscribed rectangles or the glb of sums of areas of the circumscribed rectangles, provided these two numbers are equal. In this section we shall prove that these two numbers are, in fact, equal.

Furthermore we shall do the following:

1. Remove the restriction that f be nonnegative. Thus, in place of sums of areas of inscribed or circumscribed rectangles we shall introduce *lower Riemann sums* $L_f(P)$ and *upper Riemann sums* $U_f(P)$.
2. Show that, when f is *continuous* on the closed bounded interval $[a, b]$, there exists a unique number I which has the property that

$$L_f(P) \leq I \leq U_f(P) \tag{2}$$

for all partitions P of $[a, b]$.
3. Define the Riemann integral and show that it is the number I described above (when f is continuous).

Notation. Suppose P is the partition

$$P: \{a = a_0 < a_1 < a_2 < \cdots < a_n = b\}. \tag{3}$$

By the *norm of P*, which we denote by $\|P\|$, we mean the maximum of the lengths of the subintervals $[a_{j-1}, a_j]$, $j = 1, 2, \ldots, n$:

$$\|P\| = \max_{1 \leq j \leq n} \delta_j, \tag{4a}$$

where

$$\begin{aligned}\delta_j &= (a_j - a_{j-1}) \\ &= \text{length of the } j\text{th subinterval } [a_{j-1}, a_j].\end{aligned} \tag{4b}$$

Suppose, next, that we choose an arbitrary number x_1 in the first subinterval $[a_0, a_1]$, x_2 in the second subinterval, and so on:

$$a_0 \leq x_1 \leq a_1, \quad a_1 \leq x_2 \leq a_2, \quad \ldots, \quad a_{n-1} \leq x_n \leq a_n. \tag{5}$$

Let

$$Q = \{x_1, x_2, \ldots, x_n\}, \tag{6}$$

and form the sum

$$\begin{aligned}S_f(P, Q) &= f(x_1)\,\delta_1 + f(x_2)\,\delta_2 + \cdots + f(x_n)\,\delta_n \\ &= \sum_{j=1}^{n} f(x_j)\,\delta_j.\end{aligned} \tag{7}$$

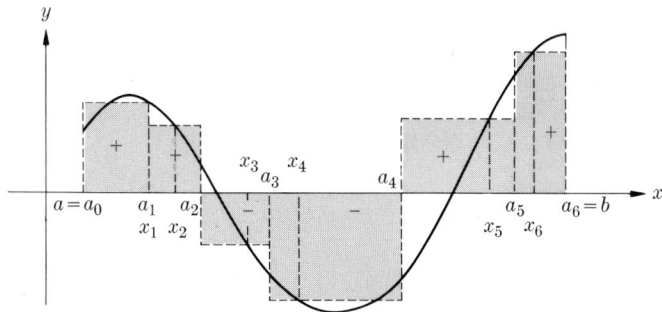

FIG. 12–4. Geometric representation of $S_f(P, Q)$ for the special case $n = 6$. Areas of rectangles marked $+$ are added, those marked $-$ are subtracted.

Whenever $f(x_j)$ is positive, the product $f(x_j)\, \delta_j$ represents the area of a rectangle of altitude

$$y_j = f(x_j)$$

and base

$$\delta_j = \text{length of the } j\text{th subinterval.}$$

If $f(x_j)$ is negative, the corresponding product is the negative of the area of the associated rectangle. The sum, (7), depends on the function f, the partition P, and the specific choices of points Q in the subintervals. It is called a *Riemann* sum for f on $[a, b]$. Geometrically it represents the algebraic sum of

(areas of rectangles above the x-axis)
minus
(areas of rectangles below the x-axis).

See Fig. 12–4.

If the points of Q are chosen at places where f takes its maximum values, in each subinterval, the corresponding Riemann sum is called an *upper Riemann sum*, and will be denoted by $U_f(P)$. Thus, if

$$M_j = \max\,\{f(x)\colon a_{j-1} \le x \le a_j\}, \qquad j = 1, 2, \ldots, n, \tag{8}$$

then the upper Riemann sum of f, for P, is

$$U_f(P) = \sum_{j=1}^{n} M_j\, \delta_j. \tag{9}$$

Similarly, the *lower Riemann sum* $L_f(P)$ is given by

$$L_f(P) = \sum_{j=1}^{n} m_j\, \delta_j, \tag{10}$$

where
$$m_j = \min \{f(x): a_{j-1} \leq x \leq a_j\}, \quad 1 \leq j \leq n. \tag{11}$$

Clearly, since points of Q are chosen so that
$$a_{j-1} \leq x_j \leq a_j,$$

Eqs. (8) and (11) show that
$$m_j \leq f(x_j) \leq M_j \quad \text{for} \quad j = 1, 2, \ldots, n. \tag{12}$$

The lengths δ_j of the subintervals are positive, so (12) implies
$$m_j \, \delta_j \leq f(x_j) \, \delta_j \leq M_j \, \delta_j \quad \text{for} \quad 1 \leq j \leq n,$$

which in turn leads to
$$\sum_{j=1}^{n} m_j \, \delta_j \leq \sum_{j=1}^{n} f(x_j) \, \delta_j \leq \sum_{j=1}^{n} M_j \, \delta_j,$$

or
$$L_f(P) \leq S_f(P, Q) \leq U_f(P). \tag{13}$$

In words, inequalities (13) show that *all* Riemann sums for f, based on the same partition P, are bounded below by the lower sum and above by the upper sum.

Definition of Riemann Integral. We say that f is *Riemann integrable* on $[a, b]$ if there exists a number I such that all Riemann sums $S_f(P, Q)$ have I as limit when the norm of P approaches zero:
$$\lim_{\|P\| \to 0} S_f(P, Q) = I. \tag{14}$$

If such a limit I exists, it is called the *Riemann integral* of f on $[a, b]$, and this fact is represented by the notation
$$\int_a^b f = I \tag{15a}$$

or the more traditional notation
$$\int_a^b f(x) \, dx = I. \tag{15b}$$

In order to establish that a given function (or class of functions) is integrable on a particular interval, it is necessary (and sufficient) to

a) show the existence of the appropriate number I, and

b) show that to each $\epsilon > 0$ there corresponds $\delta > 0$ such that

$$|S_f(P, Q) - I| < \epsilon \qquad (16)$$

whenever P is any partition of norm less than δ, and the n points in Q are arbitrarily chosen in the n subintervals of $[a, b]$ determined by P.

Example 1. If $[a, b]$ is an arbitrary finite interval and $f(x) = c$ is constant on $[a, b]$, then for every partition P, and for every choice of points of Q,

$$S_f(P, Q) = \sum_{j=1}^{n} f(x_j)\, \delta_j$$

$$= \sum_{j=1}^{n} c\, \delta_j = c\, \delta_1 + c\, \delta_2 + \cdots + c\, \delta_n$$

$$= c(\delta_1 + \delta_2 + \cdots + \delta_n)$$

$$= c(b - a).$$

Hence,

$$U_f(P) = L_f(P) = S_f(P, Q) = c(b - a),$$

and for any $\epsilon > 0$, inequality (16) is satisfied in the form

$$|S_f(P, Q) - I| = |c(b - a) - I| = 0 < \epsilon$$

if we take

$$I = c(b - a).$$

Here, the norm of P doesn't much matter; the condition (16) is met if $\|P\| < \delta$ provided

$$\delta = \text{any positive number} \leq (b - a).$$

The result is: a *constant* function is integrable, and

$$\int_a^b \mathbf{c} = c(b - a), \qquad (17a)$$

or

$$\int_a^b c\, dx = c(b - a). \qquad (17b)$$

Remark 1. In (17a), $\mathbf{c}$ represents the function f whose values are $f(x) = c$ for $a \leq x \leq b$.

The chief result to be established in this section is to show that every function that is continuous on $[a, b]$ is *Riemann integrable* (or, more briefly, *integrable*) there.

Theorem 12–1a. Let a and b be real numbers, with $a < b$. Let f be a real-valued function continuous on the closed bounded interval $[a, b]$. Then $\int_a^b f$ exists.

Proof. By Theorem 6–7, f is bounded on $[a, b]$. That is, there exist real numbers m, M such that

$$m \leq f(x) \leq M \quad \text{for all } x \in [a, b].$$

Let P be a partition of $[a, b]$ and note that

$$m \leq m_j \leq M_j \leq M \quad \text{for} \quad j = 1, 2, \ldots, n.$$

Therefore, multiplying by δ_j and summing on j, we get

$$m(b - a) \leq L_f(P) \leq U_f(P) \leq M(b - a).$$

That is, for each partition P, the lower sum $L_f(P)$ and upper sum $U_f(P)$ are bounded below by $m(b - a)$ and above by $M(b - a)$. Therefore the bounded nonempty set of *upper sums* has a *greatest lower bound*, which we call I:

$$I = \text{glb } \{U_f(P) \colon P \text{ any partition of } [a, b]\}. \tag{18}$$

We shall show that this number I satisfies Eq. (14). To this end, suppose $\epsilon > 0$. Then $\epsilon/(b - a)$ is also positive, and because f is *uniformly* continuous on the closed bounded interval $[a, b]$, there is a $\delta > 0$ such that, for any x' and x'' in $[a, b]$,

$$|f(x') - f(x'')| < \frac{\epsilon}{b - a} \tag{19a}$$

whenever

$$|x' - x''| < \delta. \tag{19b}$$

Now let P be any partition of $[a, b]$ with $\|P\| < \delta$. Because f is continuous on each subinterval, there exists x_j' and x_j'' in $[a_{j-1}, a_j]$ such that

$$m_j = f(x_j') = \min \{f(x) \colon a_{j-1} \leq x \leq a_j\},$$
$$M_j = f(x_j'') = \max \{f(x) \colon a_{j-1} \leq x \leq a_j\},$$

and

$$|x_j' - x_j''| \leq (a_j - a_{j-1}) \leq \|P\| < \delta.$$

Therefore,

$$0 \leq M_j - m_j < \frac{\epsilon}{b - a} \quad \text{for} \quad 1 \leq j \leq n. \tag{20}$$

Multiplying (20) by $\delta_j > 0$ and summing, we get

$$0 \leq \sum_{j=1}^{n} (M_j - m_j) \delta_j < \sum_{j=1}^{n} \frac{\epsilon}{b-a} \delta_j = \frac{\epsilon}{b-a} \sum_{j=1}^{n} \delta_j = \epsilon,$$

or

$$0 \leq \sum_{j=1}^{n} M_j \delta_j - \sum_{j=1}^{n} m_j \delta_j = U_f(P) - L_f(P) < \epsilon. \qquad (21)$$

Inequality (21) tells us that, for any partition P with norm less than the δ of inequalities (19a) and (19b), the upper and lower Riemann sums differ by less than ϵ. Certainly we can find partitions of norm less than δ: we simply take n large enough so that

$$\Delta x = \frac{b-a}{n} < \delta,$$

or

$$n > \frac{b-a}{\delta},$$

and divide the interval $[a, b]$ into n subintervals of equal length. The corresponding regular partition P has $\|P\| = \Delta x < \delta$.

Now, the number I of Eq. (18) is a *lower bound* for all upper sums for f on $[a, b]$, so

$$I \leq U_f(P) \qquad \text{for all } P. \qquad (22)$$

If we could also be sure that

$$L_f(P) \leq I \qquad \text{for all } P \qquad (23)$$

(which, in fact, we *shall prove* after the lemma at the conclusion of this proof), we could complete our present proof as follows:

Let P be any partition of norm $< \delta$, let $S_f(P, Q)$ be an arbitrary associated Riemann sum. Then, from (13),

$$L_f(P) \leq S_f(P, Q) \leq U_f(P) \qquad (24a)$$

and, from (22) and (23),

$$L_f(P) \leq I \leq U_f(P). \qquad (24b)$$

Therefore, I and $S_f(P, Q)$ are both in the closed interval of numbers $[L_f(P), U_f(P)]$, so

$$|S_f(P, Q) - I| \leq |U_f(P) - L_f(P)| \qquad (25)$$

and the right-hand side of (25) is less than ϵ by inequality (21); hence $|S_f(P, Q) - I| < \epsilon$ whenever $\|P\| < \delta$. Therefore $\int_a^b f = I$. Q.E.D.

In the exercises, the reader is asked to complete the proof of the following lemma.

Lemma. If P_1 and P_2 are any partitions of $[a, b]$, then

$$L_f(P_1) \leq U_f(P_2). \tag{26}$$

Partial Proof. Suppose P_1' is a *refinement* of the partition P_1. By this we mean that P_1' is obtained by using all of the partition points used in P_1 and, possibly, a finite number of additional points. How does $L_f(P_1')$ compare with $L_f(P_1)$? [First, look at the simplest possible case where $P_1 = \{a, b\}$ and $P_1' = \{a, c, b\}$. See inequality (11), Section 12-1. The general case behaves the same way and can be handled by inserting points, one at a time, to go from P_1 to P_1'.] You should reach the conclusion that

$$L_f(P_1) \leq L_f(P_1'). \tag{27}$$

Similarly, if P_2' is a refinement of P_2, then

$$U_f(P_2') \leq U_f(P_2). \tag{28}$$

Now, let P_3 be a refinement of both P_1 and P_2. For example, one could take P_3 to be their union. Then inequalities (27) and (28) would apply with P_3 in place of P_1' and P_2'. Moreover, with

$$P_1' = P_2' = P_3,$$

we would be able to conclude that

$$L_f(P_1) \leq L_f(P_3) \leq U_f(P_3) \leq U_f(P_2), \tag{29}$$

which establishes inequality (26). [Why is the middle inequality in (29) correct?]

Remark 2. Inequality (26) enables us to complete the proof of Theorem 12-1a by establishing (23). Recall that I, in (23), is the *greatest lower bound* of *all* the upper sums. By inequality (26), if P_1 is an arbitrary partition, then $L_f(P_1)$ is a lower bound for *all* upper sums, because P_2 is an arbitrary partition. But I is the *greatest* lower bound. Therefore, $L_f(P_1) \leq I$. Since P_1 is arbitrary, this is the same as (23).

Corollary 12-1.1. Let f be continuous on $[a, b]$ with $f(x) \geq 0$ for $a \leq x \leq b$. Let

$$R = \{(x, y): a \leq x \leq b, 0 \leq y \leq f(x)\}.$$

Then the area of R exists, and

$$A(R) = \underline{A}(R) = \overline{A}(R) = \int_a^b f. \tag{30}$$

Proof. Equation (18) in the proof of Theorem 12–1a defines I to be the glb of upper Riemann sums. This is the same as the definition of outer area. Hence

$$\overline{A}(R) = I. \tag{31}$$

Inequality (23) implies that I is an upper bound of all lower sums $L_f(P)$, and therefore I is greater than or equal to their *least* upper bound $\underline{A}(R)$:

$$\underline{A}(R) \leq I. \tag{32}$$

We want to prove that equality must hold in (32). We assume the contrary, arrive at a contradiction, and thus complete the proof. Thus, suppose that

$$\underline{A}(R) < I = \overline{A}(R). \tag{33a}$$

Then $\overline{A}(R) - \underline{A}(R)$ is positive, and we let

$$\epsilon = \tfrac{1}{2}[\overline{A}(R) - \underline{A}(R)]. \tag{33b}$$

By inequality (21), there exists a partition P_0 such that

$$U_f(P_0) < L_f(P_0) + \epsilon. \tag{34a}$$

But, for every partition P,

$$L_f(P) \leq \underline{A}(R),$$

so that from (34a) we deduce that

$$U_f(P_0) < \underline{A}(R) + \epsilon. \tag{34b}$$

On the other hand, all upper sums are greater than or equal to $\overline{A}(R)$ because

$$\overline{A}(R) = I = \text{glb } U_f(P).$$

Hence,

$$U_f(P_0) \geq \overline{A}(R),$$

so that (34b) implies

$$\overline{A}(R) \leq U_f(P_0) < \underline{A}(R) + \epsilon$$

or

$$\overline{A}(R) < \underline{A}(R) + \epsilon. \tag{35a}$$

Therefore,
$$\overline{A}(R) - \underline{A}(R) < \epsilon \tag{35b}$$
or, from (33b),
$$2\epsilon < \epsilon.$$

Since $\epsilon > 0$, the last inequality is false. Therefore, the assumption (33a) is false. Statement (32) is therefore satisfied only in the form
$$\underline{A}(R) = I. \qquad \text{Q.E.D.}$$

Remark 3. If f is not restricted to positive values, we replace outer area $\overline{A}(R)$ by the "*upper* Riemann integral of f," written $\overline{\int}_a^b f$, and defined as follows:
$$\overline{\int}_a^b f = \underset{P}{\text{glb}}\ U_f(P). \tag{36a}$$

Similarly, the *lower* Riemann integral is given by
$$\underline{\int}_a^b f = \text{lub}\ L_f(P). \tag{36b}$$

When f is continuous on $[a, b]$, *but is not restricted to nonnegative values*, then Corollary 12–1.1 is replaced by the following.

Corollary 12–1.2. Let f be continuous on $[a, b]$. Then
$$\overline{\int}_a^b f = \underline{\int}_a^b f = \int_a^b f.$$

The proof is left as an exercise. The argument is almost exactly that given above with upper and lower Riemann integrals replacing $\overline{A}(R)$ and $\underline{A}(R)$, respectively.

Remark 4. For any *bounded* function f, the lower and upper Riemann integrals always exist, and
$$\underline{\int}_a^b f \leq \overline{\int}_a^b f.$$

A necessary and sufficient condition that f be *integrable* on $[a, b]$ is that these lower and upper integrals be equal. By Theorem 12–1a and Corollary 12–1.2, a *sufficient* condition is that f be *continuous* on $[a, b]$. However, continuity of f is not necessary. The following theorem will not be proved here. (See, for example, J. M. H. Olmsted, *Real Variables*, Appleton-Century-Crofts, 1959, p. 153, Exercise 54, for an outline of the proof.)

Theorem 12–1b. A function f defined on a closed interval $[a, b]$ is integrable there if and only if it is bounded and its discontinuities form a set of measure zero.

[A set of real numbers has *measure zero* if and only if to every $\epsilon > 0$ corresponds a covering of the set by a finite or countably infinite collection of open intervals the sum of whose lengths is less than ϵ. Thus, for example, any countably infinite set has measure zero. For, if the points of the set are

$$x_1, x_2, \ldots$$

and $\epsilon > 0$, we cover

x_1 by an interval of length $\epsilon/4$
x_2 by an interval of length $\epsilon/8$
$\vdots$
x_n by an interval of length $\epsilon/2^{n+1}$.

The sum of lengths is

$$\frac{\epsilon}{4} + \frac{\epsilon}{8} + \frac{\epsilon}{16} + \cdots = \frac{\epsilon}{4}\left(1 + \frac{1}{2} + \frac{1}{4} + \cdots\right)$$
$$= \frac{\epsilon}{2} < \epsilon.]$$

The *ruler function* defined on [0, 1] by

$$f(x) = \begin{cases} 0 & \text{if } x \text{ is irrational,} \\ 1/q & \text{if } x = p/q, \, p \text{ and } q \text{ are relatively prime positive integers,} \end{cases}$$

is continuous at the irrationals and discontinuous at the rationals. Since the rationals are countably infinite, and the ruler function is bounded, it is integrable on [0, 1]. In fact,

$$\int_0^1 f = 0.$$

Many functions we study in calculus are either monotonic, or piecewise monotonic. Such functions are integrable on any interval on which they are bounded. This is a consequence of the following theorem.

Theorem 12–1c. *If f is monotonic and bounded on $[a, b]$, then*

$$\underline{\int_a^b} f = \overline{\int_a^b} f.$$

Proof (for monotonic increasing functions). Suppose that f is monotone increasing on $[a, b]$ and

$$f(a) = m, \quad f(b) = M.$$

Let ϵ be an arbitrary positive number. Let P be a partition of $[a, b]$ with
$$\|P\| < \frac{\epsilon}{1 + |M - m|}.$$
Then, in the notations of Eqs. (3) and (4b),
$$U_f(P) = f(a_1)\,\delta_1 + f(a_2)\,\delta_2 + \cdots + f(a_n)\,\delta_n, \tag{37a}$$
$$L_f(P) = f(a_0)\,\delta_1 + f(a_1)\,\delta_2 + \cdots + f(a_{n-1})\,\delta_n. \tag{37b}$$

(If f were monotone decreasing, we would have to interchange the two expressions above.) Therefore,
$$\begin{aligned}U_f(P) - L_f(P) &= [f(a_1) - f(a_0)]\,\delta_1 + [f(a_2) - f(a_1)]\,\delta_2 \\ &\quad + \cdots + [f(a_n) - f(a_{n-1})]\,\delta_n \\ &\leq [f(a_1) - f(a_0)]\|P\| + [f(a_2) - f(a_1)]\|P\| \\ &\quad + \cdots + [f(a_n) - f(a_{n-1})]\|P\|.\end{aligned} \tag{38}$$

In this inequality, $\|P\|$ is a common factor on the right-hand side, and the other factors have a sum which telescopes as follows:
$$\begin{aligned}[f(a_1) - f(a_0)] + [f(a_2) - f(a_1)] + \cdots &+ [f(a_n) - f(a_{n-1})] \\ &= f(a_n) - f(a_0) = M - m.\end{aligned} \tag{39}$$

Therefore, if $\|P\| < \epsilon/(1 + |M - m|)$, from (38) and (39), we deduce that
$$|U_f(P) - L_f(P)| \leq |M - m| \frac{\epsilon}{1 + |M - m|} < \epsilon. \tag{40}$$

Now for any *bounded* function, both the lower and upper integrals exist (by the completeness postulate for the real numbers):
$$\underline{I} = \int_{\underline{a}}^b f = \operatorname*{lub}_P L_f(P)$$
$$\overline{I} = \int_a^{\overline{b}} f = \operatorname*{glb}_P U_f(P).$$

By definition of lub and glb, this also implies that, for any bounded function f and any partition P,
$$L_f(P) \leq \underline{I} \leq \overline{I} \leq U_f(P). \tag{41}$$

To complete our proof, we now show that inequality (40) implies that $\underline{I} = \overline{I}$. For, if this were not so, we could define the positive number ϵ_0 to

AREA AND INTEGRATION

be $\epsilon_0 = \overline{I} - \underline{I}$. Then (41) shows that

$$|U_f(P) - L_f(P)| \geq \overline{I} - \underline{I} = \epsilon_0$$

for *all* partitions P. But this is false because inequality (40) holds with $\epsilon_0 = \epsilon$ when $\|P\|$ is small enough. Since the assumption that the lower and upper integrals are not equal leads to this contradiction, it follows that

$$\overline{I} = \underline{I}. \qquad \text{Q.E.D.}$$

Remark 5. The above result applies, for example, to the greatest-integer function, $f(x) = [x]$. Although this function is discontinuous at every integral value of x, it is bounded and monotonic on any bounded interval, and is therefore integrable there. [Inequalities (40) and (41) show that any Riemann sum $S_f(P, Q)$ for a bounded monotonic function differs from the common value of upper and lower integrals by less than ϵ when $\|P\|$ is small enough, so that

$$\lim_{\|P\| \to 0} S_f(P, Q) = \overline{I} = \underline{I}.]$$

EXERCISES 12–2

1. Verify that the equations below are valid:

a) $\sum_{k=1}^{5} (7a_k) = 7 \left(\sum_{k=1}^{5} a_k \right)$ b) $\sum_{k=1}^{4} (a_k + b_k) = \left(\sum_{k=1}^{4} a_k \right) + \left(\sum_{k=1}^{4} b_k \right)$

c) $\sum_{k=1}^{n} (a_{k+1} - a_k) = a_{n+1} - a_1$

d) $\sum_{k=1}^{n} [(k+1)^2 - k^2] = (n+1)^2 - 1^2 = n^2 + 2n$

e) $\sum_{k=1}^{n} [(k+1)^2 - k^2] = \sum_{k=1}^{n} (2k+1)$

f) $\sum_{k=1}^{n} (2k+1) = 2 \left(\sum_{k=1}^{n} k \right) + n$

2. Combine the results of parts (d), (e), and (f) in the exercise above, and show that

$$2 \left(\sum_{k=1}^{n} k \right) = n^2 + n \quad \text{or} \quad 1 + 2 + 3 + \cdots + n = \frac{n(n+1)}{2}.$$

3. Show that

$$\sum_{k=1}^{n}[(k+1)^3 - k^3] = (n+1)^3 - 1,$$

and from this derive the result

$$n^3 + 3n^2 + 3n = \sum_{k=1}^{n}(3k^2 + 3k + 1)$$

$$= 3\left(\sum_{k=1}^{n}k^2\right) + 3\left(\sum_{k=1}^{n}k\right) + n$$

or

$$\sum_{k=1}^{n}k^2 = \frac{n^3 + 3n^2 + 2n}{3} - \sum_{k=1}^{n}k.$$

Substitute

$$\sum_{k=1}^{n}k = \frac{n(n+1)}{2}$$

from the preceding exercise and show that

$$1^2 + 2^2 + 3^2 + \cdots + n^2 = \frac{n(n+1)(2n+1)}{6}.$$

4. Divide the interval $0 \leq x \leq b$ into n equal subintervals, each of length $\Delta x = b/n$, and calculate $L_f(P_n)$, $U_f(P_n)$, and $E_n = U_f(P_n) - L_f(P_n)$ for the function $f(x) = x^2$. The formula

$$1^2 + 2^2 + 3^2 + \cdots + m^2 = \frac{m(m+1)(2m+1)}{6}$$

may be assumed for the purposes of this exercise.

5. Divide the interval $0 \leq x \leq \pi/2$ into n equal subintervals, each of length $\Delta x = \pi/(2n)$, and calculate $L_f(P_n)$, $U_f(P_n)$, and $U_f(P_n) - L_f(P_n)$ for the function $f(x) = \sin x$. The formula

$$\sin \alpha + \sin 2\alpha + \sin 3\alpha + \cdots + \sin m\alpha = \frac{\cos (\alpha/2) - \cos (m + \frac{1}{2})\alpha}{2 \sin (\alpha/2)}$$

may be assumed for the purposes of this exercise.

6. Let $L_f(P_n)$ be a lower sum, as in Eq. (10), and form another lower sum $L_f(P_{2n})$ corresponding to a finer subdivision of the interval $a \leq x \leq b$ obtained by halving each of the original subintervals. Is $L_f(P_n)$ the same as, greater than, or less than $L_f(P_{2n})$?

7. Replace $L_f(P_n)$ and $L_f(P_{2n})$ of Exercise 6 by $U_f(P_n)$ and $U_f(P_{2n})$, respectively, and answer the corresponding question.

AREA AND INTEGRATION

8. Suppose the graph of $y = f(x)$ is always rising to the right between $x = a$ and $x = b$. Take all the Δx_k's equal and call their common value simply Δx. Show by reference to Fig. 12-4(a), that the error $U_f(P_n) - L_f(P_n)$ is representable graphically as the area

$$[f(b) - f(a)]\Delta x$$

of the rectangle R. [*Hint:* The error is the sum of rectangles with diagonals P_0P_1, P_1P_2, etc., along the curve, and there is no overlapping when these are all displaced horizontally into the rectangle R.]

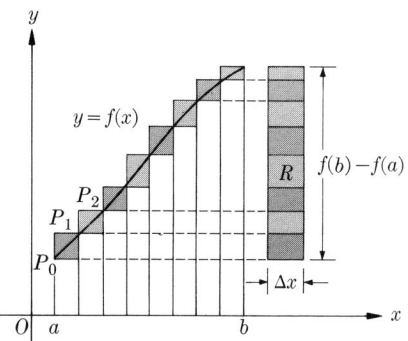

9. Draw a figure to represent a continuous curve $y = f(x)$ which is always falling to the right between $x = a$ and $x = b$. Suppose all the subdivisions Δx_k are again taken to be equal, $\Delta x_k = \Delta x = (b-a)/n$. Obtain an expression for the error analogous to the expression in Exercise 8.

10. In Exercise 8 or 9, if the Δx_k's are not all equal, show that

$$U_f(P_n) - L_f(P_n) \leq |f(b) - f(a)|(\Delta x_{\max}),$$

where $\Delta x_{\max}$ is the largest of the Δx_k's for $k = 1, 2, \ldots, n$.

11. Definition of *refinement*: a partition of $[a, b]$ is a *refinement* of the partition P if it contains P as a subset. If f is continuous on $[a, b]$ and P' is a refinement of P, show that:
   a) $L_f(P') \geq L_f(P)$,
   b) $U_f(P') \leq U_f(P)$.

12. Give the complete proof of Eq. (26) (lemma).

13. Prove Corollary 12–1.2.

14. Compute the lower Riemann sums for the function $f(x) = x$, on $[0, 1]$, corresponding to the following partitions:
   a) $P_1 : a_0 = 0, a_1 = \frac{1}{2}, a_2 = 1$
   b) $P_2 = \{0, \frac{1}{4}, \frac{1}{2}, 1\}$
   c) $P_3 = \{0, 1/n, 2/n, 3/n, \ldots (n-1)/n, n/n\}$, where n represents an arbitrary positive integer. The formula $1 + 2 + 3 + \cdots + m = m(m+1)/2$ may be assumed as known.

15. Suppose that $a < b < c$, where a, b, c are three real numbers. Let f be a real-valued function that is continuous on $[a, c]$. Let P_1 be a partition of $[a, b]$, P_2 a partition of $[b, c]$, and $P_3 = P_1 \cup P_2$.
   a) Show that P_3 is a partition of $[a, c]$.

b) Is it true that $U_f(P_3) = U_f(P_1) + U_f(P_2)$? Give the reasons for your answer.

c) Is it true that $L_f(P_3) = L_f(P_1) + L_f(P_2)$? Give the reasons for your answer.

16. Suppose that a, b, c are three real numbers with $a < b < c$. Let P be a partition of $[a, c]$ that does *not* include b as a partition point. Let P' be the partition of $[a, c]$ obtained by including b along with all points of P:

$$P' = P \cup \{b\}.$$

Let P_1 be the intersection of P' with $[a, b]$ and P_2 the intersection of P' with $[b, c]$.

a) Is it true that $U_f(P) = U_f(P_1) + U_f(P_2)$, assuming that f is an arbitrary real-valued function that is continuous on $[a, c]$? Give reasons for your answer.

b) Discuss the relation between $L_f(P)$ and the sum $L_f(P_1) + L_f(P_2)$.

12–3 INTEGRALS EVALUATED DIRECTLY FROM THE DEFINITION

In Section 12–2 we saw that, associated with any function f that is *continuous* on the interval $[a, b]$ there is a number

$$I = \int_a^b f \tag{1}$$

with these properties:

a) Given any $\epsilon > 0$, there exists $\delta > 0$ such that every Riemann sum $S_f(P, Q)$ with $\|P\| < \delta$ is within ϵ-distance of I:

$$|S_f(P, Q) - I| < \epsilon \quad \text{if} \quad \|P\| < \delta. \tag{2}$$

b) For any partition P of $[a, b]$, the lower and upper Riemann sums $L_f(P)$ and $U_f(P)$ bracket I:

$$L_f(P) \leq I \leq U_f(P). \tag{3}$$

c) Given $\epsilon > 0$, there is a partition P for which the lower and upper sums differ by less than ϵ. In fact, there exists $\delta > 0$ such that

$$U_f(P) - L_f(P) < \epsilon \quad \text{for all } P \text{ such that } \|P\| < \delta. \tag{4}$$

d) The number I is the greatest lower bound of the set of all upper Riemann sums:

$$I = \operatorname*{glb}_{P} U_f(P). \tag{5}$$

e) I is also *an* upper bound for all lower Riemann sums. It is easy to deduce from (3) and (4) that I is, in fact, the *least* upper bound of the set of all lower Riemann sums:

$$I = \operatorname*{lub}_{P} L_f(P). \tag{6}$$

Example 1. Evaluate the integral from 0 to b, $b > 0$, of the identity function

$$f(x) = x.$$

Solution. By Theorem 12–1a of Section 12–2, we know that the continuous function $f(x) = x$ is integrable on $[0, b]$, and there is a number I such that

$$\int_0^b x \, dx = I.$$

Figure 12–5 shows a portion of the graph of $f(x) = x$ and a partition of $[0, b]$ into $n = 6$ subintervals of unequal lengths. For

$$P = \{a_0, a_1, \ldots, a_6\}$$

and $f(x) = x$, the minimum value of $f(x)$ for $a_{i-1} \leq x \leq a_i$ is $m_i = f(a_{i-1}) = a_{i-1}$. Therefore,

$$L_f(P) = f(a_0)(a_1 - a_0) + f(a_1)(a_2 - a_1) + \cdots + f(a_5)(a_6 - a_5)$$

$$= \sum_{i=1}^{6} f(a_{i-1})(a_i - a_{i-1}) = \sum_{i=1}^{6} a_{i-1}(a_i - a_{i-1})$$

$$= (a_0 a_1 - a_0^2) + (a_1 a_2 - a_1^2) + \cdots + (a_5 a_6 - a_5^2). \tag{7}$$

The sum (7) is not particularly attractive, but we can improve things by computing the corresponding upper sum $U_f(P)$, and the average of $L_f(P)$ and $U_f(P)$.

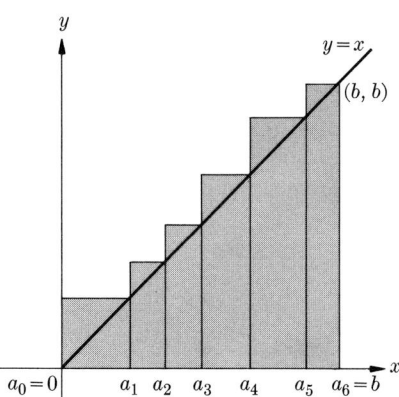

FIG. 12–5. The area of the shaded rectangles is $U_f(P)$.

For the same partition P, the maximum value of $f(x)$ for $a_{i-1} \leq x \leq a_i$ is $M_i = f(a_i) = a_i$. Hence

$$U_f(P) = f(a_1)(a_1 - a_0) + f(a_2)(a_2 - a_1) + \cdots + f(a_6)(a_6 - a_5)$$
$$= \sum_{i=1}^{6} f(a_i)(a_i - a_{i-1}) = \sum_{i=1}^{6} a_i(a_i - a_{i-1})$$
$$= (a_1^2 - a_1 a_0) + (a_2^2 - a_2 a_1) + \cdots + (a_6^2 - a_6 a_5). \tag{8}$$

Combining Eqs. (7) and (8), we get

$$L_f(P) + U_f(P) = (a_1^2 - a_0^2) + (a_2^2 - a_1^2) + \cdots + (a_6^2 - a_5^2)$$
$$= (a_1^2 + a_2^2 + \cdots + a_6^2) - (a_0^2 + a_1^2 + \cdots + a_5^2)$$
$$= a_6^2 - a_0^2 = b^2 - 0^2 = b^2. \tag{9}$$

The average of $L_f(P)$ and $U_f(P)$ is a number, $b^2/2$, midway between them, so

$$L_f(P) < \frac{b^2}{2} < U_f(P). \tag{10}$$

The same result would be true for every partition P, because Eqs. (7), (8), and (9) would simply be replaced, in the general case, by

$$L_f(P) = \sum_{i=1}^{n} (a_{i-1} a_i - a_{i-1}^2), \tag{7'}$$

$$U_f(P) = \sum_{i=1}^{n} (a_i^2 - a_i a_{i-1}), \tag{8'}$$

and

$$L_f(P) + U_f(P) = \sum_{i=1}^{n} (a_i^2 - a_{i-1}^2) = a_n^2 - a_0^2 = b^2. \tag{9'}$$

Of course we now claim that

$$\int_0^b x \, dx = b^2/2. \tag{11}$$

How does this result follow from inequalities (10)? In the following way: we know from (3) that the number I that we want is always in the interval $[L_f(P), U_f(P)]$, for any partition P; and $b^2/2$ is also in that interval by inequalities (10). Can we prove that $b^2/2$ and I are the same number? Yes, since the given function $f(x) = x$ is continuous on $[0, b]$, the difference between upper and lower sums approaches zero as the norm of P does. Because $b^2/2$ and I are fixed numbers between $L_f(P)$ and $U_f(P)$, we must have

$$|b^2/2 - I| \leq U_f(P) - L_f(P). \tag{12}$$

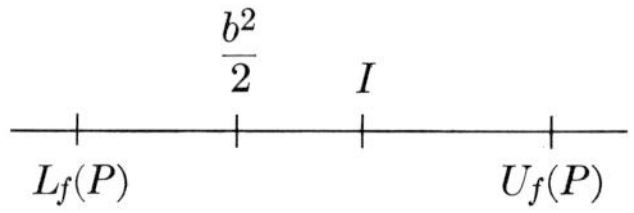

AREA AND INTEGRATION

```
         b²
         ─
         2       I
    +────+───────+────+
   L_f(P)       U_f(P)
```
FIG. 12-6. This situation is impossible.

The left-hand side of (12) remains constant, and the right-hand side can be made as small as we please by taking $\|P\|$ sufficiently small, by property (4). Therefore,
$$b^2/2 - I = 0, \tag{13}$$
for, otherwise, we could take
$$\epsilon = \tfrac{1}{2}|b^2/2 - I|$$
and arrive at a contradiction (Fig. 12-6). Hence
$$\int_0^b x\, dx = I = b^2/2.$$

We could have evaluated $\int_0^b x\, dx$ in other ways. A familiar method is to use partitions of $[0, b]$ into subintervals of equal lengths. Thus, if n is a positive integer, we let P_n be the regular partition
$$P_n = \{a_0, a_1, a_2, \ldots, a_i, \ldots, a_n = b\}$$
$$= \left\{0, \frac{b}{n}, \frac{2b}{n}, \ldots, \frac{ib}{n}, \ldots, \frac{nb}{n} = b\right\}.$$

Then, Eq. (7′) becomes
$$L_f(P_n) = \sum_{i=1}^{n}\left[\frac{(i-1)b}{n}\frac{ib}{n} - \frac{(i-1)^2 b^2}{n^2}\right]$$
$$= \frac{b^2}{n^2}\sum_{i=1}^{n}[(i^2 - i) - (i^2 - 2i + 1)]$$
$$= \frac{b^2}{n^2}\sum_{i=1}^{n}(i - 1) = \frac{b^2}{n^2}[0 + 1 + 2 + \cdots + (n-1)]$$
$$= \frac{b^2}{n^2}\frac{(n-1)(n)}{2} = \frac{b^2}{2}\left(1 - \frac{1}{n}\right). \tag{14}$$

Similarly, Eq. (8′) yields
$$U_f(P_n) = \frac{b^2}{2}\left(1 + \frac{1}{n}\right). \tag{15}$$

Finally,
$$L_f(P_n) \le I \le U_f(P_n)$$

becomes
$$\frac{b^2}{2}\left(1 - \frac{1}{n}\right) \leq I \leq \frac{b^2}{2}\left(1 + \frac{1}{n}\right). \tag{16}$$

The only number I that satisfies the inequalities (16) for every positive integer n is
$$I = b^2/2.$$

Still another method for evaluating $I = \int_0^b x\, dx$ is to compute the Riemann sum $S_f(P, Q)$, where Q consists of the midpoints of the subintervals $[a_{i-1}, a_i]$. That is,
$$Q = \{x_1, x_2, \ldots, x_n\}$$
with
$$x_i = \tfrac{1}{2}(a_i + a_{i-1}), \qquad i = 1, 2, \ldots, n.$$
Then
$$\begin{aligned}
S_f(P, Q) &= \sum_{i=1}^n f(x_i)(a_i - a_{i-1}) = \sum_{i=1}^n \tfrac{1}{2}(a_i^2 - a_{i-1}^2) \\
&= \tfrac{1}{2}[(a_1^2 + a_2^2 + \cdots + a_n^2) - (a_0^2 + a_1^2 + \cdots + a_{n-1}^2)] \\
&= \tfrac{1}{2}(a_n^2 - a_0^2) = \tfrac{1}{2}b^2. \tag{17}
\end{aligned}$$

But, for any partition P and any Riemann sum $S_f(P, Q)$, we know that
$$L_f(P) \leq S_f(P, Q) \leq U_f(P).$$
Then, as before, we have
$$L_f(P) \leq b^2/2 \leq U_f(P)$$
and
$$L_f(P) \leq I \leq U_f(P).$$

The conclusion that $I = b^2/2$ follows from the fact that both $U_f(P)$ and $L_f(P)$ approach I when $\|P\|$ tends to zero.

Equation (17) is a special case of the following. Suppose that F is a continuous function whose domain includes $[a, b]$ and that
$$F'(x) = f(x) \qquad \text{for all } x \in (a, b).$$

Then, by the mean-value theorem, in each subinterval (a_{i-1}, a_i) there is an x_i such that
$$\frac{F(a_i) - F(a_{i-1})}{a_i - a_{i-1}} = F'(x_i) = f(x_i)$$

or, equivalently,

$$f(x_i)(a_i - a_{i-1}) = F(a_i) - F(a_{i-1}). \tag{18}$$

The left-hand side of (18) is a typical term in $S_f(P, Q)$, provided the points x_i in Q are chosen in the way just described. Hence, for such a choice of Q, we have

$$\begin{aligned} S_f(P, Q) &= \sum_{i=1}^{n} f(x_i)(a_i - a_{i-1}) = \sum_{i=1}^{n} [F(a_i) - F(a_{i-1})] \\ &= [F(a_1) + F(a_2) + \cdots + F(a_n)] \\ &\quad - [F(a_0) + F(a_1) + \cdots + F(a_{n-1})] \\ &= F(a_n) - F(a_0) = F(b) - F(a). \end{aligned} \tag{19}$$

Now, if f is integrable on $[a, b]$, we know that

$$\lim_{\|P\| \to 0} S_f(P, Q) = \int_a^b f(x)\, dx. \tag{20}$$

Since $F(b) - F(a)$ remains fixed as $\|P\| \to 0$, if we combine (19) and (20) we get

$$\int_a^b f(x)\, dx = F(b) - F(a). \tag{21}$$

We have therefore proved the following theorem, which is known as the fundamental theorem of calculus:

Theorem 12-2. Fundamental Theorem of Calculus. Let f be integrable on $[a, b]$. Suppose F is continuous on the closed interval $[a, b]$ and differentiable on the open interval (a, b) and

$$F'(x) = f(x) \quad \text{for} \quad a < x < b. \tag{22a}$$

Then

$$\int_a^b f = F(b) - F(a). \tag{22b}$$

Remark 1. When two functions f and F satisfy the relation $F'(x) = f(x)$ over some domain, we say that

f is the derivative of F

and that

F is a primitive of f.

Note that the derivative is unique (when it exists), but a primitive is not unique, because an arbitrary constant function can be added to it and the result is another primitive of f. Hence, briefly, the fundamental theorem says, if f is integrable and F is a primitive of f, then Eq. (22b) holds.

Remark 2. It is helpful, in applying the fundamental theorem to the evaluation of integrals, to introduce the notation

$$F(x)]_a^b = F(b) - F(a). \tag{23}$$

Example 2. Since

$$\frac{d}{dx}\left(\frac{x^3}{3}\right) = x^2,$$

we have

$$\int_1^4 x^2 \, dx = \frac{x^3}{3}\bigg]_1^4 = \frac{64}{3} - \frac{1}{3} = 21.$$

Remark 3. In each application of the fundamental theorem we have to guess a primitive of the function we want to integrate. If we can't guess such a primitive, it doesn't mean that the integral fails to exist, but only that we can't evaluate it by that particular method. However, if the integrand function f is continuous, we can approximate

$$\int_a^b f = I$$

to any desired accuracy, $\epsilon > 0$, by using a partition P such that

$$U_f(P) - L_f(P) < \epsilon. \tag{24}$$

Then, for any Riemann sum $S_f(P, Q)$ we exploit the fact that

$$L_f(P) \leq S_f(P, Q) \leq U_f(P)$$

and

$$L_f(P) \leq I \leq U_f(P)$$

to conclude that

$$|S_f(P, Q) - I| < \epsilon.$$

Example 3. Estimate

$$\int_1^2 \frac{1}{x} \, dx \tag{25}$$

to an accuracy of $\epsilon = \frac{1}{4}$.

Solution. Meeting the condition (24) is our goal. (Here we assume that the natural logarithm function has not yet been introduced, so we can't apply the fundamental theorem but must estimate the integral (25) by Riemann sums.) Let P be a partition

$$P = \{a_0 = 1 < a_1 < a_2 < \cdots < a_n = 2\}$$

with

$$a_i - a_{i-1} = \frac{2-1}{n} = \frac{1}{n} \quad \text{for} \quad i = 1, 2, \ldots, n.$$

If

$$m_i = \min(1/x), \quad M_i = \max(1/x)$$

for $a_{i-1} \leq x \leq a_i$, then

$$m_i = 1/a_i, \quad M_i = 1/a_{i-1}$$

and

$$M_i - m_i = \frac{a_i - a_{i-1}}{a_{i-1} \cdot a_i} = \frac{1/n}{a_{i-1} a_i}. \tag{26}$$

Since

$$a_i > a_{i-1} \geq a_0 = 1,$$

the denominator on the right-hand side of (26) is >1, so

$$M_i - m_i < 1/n \quad \text{for} \quad 1 \leq i \leq n. \tag{27}$$

Hence

$$U_f(P) - L_f(P) = \sum_{i=1}^{n} (M_i - m_i)(a_i - a_{i-1})$$

$$< \frac{1}{n} \sum_{i=1}^{n} (a_i - a_{i-1}) = \frac{1}{n}(2 - 1).$$

Therefore, given $\epsilon = \frac{1}{4}$, the condition

$$U_f(P) - L_f(P) < \epsilon$$

is satisfied if $n \geq 4$. Using $n = 4$ and $P = \{1, \frac{5}{4}, \frac{6}{4}, \frac{7}{4}, 2\}$, we compute

$$L_f(P) = \sum_{i=1}^{4} m_i(a_i - a_{i-1}) = \frac{1}{4} \sum_{i=1}^{4} m_i$$

$$= \frac{1}{4}[\frac{4}{5} + \frac{2}{3} + \frac{4}{7} + \frac{1}{2}] = \frac{1}{5} + \frac{1}{6} + \frac{1}{7} + \frac{1}{8} \approx 0.634524$$

and

$$U_f(P) = \frac{1}{4}[1 + \frac{4}{5} + \frac{2}{3} + \frac{4}{7}] = \frac{1}{4} + \frac{1}{5} + \frac{1}{6} + \frac{1}{7} \approx 0.759524.$$

[Note that $U_f(P) - L_f(P) = \frac{1}{4} - \frac{1}{8} = \frac{1}{8}$.] Therefore

$$0.634 < \int_1^2 \frac{1}{x}\,dx < 0.760,$$

and the approximation we get by averaging the upper and lower sums is

$$I = \int_1^2 \frac{1}{x}\,dx \approx 0.697.$$

Since I is in the interval $[L_f(P), U_f(P)]$ of length $\frac{1}{8}$, the midpoint of this interval differs from the true value of I by at most $\frac{1}{16}$, or 0.0625. (The value of I to 5 decimals is 0.69315, so our estimate of 0.697 is off by approximately 0.004. In other words, I also lies fairly near the midpoint of the interval [0.634, 0.760].

The fundamental theorem is sometimes used as a tool for approximating sums. This is the reverse of what we did in Section 12–2. There we found formulas for sums of first powers, and of squares, of the positive integers 1 through n. Then we used these formulas to compute limits of sums of areas of inscribed rectangles for the graphs of $y = mx$, $a \leq x \leq b$, and of $y = x^2$, $0 \leq x \leq b$. In the following example, we work back from the definite integral (or, what amounts to the same thing, the area under the graph) to an approximation for the sum of the square roots of the integers 1 through n. The process is not completely reversible. We can go from exact formulas for sums to definite integrals by way of limits, but we cannot go back from the definite integral to an exact formula for the sum, because we don't know which terms went to zero in the limit process, and hence we cannot recover them.

Example 4. Consider the function f defined by

$$f(x) = \sqrt{x}, \quad 0 \leq x \leq 1.$$

Let n be a positive integer greater than 1, and let $\Delta x = 1/n$. Let P be a partition of $[0, 1]$ into n subintervals of equal lengths Δx. Since the function f is monotone increasing, the area of a rectangle inscribed under the graph of f over the subinterval from $(j - 1)\,\Delta x$ to $j\,\Delta x$ is

$$f(x_{j-1})\,\Delta x = \sqrt{x_{j-1}}\,\Delta x = \sqrt{(j-1)\,\Delta x}\,\Delta x = \sqrt{(j-1)}\,(\Delta x)^{3/2} \quad (28)$$

for $j = 1, 2, \ldots, n$. We get the lower Riemann sum $L_f(P)$ by successively substituting the values of j from 1 through n in Eq. (28) and summing. Because $\Delta x = 1/n$ does not vary with j, the common factor

$$(\Delta x)^{3/2} = n^{-3/2}$$

appears in each summand, and we get

$$L_f(P) = \sum_{j=1}^{n} \sqrt{(j-1)}\, n^{-3/2} = \frac{\sqrt{0} + \sqrt{1} + \sqrt{2} + \cdots + \sqrt{n-1}}{n^{3/2}}.$$
(29)

In order to get the upper Riemann sum $U_f(P)$, we replace $j - 1$ by j in Eqs. (28) and (29) and get

$$U_f(P) = \sum_{j=1}^{n} \sqrt{j}\, n^{-3/2} = \frac{\sqrt{1} + \sqrt{2} + \sqrt{3} + \cdots + \sqrt{n}}{n^{3/2}}.$$
(30)

Now suppose that we let $n \to \infty$ in Eqs. (29) and (30). Since the function f is continuous on $[0, 1]$, it is integrable there and

$$\lim_{n \to \infty} L_f(P) = \lim_{n \to \infty} U_f(P) = \int_0^1 \sqrt{x}\, dx = \frac{x^{3/2}}{\frac{3}{2}}\bigg]_0^1 = \frac{2}{3}.$$
(31)

When n is large, both $L_f(P)$ and $U_f(P)$ are close to their common limit, $\frac{2}{3}$. For the upper sum, this means that the numerator on the right-hand side of Eq. (30) is approximately equal to $\frac{2}{3}n^{3/2}$:

$$\sqrt{1} + \sqrt{2} + \sqrt{3} + \cdots + \sqrt{n} \approx \frac{2}{3}n^{3/2}.$$
(32)

For $n = 10$, the sum of the square roots is found to be 22.5^-, while $\frac{2}{3}(10)^{3/2}$ is 21.1^-, so the approximation (32) is in error by about 6%. We could improve our estimate by taking advantage of the relation

$$L_f(P) \leq \tfrac{2}{3} \leq U_f(P).$$

For example, take $n = 11$ in Eq. (29) and find that the sum of the square roots of the integers from 1 through 10 is less than $\frac{2}{3}(11)^{3/2}$, which is 24.32, to two decimals. The average of the two estimates, 21.1 and 24.3, is 22.7, and this differs from 22.5 by less than 1%.

EXERCISES 12–3

1. For each of the following cases, integrate the given function f to find a new function F, defined by $F(x) = \int f(x)\, dx$. Apply the mean-value theorem to this new function F to find an expression for c_k in terms of x_k and x_{k-1} such that
$$F(x_k) - F(x_{k-1}) = F'(c_k)(x_k - x_{k-1}).$$

   a) $f(x) = x$
   c) $f(x) = x^3$
   b) $f(x) = x^2$
   d) $f(x) = 1/\sqrt{x}$

2. Consider the function defined by $f(x) = x$ and take $a = 0$, $b > 0$. Let $S_n = \sum_{k=1}^n f(c_k)\,\Delta x$. Show that by taking
$$c_k = \tfrac{1}{2}(x_k + x_{k-1}),$$
the resulting sum has the constant value
$$S_n = \tfrac{1}{2}b^2,$$
independent of the value of n. (Let $a = x_0$ and $b = x_n$.)

3. Take $f(x) = x^2$, $a = 0$, $b > 0$, and form S_n as in Exercise 2, using
$$c_k = \sqrt{\tfrac{1}{3}(x_k^2 + x_k x_{k-1} + x_{k-1}^2)}.$$
Show that no matter what value n has, the resulting sum S_n has the constant value
$$S_n = \tfrac{1}{3}b^3.$$
(Let $a = x_0$ and $b = x_n$.) What is the limit of S_n as $n \to \infty$? Note that one should substitute $x_k - x_{k-1}$ for Δx and recognize that
$$\left(\frac{x_k^2 + x_k x_{k-1} + x_{k-1}^2}{3}\right)\Delta x = \frac{x_k^3 - x_{k-1}^3}{3}.$$

4. Take $f(x) = x^3$, $a = 0$, $b > 0$, and calculate S_n as given in Exercise 2, taking
$$c_k = \sqrt[3]{\tfrac{1}{4}(x_k^3 + x_k^2 x_{k-1} + x_k x_{k-1}^2 + x_{k-1}^3)}.$$
Express the result in a form which is independent of the number of subdivisions n. (Let $a = x_0$ and $b = x_n$.)

5. Take $f(x) = 1/\sqrt{x}$, $a = 1$, $b > 1$. Use the intermediate values
$$c_k = \left(\frac{\sqrt{x_{k-1}} + \sqrt{x_k}}{2}\right)^2$$
and calculate S_n, where $S_n = \sum_{k=1}^n f(c_k)\,\Delta x$. Express the result in a form which is independent of the number of subdivisions. (Let $a = x_0$, and $b = x_n$.)

Evaluate each of the following definite integrals (6 through 16).

6. $\displaystyle\int_1^2 (2x + 5)\,dx$

7. $\displaystyle\int_0^1 (x^2 - 2x + 3)\,dx$

8. $\displaystyle\int_{-1}^1 (x+1)^2\,dx$

9. $\displaystyle\int_0^2 \sqrt{4x+1}\,dx$

10. $\displaystyle\int_0^\pi \sin x\,dx$

11. $\displaystyle\int_0^\pi \cos x\,dx$

AREA AND INTEGRATION

12. $\displaystyle\int_{\pi/4}^{\pi/2} \frac{\cos x \, dx}{\sin^2 x}$ (Let $u = \sin x$.)   13. $\displaystyle\int_0^{\pi/6} \frac{\sin 2x}{\cos^2 2x} \, dx$

14. $\displaystyle\int_0^{\pi} \sin^2 x \, dx$ 15. $\displaystyle\int_0^{2\pi/\omega} \cos^2 (\omega t) \, dt$ (ω constant)   16. $\displaystyle\int_0^1 \frac{dx}{(2x+1)^3}$

17. a) Express the area between the curves $y = x^2$, $y = 18 - x^2$ as a limit of a sum of areas of rectangles. b) Evaluate the area.

18. Approximate the area under the curve $y = 1/x$ between $x = 1$ and $x = 2$ by five rectangles, each of base $\Delta x = 0.2$. Use the ordinate of the curve at the midpoint of each subinterval as the altitude of the approximating rectangle. The following short table of reciprocals may be used for convenience:

x	1.1	1.3	1.5	1.7	1.9
1/x	0.909	0.769	0.667	0.588	0.526

(Compare your answer with 0.693, which is the value of $\int_1^2 dx/x$ to 3 decimals.)

19. Let $f(x)$ be a continuous function. Express

$$\lim_{n \to \infty} \frac{1}{n}\left[f\left(\frac{1}{n}\right) + f\left(\frac{2}{n}\right) + \cdots + f\left(\frac{n}{n}\right) \right]$$

as a definite integral.

20. Use the result of Exercise 19 to evaluate:

a) $\displaystyle\lim_{n \to \infty} \frac{1}{n^{16}} [1^{15} + 2^{15} + 3^{15} + \cdots + n^{15}]$,

b) $\displaystyle\lim_{n \to \infty} \frac{\sqrt{1} + \sqrt{2} + \sqrt{3} + \cdots + \sqrt{n}}{n^{3/2}}$,

c) $\displaystyle\lim_{n \to \infty} \frac{1}{n}\left[\sin\frac{\pi}{n} + \sin\frac{2\pi}{n} + \sin\frac{3\pi}{n} + \cdots + \sin\frac{n\pi}{n} \right]$.

21. Compute $U_f(P) - L_f(P)$ from (7') and (8'). Show that

$$U_f(P) - L_f(P) \le b\|P\|.$$

[Hint: $\delta_1^2 + \delta_2^2 + \cdots + \delta_n^2 \le (\delta_1 + \delta_2 + \cdots + \delta_n)\|P\|.$]

22. If

$$\frac{b^2}{2}\left(1 - \frac{1}{n}\right) \le I \le \frac{b^2}{2}\left(1 + \frac{1}{n}\right),$$

why is $b^2/2$ the only value for I which holds for every positive integer n?

23. Estimate
$$\int_1^3 \frac{1}{x}\,dx$$
to an accuracy of $\epsilon = \frac{1}{8}$.

24. Estimate
$$\int_{1/2}^{5/2} \frac{1}{x}\,dx$$
to an accuracy of $\epsilon = \frac{1}{8}$.

25. Use Eq. (29) with $n = 100$ and Eq. (30) with $n = 99$ to show that $\sqrt{1} + \sqrt{2} + \cdots + \sqrt{99}$ lies between $\frac{2}{3}(99)^{3/2}$ and $\frac{2}{3}(100)^{3/2}$. Express these two numbers and their average in decimal form, using 9.950 as the three-decimal approximation for $\sqrt{99}$. Is it true or false that the sum differs from 662 by less than 1%? Why?

12–4 TRAPEZOIDAL RULE FOR APPROXIMATING INTEGRALS

There are several ways of approximating an integral. The technique used in Example 3 of Section 12–3 was to divide $[a, b]$ into n congruent subintervals, study $U_f(P) - L_f(P)$ to see how large to choose n, and then use the average of the lower sum and upper sum:

$$\int_a^b f \approx \tfrac{1}{2}[L_f(P) + U_f(P)]. \tag{1a}$$

In the particular example $f(x) = 1/x$, the integrand is monotone decreasing so that m_i and M_i are values of f at opposite endpoints of the ith subinterval. The right-hand side of Eq. (1a) is therefore the same as T, where T is the *trapezoidal approximation* defined by

$$T = \sum_{i=1}^{n} \tfrac{1}{2}[f(a_{i-1}) + f(a_i)](a_i - a_{i-1}). \tag{1b}$$

It is customary to let

$$a_i - a_{i-1} = \frac{b-a}{n} = h \tag{2a}$$

and write T in the equivalent form

$$T = h[\tfrac{1}{2}y_0 + y_1 + y_2 + \cdots + y_{n-1} + \tfrac{1}{2}y_n], \tag{2b}$$

where

$$y_0 = f(a_0), \quad y_1 = f(a_1), \quad \ldots, \quad y_n = f(a_n). \tag{2c}$$

Example 1. Use Eqs. (2) with $n = 4$ to compute the trapezoidal approximation T for $\int_1^2 dx/x$.

Solution. Here $h = (2-1)/4 = \frac{1}{4}$, $a_0 = 1$, $a_1 = \frac{5}{4}$, $a_2 = \frac{6}{4}$, $a_3 = \frac{7}{4}$, $a_4 = \frac{8}{4}$, and $f(x) = 1/x$. Hence $y_0 = 1$, $y_1 = \frac{4}{5}$, $y_2 = \frac{4}{6}$, $y_3 = \frac{4}{7}$, $y_4 = \frac{4}{8}$, and Eq. (2b) gives

$$T = \tfrac{1}{4}(\tfrac{1}{2} + \tfrac{4}{5} + \tfrac{4}{6} + \tfrac{4}{7} + \tfrac{2}{8}) = \tfrac{1}{8} + \tfrac{1}{5} + \tfrac{1}{6} + \tfrac{1}{7} + \tfrac{1}{16}$$
$$= 0.69702\ldots$$

Example 2. Compare $\int_1^4 x^2 \, dx$ with its trapezoidal approximation for $n = 3$.

Solution. The fundamental theorem gives

$$\int_1^4 x^2 \, dx = \left.\frac{x^3}{3}\right]_1^4 = \frac{64}{3} - \frac{1}{3} = 21.$$

With $n = 3$, and $f(x) = x^2$, Eqs. (2) give

$$h = \frac{4-1}{3} = 1; \quad a_0 = 1, \quad a_1 = 2, \quad a_2 = 3, \quad a_3 = 4,$$

and

$$T = 1(\tfrac{1}{2} + 4 + 9 + 8) = 21.5.$$

Error in trapezoidal approximation

Is there any way we can estimate the difference between an integral and its trapezoidal approximation? There is, and we refer the reader to Olmsted's *Advanced Calculus*, p. 118, Exercise 16, for the derivation. The result is that

$$\int_a^b f(x) \, dx - h(\tfrac{1}{2}y_0 + y_1 + \cdots + \tfrac{1}{2}y_n) = -\frac{b-a}{12} f''(c) h^2 \quad (3)$$

for some c between a and b, assuming that f and f' are continuous on $[a, b]$ and f'' exists on the open interval (a, b).

Example 3. For $f(x) = x^2$ and $n = 3$ on $[1, 4]$ we found in Example 2 that

$$I = 21, \quad T = 21.5.$$

Here the second derivative is

$$f''(x) = 2,$$

which is constant. Thus, the right-hand side of Eq. (3) is

$$-\frac{b-a}{12} f''(c) h^2 = -\frac{4-1}{12} (2)(1)^2 = -\frac{1}{2},$$

and Eq. (3) is satisfied because the left-hand side is

$$I - T = 21 - 21.5 = -0.5.$$

EXERCISES 12–4

1. For any continuous function $f(x)$, the volume generated by rotating the area bounded by $x = a$, $x = b$, $y = 0$, $y = f(x)$, is found by evaluating the definite integral

$$\pi \int_a^b [f(x)]^2 \, dx.$$

Using this and the trapezoidal rule stated in (1b), compute the approximate volume generated by rotating the area, bounded by the lines $x = 0$, $y = 0$, $2x + 3y = 6$, around the x-axis. Use a regular partition with $a_i - a_{i-1} = \frac{1}{2}$.

2. Repeat Exercise 1, with $a_i - a_{i-1} = \frac{1}{4}$.

3. Use the definite integral, as defined in Eq. (22b) of Section 12–3, to compute the volume described in Exercise 1. Compare the results of Exercises 1, 2, 3.

4. Using the method of Exercise 1, with the curve $y = x^2$ being rotated around the x-axis, and with $a_i - a_{i-1} = \frac{1}{4}$, compute the approximate volume generated by the rotation of the area bounded by the lines $y = x^2$, $y = 0$, and $x = 2$.

5. Repeat Exercise 4, with $a_i - a_{i-1} = \frac{1}{8}$.

6. Use the definite integral, as defined in Eq. (22b) of Section 12–3, to compute the volume described in Exercise 4. Compare the results of Exercises 4, 5, 6.

7. For any function f, with continuous derivative f', the length of that portion of the graph between the points $(a, f(a))$ and $(b, f(b))$ is found by evaluating the definite integral

$$\int_a^b \sqrt{1 + f'^2(x)} \, dx,$$

where $y = f(x)$. Use the trapezoidal rule given in (1b) to approximate the length of the segment of $2x + 3y = 6$ included between the x- and y-axes, taking

$$\frac{a_i - a_{i-1}}{n} = \Delta x = \frac{1}{2}.$$

8. Repeat Exercise 7, doubling the value of n. Compare the results in Exercises 7 and 8 with the actual length of the segment.

9. Using the method of Exercise 7, find the length of the curve $y = x^2$ from the point (1, 4) to the point (3, 9), letting $\Delta x = \frac{1}{2}$.

10. Find the approximate length of one arch of the curve $y = \sin x$, letting
$$\frac{a_i - a_{i-1}}{n} = \Delta x = \frac{\pi}{8},$$
and using the method of Exercise 7.

11. A rectangular swimming pool is 30 feet wide and 50 feet long. The depth of the water h (ft) at distance x (ft) from one end of the pool is measured at five-feet intervals and found to be as follows:

x ft	0	5	10	15	20	25	30	35	40	45	50
h ft	6.0	8.2	9.1	9.9	10.5	11.0	11.5	11.9	12.3	12.7	13.0

Use the trapezoidal approximation, based on Eq. (1b), to compute the approximate volume of water in the pool.

12. Use the trapezoidal rule to find the approximate distance covered from $t = 0$ to $t = 4$ by a particle moving with velocity
$$\frac{ds}{dt} = 3t + t^2 + 4, \quad \text{where} \quad n = \frac{a_i - a_{i-1}}{\Delta t} = 4.$$

13. Solve Exercise 12 taking $n = 8$.

14. Use a definite integral to solve Exercise 12. Compare the results of Exercises 12, 13, 14.

15. Another method of approximating the value of a definite integral is given by Simpson's rule (see *Calculus and Analytic Geometry*, 3rd edition, G. B. Thomas, Jr., pp. 385–386). The rule states:
$$S = (f_0 + 4f_1 + 2f_2 + 4f_3 + 2f_4 + \cdots + 2f_{n-2} + 4f_{n-1} + f_n)\frac{\Delta x}{3}.$$
Find the length of one arch of the curve $y = \sin x$ by Simpson's rule, letting $\Delta x = \pi/8$. Compare your result with that of Exercise 10.

16. Use Simpson's rule to solve Exercise 12, taking $\Delta x = \frac{1}{2}$. Compare your results with those of Exercise 14.

17. Use Simpson's rule to solve Exercise 4, letting $\Delta x = \frac{1}{4}$. Compare your results with those of Exercise 6.

18. Consider the definite integral:
$$\int_1^2 \frac{1}{x}\, dx.$$

a) Approximate its value by the trapezoidal rule, taking $\Delta x = \frac{1}{4}$.

b) Approximate its value by Simpson's rule, taking $\Delta x = \frac{1}{4}$.

c) Approximate its value by letting $x = 1 + u$, hence

$$\int_1^2 \frac{1}{x}\,dx = \int_0^1 \left(\frac{1}{1+u}\right) du$$

$$= \int_0^1 [1 - u + u^2 - u^3 + \cdots + (-1)^n u^n + R_n]\,du,$$

where

$$R_n = (-1)^{n+1} \frac{u^{n+1}}{1+u} \leq u^{n+1} \quad \text{for} \quad 0 \leq u \leq 1.$$

Evaluate the integral

$$\int_0^1 \left(\frac{1}{1+u}\right) du$$

to five terms, eight terms, ten terms. (See Remark 2, Section 13–3.)

19. Construct a function that is defined on [0, 1], has infinitely many discontinuities there, and yet is integrable there. Evaluate the integral.

20. a) Use the first four terms of the infinite series

$$\sin x = \sum_{i=1}^{\infty} (-1)^{i+1} \frac{x^{2i-1}}{(2i-1)!}$$

to approximate the sine of 2 radians. How close to sin 2 might you expect your approximation to be?

b) Use a sufficient number of terms of the infinite series given in (a) to approximate the value of the definite integral

$$\int_0^2 \sin x\,dx$$

to the nearest hundredth.

12–5 PROPERTIES OF THE RIEMANN INTEGRAL

So far our discussion of integration has been concerned mainly with continuous functions. A function f that is continuous on $[a, b]$ and on $[b, c]$ is integrable on both intervals, and we naturally hope that f is also integrable on $[a, c]$ and that

$$\int_a^c f = \int_a^b f + \int_b^c f. \tag{1}$$

Equation (1) is correct, but we can't base the existence of $\int_a^c f$ on Theorem 12–1a, because f may be discontinuous at b, as in Fig. 12–7.

AREA AND INTEGRATION

Suppose we have two functions f_1 and f_2, such that

f_1 is continuous on $[a, b]$,

f_2 is continuous on $[b, c]$.

We define f so that

$$f(x) = \begin{cases} f_1(x) & \text{for } a \leq x < b, \\ f_2(x) & \text{for } b < x \leq c. \end{cases}$$

Then we can define $f(b)$ to be either $f_1(b)$ or $f_2(b)$ and have

$$\int_a^b f_1 = \int_a^b f, \quad \int_b^c f_2 = \int_b^c f,$$

because changing the value of a function at an endpoint changes at most one term in a Riemann sum $S_f(P, Q)$ and does not affect the limit.

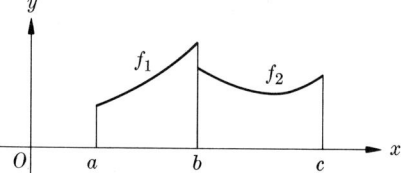

FIG. 12-7. Functions f_1 and f_2 are continuous on the domains $[a, b]$ and $[b, c]$, respectively.

If we define $f(b)$ to be any finite number, Eq. (1) will be satisfied in the sense that f is integrable on $[a, c]$ and

$$\int_a^c f = \int_a^b f_1 + \int_b^c f_2.$$

Stated formally, we have the following theorem:

Theorem 12-3. Let f_1 be integrable on $[a, b]$ and let f_2 be integrable on $[b, c]$. Define f to have the same values as f_1 on $[a, b)$ and as f_2 on $(b, c]$:

$$f(x) = \begin{cases} f_1(x) & \text{for } a \leq x < b, \\ f_2(x) & \text{for } b < x \leq c. \end{cases} \tag{2a}$$

Let

$$f(b) = k, \text{ be any real number.} \tag{2b}$$

Then f is integrable on $[a, c]$ and

$$\int_a^c f = \int_a^b f_1 + \int_b^c f_2 \tag{3a}$$

$$= \int_a^b f + \int_b^c f. \tag{3b}$$

Proof. Suppose P_1 is a partition of $[a, b]$ and $S_{f_1}(P_1, Q_1)$ is a Riemann sum for f_1 associated with this partition. By hypothesis,

$$\lim_{\|P_1\| \to 0} S_{f_1}(P_1, Q_1) = \int_a^b f_1.$$

Consider the Riemann sum $S_f(P_1, Q_1)$ for f. By definition, f and f_1 have the same values for all points in Q_1 except possibly x_n:

$$f(x_n) = k, \quad f_1(x_n) = f_1(b) \text{ may not equal } k$$

if $x_n = b$. However,

$$|S_f(P_1, Q_1) - S_{f_1}(P_1, Q_1)| \le (|k| + |f_1(b)|)\|P_1\|.$$

If $\epsilon > 0$, we can make the right-hand side of this inequality less than $\epsilon/2$ by making

$$\|P_1\| < \frac{\epsilon/2}{1 + |k| + |f_1(b)|}.$$

If also

$$\left| S_{f_1}(P_1, Q_1) - \int_a^b f_1 \right| < \epsilon/2$$

when $\|P_1\| < \delta_1$, we can let

$$\delta = \min\left\{ \delta_1, \frac{\epsilon/2}{1 + |k| + |f_1(b)|} \right\}$$

and have

$$\left| S_f(P_1, Q_1) - \int_a^b f_1 \right| < \epsilon$$

when $\|P_1\| < \delta$. This establishes the integrability of f on $[a, b]$ with

$$\int_a^b f = \int_a^b f_1.$$

The same kind of argument leads to the conclusion that

$$\int_b^c f = \int_b^c f_2.$$

Now let P be any partition of $[a, c]$. It may not contain the point b, so we construct a related partition P_3 that does, as follows. Let P_1 be the partition of $[a, b]$ consisting of b plus those points of P that are in $[a, b]$. Similarly, let P_2 be the points of P that are in $[b, c]$ together with b. Then

$$P_3 = P_1 \cup P_2 = P \cup \{b\}$$

is a partition of $[a, c]$ that consists of P and the point b (which might already have been a partition point in P, but need not have been). See Fig. 12–8.

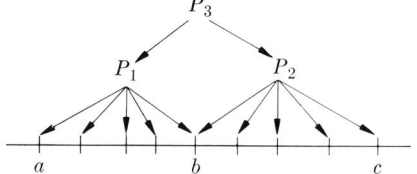

FIG. 12–8. $P_3 = P_1 \cup P_2$ is obtained from P by including the partition point b.

Next, let Q_3 be a set of base points $x_1, x_2, \ldots, x_n$ in the n subintervals $[a_{i-1}, a_i]$, $i = 1, \ldots, n$, determined by P_3. Suppose Q_1 are those points of Q_3 that are in $[a, b]$ and Q_2 those that are in $[b, c]$. Then

$$S_f(P_3, Q_3) = S_f(P_1, Q_1) + S_f(P_2, Q_2), \qquad (4)$$

and the right-hand side of (4) approaches

$$\int_a^b f + \int_b^c f$$

as $\|P\| \to 0$, because

$$(\|P_1\| \text{ and } \|P_2\|) \leq \|P_3\| \leq \|P\|.$$

Therefore,

$$\lim_{\|P\| \to 0} S_f(P_3, Q_3) = \int_a^b f + \int_b^c f. \qquad (5)$$

Equation (5) is almost, but not quite, the result we want. What we need is to show that

$$\lim_{\|P\| \to 0} S_f(P, Q) = \lim_{\|P\| \to 0} S_f(P_3, Q_3),$$

where P, P_3, Q_3 were defined above and Q is a set of base points in the subintervals of $[a, c]$ determined by P. If P happened to contain the point b, we could take

$$P = P_3, \qquad Q = Q_3$$

and have identical Riemann sums

$$S_f(P, Q) = S_f(P_3, Q_3).$$

However, suppose P does not contain b, but, for some j,

$$a_{j-1} < b < a_j.$$

Then there are three possible places where the corresponding point x_j in Q may be:

i) $a_{j-1} \leq x_j < b$,

ii) $x_j = b$,

iii) $b < x_j \leq a_j$.

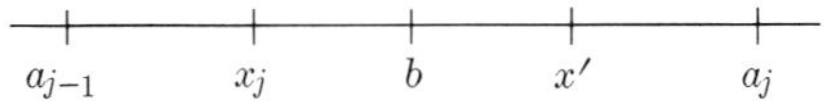

FIG. 12-9. Illustrating case (i).

In case (ii) we can take $Q_3 = Q$ and apply Eq. (4). Cases (i) and (iii) are much alike, so we consider only case (i) in detail. See Fig. 12-9. In order to obtain a suitable base set Q_3, we adjoin to Q a point x' in $(b, a_j]$. Then

$$\begin{aligned}S_f(P_3, Q_3) &- S_f(P, Q) \\ &= f(x_j)(b - a_{j-1}) + f(x')(a_j - b) - f(x_j)(a_j - a_{j-1}) \\ &= f(x_j)(b - a_{j-1} - a_j + a_{j-1}) + f(x')(a_j - b) \\ &= [f(x') - f(x_j)](a_j - b) \\ &= [f_2(x') - f_1(x_j)](a_j - b).\end{aligned}$$

Since only bounded functions are Riemann integrable, there exist numbers A and B such that

$$|f_1(x)| < A \quad \text{for} \quad a \leq x \leq b,$$
$$|f_2(x)| < B \quad \text{for} \quad b \leq x \leq c,$$

so we may conclude that

$$|S_f(P_3, Q_3) - S_f(P, Q)| < (A + B)(a_j - b). \tag{6}$$

Moreover,

$$a_j - b < a_j - a_{j-1} \leq \|P\|,$$

so

$$\lim_{\|P\| \to 0} |S_f(P_3, Q_3) - S_f(P, Q)| = 0. \tag{7}$$

When we combine Eqs. (5) and (7) we have the result

$$\lim_{\|P\| \to 0} S_f(P, Q) = \int_a^b f + \int_b^c f.$$

This means that f is integrable on $[a, c]$ and

$$\int_a^c f = \int_a^b f + \int_b^c f. \qquad \text{Q.E.D.} \tag{8}$$

Additional properties of Riemann integrals include the following:

If f and g are integrable on $[a, b]$, then $f + g$ is integrable there and

$$\int_a^b (f + g) = \int_a^b f + \int_a^b g. \tag{9}$$

If c is a constant, and f is integrable on $[a, b]$, then cf is integrable there and

$$\int_a^b (cf) = c \int_a^b f. \tag{10}$$

If f and g are integrable on $[a, b]$ and $f(x) \leq g(x)$ for $a \leq x \leq b$, then

$$\int_a^b f \leq \int_a^b g. \tag{11}$$

These properties follow from the identities:

$$S_{f+g}(P, Q) = S_f(P, Q) + S_g(P, Q), \qquad S_{cf}(P, Q) = cS_f(P, Q),$$

and

$$S_g(P, Q) - S_f(P, Q) = S_{g-f}(P, Q).$$

Up to this point we have assumed that $a < b$, but most of our results still hold for

$$b = a \quad \text{or} \quad b < a,$$

provided we adopt the usual definitions:

$$\int_a^a f = 0, \tag{12a}$$

$$\int_b^a f = -\int_a^b f. \tag{12b}$$

The intermediate-value theorem, Theorem C-II of Section 6-1, leads to the following result, which is called the intermediate-value theorem for integrals.

Theorem 12–4. Let f be continuous on $[a, b]$. Then

$$\int_a^b f = f(c)(b - a) \qquad \text{for some } c \in (a, b). \tag{13}$$

Proof. In order to simplify the proof we assume that $a < b$. Since f is continuous and $[a, b]$ is a closed bounded interval, there exist m and M such that

$$m = \min_{a \leq x \leq b} f(x), \qquad M = \max_{a \leq x \leq b} f(x).$$

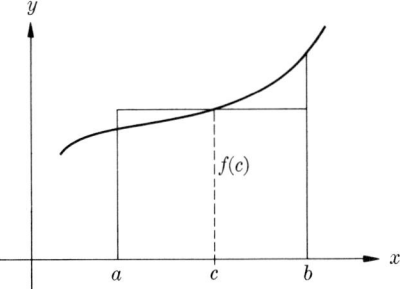

FIG. 12–10. Illustrating Theorem 12–4.

Let **m** and **M** denote the constant functions whose values on $[a, b]$ are m and M, respectively. Then, by inequality (11),

$$\int_a^b \mathbf{m} \leq \int_a^b f \leq \int_a^b \mathbf{M}$$

or

$$m(b - a) \leq \int_a^b f \leq M(b - a). \tag{14}$$

If we divide by $b - a$, we get

$$m \leq \frac{1}{b - a} \int_a^b f \leq M. \tag{15}$$

The number

$$\frac{1}{b - a} \int_a^b f$$

is contained in the range of f, since (15) says it is between m and M. Hence, by Theorem C-II of Section 6–1, there exists c between a and b such that

$$f(c) = \frac{1}{b - a} \int_a^b f. \qquad \text{Q.E.D.}$$

See Fig. 12–10.

Remark 1. Equation (13) has the following geometric interpretation. The left-hand side represents the area under the graph of f from a to b. The right-hand side represents the area of a rectangle of altitude $f(c)$ based on $[a, b]$. The equality says that for a suitable choice of c these two areas are equal. Inequalities (14) and (15) suggest that if we start with a rectangle of altitude m (too small) and move the upper boundary upward to M (too large) there is a position between m and M that is just right: the corresponding rectangle has area equal to the area under the graph of f.

AREA AND INTEGRATION

Remark 2. The fundamental theorem can be written briefly as

$$\int_a^b F' = F(b) - F(a),$$

or

$$\int_a^x F' = F(x) - F(a). \tag{16}$$

Equation (16) says that integration is [except for the constant $-F(a)$] the inverse of differentiation. Is the converse also true? That is, if f is continuous on $[a, b]$ and $a < x < b$, is it true that

$$\frac{d}{dx} \int_a^x f = f(x)? \tag{17}$$

The following theorem says that the answer is "yes."

Theorem 12–5. Let f be continuous on $[a, b]$. Let $x \in (a, b)$. Then

$$\frac{d}{dx} \int_a^x f = f(x). \tag{18}$$

Proof. Since f is continuous on $[a, x]$ for each $x \in (a, b)$, the integral in Eq. (18) exists and defines a function G with domain (a, b) and values

$$G(x) = \int_a^x f. \tag{19}$$

For any $h \neq 0$ such that

$$a < x + h < b,$$

we have

$$\frac{G(x + h) - G(x)}{h} = \frac{1}{h} \left(\int_a^{x+h} f - \int_a^x f \right) = \frac{1}{h} \left(\int_x^{x+h} f \right)$$

$$= \frac{1}{h} [(x + h - x)f(c)] = f(c)$$

for some c between x and $x + h$. As $h \to 0$, $c \to x$ and $f(c) \to f(x)$, because f is continuous at x. Therefore,

$$G'(x) = \lim_{h \to 0} \frac{G(x + h) - G(x)}{h} = \lim_{h \to 0} f(c) = f(x). \quad \text{Q.E.D.}$$

Remark 3. Theorems 12–2 and 12–5 form a pair of fundamental theorems showing that integration and differentiation are essentially inverse operations. (See "The Two Fundamental Theorems of Calculus" by F. Cunningham, Jr., *American Mathematical Monthly*, **72**, No. 4, April 1965, p. 406.)

Remark 4. The traditional notation

$$\int_a^b f(x)\, dx$$

may cause one to think that the Riemann integral of f over $[a, b]$ depends on x. However, that is not true as one soon realizes, either by thinking of the integral as a limit of Riemann sums, or by thinking of the way in which such an integral is ordinarily evaluated by applying the fundamental theorem. Thus, if F is a primitive for f,

$$\int_a^b f = F]_a^b = F(b) - F(a). \tag{20}$$

Equation (20) is written without the appearance of a single x, so the answer obviously does not depend on x.

Example 1. If f is the cosine function on $[0, \pi/2]$,

$$f(x) = \cos x, \quad a = 0, \quad b = \pi/2,$$

we can take the sine function as a primitive for f,

$$F(x) = \sin x.$$

The analogue of Eq. (20) would then be

$$\int_0^{\pi/2} \cos = \sin]_0^{\pi/2} = \sin \frac{\pi}{2} - \sin 0 = 1.$$

Example 2. Show that

$$\int_a^b (x^2 + 3x + 4)\, dx = \int_a^b (t^2 + 3t + 4)\, dt. \tag{21}$$

Solution. The left-hand side of Eq. (21) is

$$\frac{x^3}{3} + \frac{3x^2}{2} + 4x\bigg]_a^b = \left(\frac{b^3}{3} + \frac{3b^2}{2} + 4b\right) - \left(\frac{a^3}{3} + \frac{3a^2}{2} + 4a\right),$$

and the right-hand side gives the same answer. The result doesn't depend on either x or t, but only on the original function f and the limits of integration a and b.

It is almost a universal custom to write *polynomial* functions, and rational functions, in terms of x. Thus, for shorthand, we write for the integrand in Eq. (21),

$$f(x) = x^2 + 3x + 4. \tag{22}$$

Equation (22) is a rule for calculating the value of a particular polynomial function f at any x in its domain. More precisely, we might write for the function f itself,

$$f = \{(x, y) : y = x^2 + 3x + 4,\ a \leq x \leq b\},$$

or say that f maps x into $x^2 + 3x + 4$ and write

$$f : x \to x^2 + 3x + 4.$$

Professor Karl Menger has suggested the use of bold-face numerals for constant functions and the letter **j** for the identity function. Then in Menger's notation we could replace (22) by

$$f = \mathbf{j}^2 + 3\mathbf{j} + \mathbf{4} \tag{23}$$

and write the integral of Eq. (21) as

$$\int_a^b f = \int_a^b \mathbf{j}^2 + 3\mathbf{j} + \mathbf{4}.$$

The Menger notation is not widely used at present, though it does have the merit of distinguishing between a *function* f and the *value* $f(x)$ of that function at a particular x. Once we understand that the x in

$$\int_a^b f(x)\,dx$$

plays the same nonessential role as the t in

$$\int_a^b f(t)\,dt,$$

we realize that the integral of f from a to b does not depend on either x or t. Thus, for any integrable function f,

$$\int_a^b f = \int_a^b f(x)\,dx = \int_a^b f(t)\,dt = \int_a^b f(u)\,du. \tag{24}$$

The purpose of the foregoing extended remark was to pave the way for a slight change in notation. Although Eq. (24) indicates that we can use almost any letter we like as the *dummy variable* of integration, there is confusion in the interpretation of x in something like

$$\int_0^x x^2\,dx.$$

Here the x used as upper limit of the integral should represent some fixed real number (perhaps chosen arbitrarily between 0 and 10, say), but the x

used as dummy variable of integration runs over the set of real numbers from 0 to the upper limit. The Menger notation would be

$$\int_0^x \mathbf{j}^2$$

which clearly avoids such confusion. We can also avoid the confusion by merely using a different letter, for example t, as the variable of integration when x is used for one of the limits. Thus, when we write

$$\int_0^x t^2 \, dt,$$

it is clear that t varies from 0 to x. (Contrast this with the nonsensical statement "x varies from 0 to x.")

EXERCISES 12–5

1. Suppose, in Fig. 12–9, that x_j lies in the interval (b, a_j); that is, $b < x_j < a_j$. Deduce Eq. (8) using this x_j.

In Exercises 2 through 8 find $F'(x)$ for the given functions F.

2. $F(x) = \int_0^x \sqrt{1 + t^2} \, dt$

3. $F(x) = \int_1^x dt/t$

4. $F(x) = \int_x^1 \sqrt{1 - t^2} \, dt$

5. $F(x) = \int_0^x \dfrac{dt}{1 + t^2}$

6. $F(x) = \int_1^{2x} \cos t^2 \, dt$ [Hint: Chain rule.]

7. $F(x) = \int_1^{x^2} \dfrac{dt}{1 + \sqrt{1 - t}}$

8. $F(x) = \int_{\sin x}^0 \dfrac{dt}{2 + t}$

9. Find $\lim\limits_{h \to 0} \dfrac{1}{h} \int_x^{x+h} \dfrac{du}{u + \sqrt{u^2 + 1}}$.

10. Find $\lim\limits_{x \to x_1} \left[\dfrac{x}{x - x_1} \int_{x_1}^x f(t) \, dt \right]$.

11. Variables x and y are related by the equation

$$x = \int_0^y \dfrac{1}{\sqrt{1 + 4t^2}} \, dt.$$

Show that d^2y/dx^2 is proportional to y, and find the constant of proportionality.

12. The area bounded by the x-axis, the curve $y = f(x)$, and the lines $x = 1$ and $x = b > 1$ is equal to $\sqrt{b} - 1$. Find $f(x)$.

13. Given $f(x) = x^2$, and $[a, b] = [1, 3]$. Using a regular partition with $\|P\| = \frac{1}{2}$, form three Riemann sums as follows
   a) $L_f(P)$
   b) $S_f(P, Q)$ such that $x_j = \frac{1}{2}(a_{i-1} + a_i)$
   c) $U_f(P)$

   What is the average (mean) of results (a) and (c)? Compare this, as well as the result (b), with the area found by evaluating the definite integral.

14. Redefine the function of Exercise 13 over the interval $[1, 3]$ as follows:

$$f(x) = \begin{cases} \frac{1}{2} & \text{if } x = 1, \\ x^2 & \text{if } 1 < x < 3, \\ 9.5 & \text{if } x = 3. \end{cases}$$

   Form the three Riemann sums corresponding to the sums (a), (b), and (c) of Exercise 13. Compare the results. In each case, how many terms in each of the Riemann sums are affected by the change? (See Fig. 12–7 and accompanying text.)

15. Remark 4 in the preceding text points out that a definite integral is, in fact, a function of its limits, and that the variable in the integrand is a "dummy" variable. With this in mind, show that

$$\int_0^{\pi/2} \sqrt{\sin x} \, dx = \int_0^{\pi/2} \sqrt{\cos x} \, dx.$$

   [*Hint:* Let $x = \pi/2 - t$.]

13
LOGARITHMIC AND EXPONENTIAL FUNCTIONS

13-1 INTRODUCTION

In a first introduction to logarithms, the base 10 is ordinarily used, and the following definition is given:
If
$$x = 10^y,$$
then we say that y is the common *logarithm* of x. Other bases are also introduced. Thus, since $8 = 2^3$, we say that 3 is the logarithm of 8 to the base 2 and write $\log_2 8 = 3$. In general, we say that y is the logarithm of x to the base a, $a > 0$, $a \neq 1$, if y is the power to which a must be raised to give x:
$$a^y = x \Leftrightarrow y = \log_a x. \tag{1}$$

If we are willing to *assume* the validity of the various laws of exponents, for arbitrary *real* exponents (irrational as well as rational), we can use Eq. (1) to study the logarithmic function corresponding to a fixed base a, say $a = 10$ or $a = 2$. The chief difficulties in putting such a program on a firm basis are:

a) properly defining things like $10^{\sqrt{2}}$, 10^π, and

b) proving that
$$10^{\sqrt{2}} \, 10^\pi = 10^{(\sqrt{2}+\pi)}.$$

The easiest way around such difficulties is first to define a particular logarithmic function, then define the inverse of that function, and use that for defining a^u. We shall not go through all of this program here because it is available in standard calculus textbooks. (For example, see G. B. Thomas, Jr., *Calculus and Analytic Geometry*, 3rd edition, p. 317; or S. Lang, *A First Course in Calculus*, p. 126.)

The *natural logarithm* is the function with domain $x > 0$ defined by the Riemann integral of $1/t$ from 1 to x:
$$\ln x = \int_1^x \frac{1}{t} \, dt, \quad x > 0. \tag{2}$$

If we start with Eq. (2), it is quite easy to derive standard properties of the logarithm, like the following:

$$\ln(ax) = \ln a + \ln x, \tag{3a}$$

$$\ln x^n = n \ln x, \quad n \text{ rational}, \tag{3b}$$

$$\ln 1 = 0, \tag{3c}$$

$$\ln(u/v) = \ln u - \ln v. \tag{3d}$$

We shall first show why one might be led to something like Eq. (2), and then derive Eq. (3a) directly from Eq. (2) using the definition of the integrals involved.

EXERCISES 13-1

1. Using Eq. (1), find: $\log_2 32$, $\log_8 32$, $\log_{1/2} 32$.

2. Using Eq. (1), find: $\log_3 81$, $\log_3 (3\sqrt{3})$, $\log_9 3$.

3. Sketch the graph of $y = 2^x$ for $-3 \leq x \leq 3$. Estimate the slope of the tangent at $x = 0$.

4. Sketch the graph of $y = 3^x$ for $-2 \leq x \leq 2$. Estimate its slope at $x = 0$.

5. Use the graph of Exercise 3 to construct a portion of the graph of $y = \log_2 x$, $\frac{1}{8} \leq x \leq 8$. Estimate its slope at $x = 1$.

6. Use the graph of Exercise 4 to construct a portion of the graph of $y = \log_3 x$, $\frac{1}{9} \leq x \leq 9$. Estimate its slope at $x = 1$.

7. On the graph of Exercise 5 consider the points (x_1, y_1) and (x_2, y_2) with $x_1 = 1.5$ and $x_2 = 2$. Compare the numbers $y_1 + y_2$ and the ordinate y_3 of the point $(3, y_3)$ on your graph.

8. On the graph of Exercise 6 consider the points (x_1, y_1) and (x_2, y_2) with $x_1 = 2$ and $x_2 = 3$. Estimate $y_1 + y_2$ and compare it with the ordinate y_3 of the point $(6, y_3)$ on your graph.

9. Sketch the graph of $y = 2^{-x}$ for $-3 \leq x \leq 3$. Estimate the slope of the tangent at $x = 0$. Compare with the results of Exercise 3.

10. Let $f(x) = a^x$. What restrictions must be placed on a so that f is a continuous function? so that its inverse is also a continuous function?

11. What are the domain and the range of the family of curves given in Exercise 10?

12. Define a function, $m_a(x)$, which will give the slope of any member of the family of curves in Exercise 11, for fixed $a > 0$, $a \neq 1$. Can you tell what the domain and range of m are? Can you find any value of a for which $m_a(0) = 1$?

13-2 HEURISTIC APPROACH: IN SEARCH OF THE LOGARITHM

In this section, we do *not* use Eq. (2) of Section 13-1. Rather, we try to show why that equation is a reasonable definition. (Of course it isn't mathematically necessary to justify a definition. Nor do we pretend to be proving anything, because one never proves a definition.)

Suppose, therefore, that L is to be a logarithmic function, with the set of all positive numbers as its domain (negative numbers have been found to have imaginary logarithms), such that

$$L(ax) = L(a) + L(x) \quad \text{for all } a > 0, x > 0. \tag{1}$$

Equation (1) says that the "logarithm of a product equals the sum of the logarithms of the factors." The first thing we can deduce from (1) is that

$$L(1) = 0, \tag{2}$$

because if $a = x = 1$, then $ax = 1$, and Eq. (1) becomes

$$L(1) = L(1) + L(1);$$

so, by subtraction,

$$0 = L(1).$$

Then Eqs. (1) and (2) lead to

$$L\left(\frac{1}{b}\right) = -L(b) \tag{3}$$

when we take $a = b$, $x = 1/b$:

$$L(1) = L\left(b \cdot \frac{1}{b}\right) = L(b) + L\left(\frac{1}{b}\right),$$

so

$$0 = L(b) + L\left(\frac{1}{b}\right).$$

If we take $a = x$ in Eq. (1), we get

$$L(x^2) = 2L(x).$$

By induction on n, by taking $a = x^{n-1}$ and assuming that

$$L(x^{n-1}) = (n-1)L(x),$$

we get, from (1), that

$$L(x^n) = (n-1)L(x) + L(x) = nL(x).$$

Therefore, for every positive integer n,

$$L(x^n) = nL(x). \tag{4}$$

Assume that we know the definition of $x^{n/m}$ for positive integers n and m:

$$x^{n/m} = \sqrt[m]{x^n}. \tag{5}$$

13-2 HEURISTIC APPROACH: IN SEARCH OF THE LOGARITHM

In particular,
$$x^{1/m} = \sqrt[m]{x}$$
so that
$$(x^{1/m})^m = x.$$

Equation (4) then gives
$$L(x) = mL(x^{1/m}),$$
which gives
$$L(x^{1/m}) = \frac{1}{m} L(x). \tag{6}$$

If we combine Eqs. (4), (5), and (6), we get
$$L(x^{n/m}) = \frac{1}{m} L(x^n) = \frac{n}{m} L(x),$$
or
$$L(x^r) = rL(x), \tag{7}$$

provided r is a positive rational number. For negative rational exponents, we use the definition
$$x^{-r} = \frac{1}{x^r},$$
then combine Eqs. (3) and (7) with $b = x^r$ and get
$$L(x^{-r}) = L(1/x^r) = -L(x^r) = -rL(x).$$

This equation is just Eq. (7) with r replaced by $-r$. Hence Eq. (7) holds for negative and positive rational exponents r [and for $r = 0$ as well, by virtue of Eq. (2)].

As the last preliminary result, we easily get
$$L(u/v) = L(u) - L(v) \tag{8}$$
by taking $a = u$, $x = 1/v$ and applying Eqs. (1) and (3).

Are the foregoing properties of L sufficient for finding the derivative
$$L'(x) = \lim_{h \to 0} \frac{L(x+h) - L(x)}{h} ? \tag{9}$$

Almost. For we can write
$$\frac{1}{h} [L(x+h) - L(x)] = \frac{1}{h} L\left(\frac{x+h}{x}\right) = \frac{1}{h} L\left(1 + \frac{h}{x}\right), \tag{10}$$

and, if h is *rational*, this is equal to

$$L\left[\left(1 + \frac{h}{x}\right)^{1/h}\right].$$

If x is *also rational*, then we can put $h/x = u$, or $h = xu$, and get

$$L\left[\left(1 + \frac{h}{x}\right)^{1/h}\right] = L[(1 + u)^{1/xu}] = \frac{1}{x} L[(1 + u)^{1/u}]. \tag{11}$$

Equations (10) and (11) show that the difference quotient on the right-hand side of Eq. (9) has a limit as $h \to 0$ through rational values, provided:

i) x is rational and positive,
ii) $x + h$ is rational and positive,
iii) $u = xh \to 0$ through rational values, and
iv) $\lim_{x \to 0} L(1 + u)^{1/u}$ exists.

One more step completes the present heuristic approach. Since

$$L(1) = 0,$$

the derivative of the L function, at 1, is

$$L'(1) = \lim_{u \to 0} \frac{L(1 + u) - L(1)}{u} = \lim_{u \to 0} \left[\frac{1}{u} L(1 + u)\right]$$
$$= \lim_{u \to 0} L[(1 + u)^{1/u}], \tag{12}$$

provided, of course, this limit exists. Equations (10), (11), and (12), applied to the right-hand side of Eq. (9), lead to the conclusion that the derivative of the logarithmic function L should satisfy the condition

$$L'(x) = \frac{1}{x} L'(1). \tag{13}$$

The coefficient of $1/x$ on the right-hand side of Eq. (13) is just the slope of the graph of L at $x = 1$. This number,

$$C = L'(1),$$

depends on the base of logarithms. When that base is chosen for which $C = 1$, we have

$$L'_e(x) = \frac{1}{x}, \qquad x > 0. \tag{14}$$

(The subscript e stands for the base of natural logarithms. It has been computed to more than 100,000 decimal places by Daniel Shanks and

John W. Wrench, Jr., using an IBM 7090 computer. The computation took 2.5 hours, as compared with 8.7 hours for π. See "Calculation of π to 100,000 Decimals," by Shanks and Wrench, *Mathematics of Computation*, 16, No. 77, Jan. 1962, p. 78.)

Our heuristic program has thus led us to Eq. (14). We have *not* proved this equation. The reader can easily see where there were gaps in the derivation. But what we have shown is that a natural logarithmic function (if one exists) should satisfy Eqs. (1) through (14). Combining these results with the fundamental theorem of calculus, we get

$$L(x) - L(1) = \int_1^x L'$$

or, since $L(1) = 0$ and we want $L'_e(t) = 1/t$, we are led to the result

$$L_e(x) = \int_1^x \frac{1}{t}\, dt, \qquad x > 0. \tag{15}$$

EXERCISES 13–2

In Exercises 1 through 6, use Eq. (15) and the trapezoidal approximation with $n = 4$ to estimate $L_e(x)$ for the given value of x:

1. 2   2. 1.5   3. 3   4. 5   5. 10   6. 0.5

7. Sketch that portion of the hyperbola $y = 1/t$ for $\frac{1}{2} \leq t \leq 10$. Let $G(a, b)$ be the area of the region $a \leq t \leq b$, $0 \leq y \leq 1/t$. Compare the following pairs:
   a) $G(1, 2)$ and $G(5, 10)$
   b) $G(\frac{1}{2}, 1)$ and $G(1, 2)$
   c) $G(1, 3)$ and $G(2, 6)$

8. If $a > 0$ and $b > a$,

$$G(a, b) = \int_a^b \frac{1}{t}\, dt$$

is the limit of Riemann sums representing areas of certain inscribed rectangles. In the next section (Section 13–3) we shall show why $G(1, x) = G(a, ax)$ when $x > 0$ and $a > 0$. Before reading that section, see if you can set up a one-to-one correspondence between rectangles with bases on the interval [1, 2] and those with bases on the interval [5, 10], inscribed under the graph of $y = 1/t$, and such that corresponding rectangles have equal areas.

9. If you worked Exercises 1, 2, and 3, compare $L_e(2) + L_e(1.5)$ with $L_e(3)$.

10. If you worked Exercises 1, 4, and 5, compare $L_e(2) + L_e(5)$ with $L_e(10)$.

11. If you worked Exercises 1 and 6, compare $L_e(2) + L_e(0.5)$ with $L_e(1) = 0$.

LOGARITHMIC AND EXPONENTIAL FUNCTIONS

12. If you worked Exercises 2, 3, and 6, compare $L_e(3) + L_e(0.5)$ with $L_e(1.5)$.

13. Assume (as will be proved in Section 13–3) that $L_e(x^m) = mL_e(x)$ for m rational and $x > 0$.

   a) Use Eq. (15) and a rectangle with altitude $1/1.05$ to estimate
   $$L_e(1.1) = \int_1^{1.1} \frac{1}{t}\, dt.$$

   b) Use the result of part (a) to approximate each of the following:
   i) $2L_e(1.1) = L_e(1.1^2) = L_e(1.21)$
   ii) $4L_e(1.1) = L_e(1.1^4) \approx L_e(1.46)$
   iii) $5L_e(1.1) = L_e(1.1^5) \approx L_e(1.61)$
   iv) $8L_e(1.1) = L_e(1.1^8) \approx L_e(2.14)$

   c) Use the results of part (b) to sketch a portion of the graph of $y = L_e(x)$. Plot the points with abscissas 1, 1.1, 1.21, 1.46, 1.61, and 2.14 and sketch a smooth curve through them.

   d) From your graph for part (c), estimate $L_e(2)$.

   e) Use the results of part (b) and linear interpolation between $L_e(1.21)$ and $L_e(1.10)$ to estimate $L_e(1.19) \approx L_e(\sqrt[4]{2})$. Compare $4L_e(1.19)$ with your answer to part (d).

13–3 THE NATURAL LOGARITHMIC FUNCTION

It is customary to start the discussion of the natural logarithm with the following *definition*.

Definition. The *natural logarithm* is the function with domain $x > 0$ and values given by
$$\ln x = \int_1^x \frac{1}{t}\, dt. \tag{1}$$

Observe that the function* $f = 1/\mathbf{j}$,
$$f(t) = \frac{1}{t},$$
is continuous on the interval $[1, x]$ (or $[x, 1]$ if $0 < x < 1$) for any $x > 0$. Therefore, the integral on the right-hand side of Eq. (1) exists for all $x > 0$ and the domain consists of *all* positive *real* numbers (not just the rationals). By Theorem 12–5,
$$\frac{d}{dx} \ln x = \frac{d}{dx}\left[\int_1^x \frac{1}{\mathbf{j}}\right] = \frac{1}{\mathbf{j}(x)} = \frac{1}{x}. \tag{2}$$

---
* Cf. Section 12–5 for reference to Menger's notation $\mathbf{j}$.

There are now two ways to show that

$$\ln(ax) = \ln a + \ln x. \tag{3}$$

First way to prove Eq. (3). Let $y = \ln(ax)$. Then, by Eq. (2) and the chain rule,

$$\frac{dy}{dx} = \frac{1}{ax}\frac{d}{dx}(ax) = \frac{a}{ax} = \frac{1}{x}.$$

Hence

$$\frac{d}{dx}\ln ax = \frac{d}{dx}\ln x.$$

Therefore,

$$\ln ax = \ln x + C,$$

and we can evaluate the constant of integration by taking $x = 1$:

$$\ln a = \ln 1 + C = 0 + C.$$

Hence

$$C = \ln a$$

and

$$\ln ax = \ln x + \ln a.$$

Second way to prove Eq. (3). The left-hand side of Eq. (3) is, by definition, the integral of $1/\mathbf{j}$ from 1 to ax, while the right-hand side is the sum of the integrals from 1 to a and from 1 to x. But we can show that the integral from 1 to x is the same as the integral from a to ax:

$$\int_1^x \frac{1}{\mathbf{j}} = \int_a^{ax} \frac{1}{\mathbf{j}}. \tag{4}$$

We prove Eq. (4) for $x > 1$ and leave the proofs for $x = 1$ and for $x < 1$ as exercises. Hence, suppose $a > 0$, $x > 1$, and P is a partition of $[1, x]$. Let Q be an associated set of base points for a Riemann sum $S_f(P, Q)$, where $f = 1/\mathbf{j}$; that is, $f(t) = 1/t$. Multiply each element in P and Q by a, and call the resulting sets P_a and Q_a. Then points in P_a are a times as far apart as corresponding points in P and

$$\|P_a\| = a\|P\|,$$

so

$$\|P_a\| \to 0 \quad \Leftrightarrow \quad \|P\| \to 0.$$

But
$$S_f(P, Q) = S_f(P_a, Q_a), \tag{5}$$

because a typical term in the sum on the left-hand side of Eq. (5) is

$$\frac{1}{x_j}(a_j - a_{j-1}) \tag{6a}$$

and the corresponding term in the sum on the right-hand side is

$$\frac{1}{ax_j}(aa_j - aa_{j-1}), \tag{6b}$$

and (6a) and (6b) are equal. Therefore,

$$\lim_{\|P\|\to 0} S_f(P, Q) = \lim_{\|P_a\|\to 0} S_f(P_a, Q_a),$$

or

$$\int_1^x \frac{1}{j} = \int_a^{ax} \frac{1}{j}. \qquad \text{Q.E.D.}$$

We saw in Section 13–2 how the usual properties of logarithms can be deduced from Eq. (3), so we do not repeat the derivation of these properties for the natural logarithm:

$$\ln ax = \ln a + \ln x, \tag{7a}$$

$$\ln 1 = 0, \tag{7b}$$

$$\ln x^n = n \ln x, \qquad n \text{ rational}, \tag{7c}$$

$$\ln (u/v) = \ln u - \ln v. \tag{7d}$$

Computation of natural logarithms

By integrating

$$\frac{1}{1+u} = 1 - u + u^2 - u^3 + \cdots + (-1)^n u^n + R_{n+1},$$

where

$$R_{n+1} = (-1)^{n+1} \frac{u^{n+1}}{1+u},$$

from $u = 0$ to $u = x$, we get the following expansion:

$$\ln (1 + x) = x - \frac{x^2}{2} + \frac{x^3}{3} - \frac{x^4}{4} + \cdots + (-1)^n \frac{x^{n+1}}{n+1} + \rho_{n+1}, \tag{8a}$$

where
$$\rho_{n+1} = (-1)^{n+1} \int_0^x \frac{u^{n+1}}{1+u}\,du. \tag{8b}$$

In (8a), we assume $1 + x > 0$, and when u goes from 0 to x, $1 + u$ goes from 1 to $1 + x$. Hence

$$1 \leq 1 + u \leq 1 + x \quad \text{if} \quad x > 0,$$
$$1 + x \leq 1 + u \leq 1 \quad \text{if} \quad -1 < x \leq 0.$$

The remainder, ρ_{n+1}, is therefore bounded as follows:

$$|\rho_{n+1}| \leq \int_0^x u^{n+1}\,du \quad \text{if} \quad x > 0,$$

and

$$|\rho_{n+1}| = \left|\int_0^x \frac{u^{n+1}}{1+u}\,du\right| \leq \frac{1}{1+x}\int_0^{|x|} u^{n+1}\,du \quad \text{if} \quad -1 < x \leq 0.$$

In other words,

$$|\rho_{n+1}| \leq \begin{cases} \dfrac{x^{n+2}}{n+2} & \text{if} \quad x > 0, \\ \dfrac{1}{1+x}\dfrac{|x|^{n+2}}{n+2} & \text{if} \quad -1 < x \leq 0. \end{cases} \tag{9}$$

If $|x| < 1$, both expressions on the right-hand side of (9) approach 0 as $n \to \infty$, so we have the series expansion

$$\ln(1+x) = \sum_{n=1}^{\infty} (-1)^{n-1} \frac{x^n}{n}, \qquad |x| < 1. \tag{10}$$

Remark 1. The series (10) obviously converges most rapidly when $|x|$ is small. For purposes of illustration we shall use this series and some algebraic, or arithmetic, identities to compute $\ln 2$ and $\ln 10$. However, in order to use Eq. (10), we don't compute either of these logarithms directly, but look for a number that can be expressed as $1 + x$ with $|x| < 1$, that is directly related to a power of 2, or a power of 10. In Example 1 below we use $\sqrt{2} = 1 + x$ and get $\ln 2$ from the relation $\ln \sqrt{2} = \frac{1}{2}\ln 2$, or $\ln 2 = 2\ln\sqrt{2}$. In Example 2 we use a modified series to compute $\ln 2$ directly to six decimal places. In a later example we shall use the relation $2^{10} = 1.024 \times 10^3$ to find $\ln 10$ from values of $\ln 2$ and $\ln 1.024$.

Example 1. If we put

$$x = \sqrt{2} - 1, \qquad x^2 = 3 - 2\sqrt{2},$$
$$x^3 = 5\sqrt{2} - 7, \qquad x^4 = 17 - 12\sqrt{2},$$

Eqs. (8a) and 8(b), with $n + 1 = 4$, yield

$$\ln \sqrt{2} = (\sqrt{2} - 1) - \tfrac{1}{2}(3 - 2\sqrt{2}) + \tfrac{1}{3}(5\sqrt{2} - 7)$$
$$- \tfrac{1}{4}(17 - 12\sqrt{2}) + \rho_4.$$

Here

$$\rho_4 = \int_0^{\sqrt{2}-1} \frac{u^4}{1+u}\, du,$$

so that

$$0 < \rho_4 \leq \frac{(\sqrt{2}-1)^5}{5} = \frac{29\sqrt{2} - 41}{5} = \frac{29}{5}\left(\sqrt{2} - \frac{41}{29}\right).$$

To six decimals,

$$\sqrt{2} = 1.414214 \quad \text{and} \quad \tfrac{41}{29} = 1.413793.$$

We therefore get

$$\ln \sqrt{2} = \frac{20\sqrt{2}}{3} - \frac{109}{12} + \rho_4 = 9.42809 - 9.08333 + \rho_4$$
$$= 0.34476 + \rho_4,$$

with

$$0 < \rho_4 < 0.0025.$$

That is,

$$0.34476 < \ln \sqrt{2} < 0.34726$$

or, since $\ln 2 = 2 \ln \sqrt{2}$,

$$0.68952 < \ln 2 < 0.69452.$$

Therefore, to two decimals,

$$\ln 2 = 0.69.$$

Remark 2. It is obvious that we could compute $\ln \sqrt{2}$ to greater accuracy by taking a larger value of n in Eqs. (8a) and (8b). We can also use the average, $\tfrac{1}{2}(0.68952 + 0.69452)$ or 0.69202, as an estimate of $\ln 2$.

Remark 3. In Eq. (10) if we replace x by $-x$, we get

$$\ln(1 - x) = \sum_{n=1}^{\infty} (-1)^{2n-1} \frac{x^n}{n}$$
$$= -x - \frac{x^2}{2} - \frac{x^3}{3} - \cdots, \qquad |x| < 1. \qquad (11)$$

13-3 THE NATURAL LOGARITHMIC FUNCTION

Finally, if we add the series for $\ln(1+x)$ and that for $-\ln(1-x)$, we get

$$\ln\left(\frac{1+x}{1-x}\right) = 2\left(x + \frac{x^3}{3} + \frac{x^5}{5} + \cdots\right), \qquad |x| < 1. \qquad (12)$$

Every positive real number y can be put in the form

$$y = \frac{1+x}{1-x} \qquad (13a)$$

by setting

$$x = \frac{y-1}{y+1}, \qquad (13b)$$

and

$$-1 < x < 1 \quad \text{if} \quad 0 < y < \infty.$$

Figure 13-1 shows part of the graph of

$$y = \frac{1+x}{1-x} \quad \text{or} \quad x = \frac{y-1}{y+1}$$

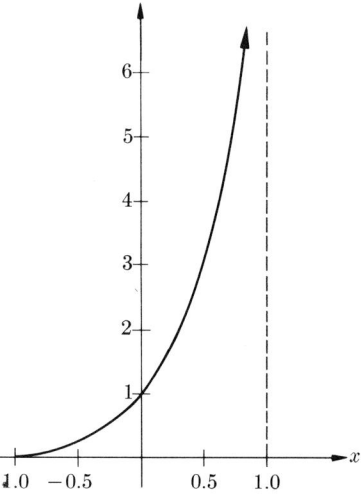

FIG. 13-1. Portion of the graph of the hyperbola $y = (1+x)/(1-x)$, $|x| < 1$.

for $-1 < x < 1$.

Substituting Eqs. (13a) and (13b) into Eq. (12), we get

$$\ln y = 2\left[\frac{y-1}{y+1} + \frac{1}{3}\left(\frac{y-1}{y+1}\right)^3 + \frac{1}{5}\left(\frac{y-1}{y+1}\right)^5 + \cdots\right], \qquad 0 < y < \infty. \qquad (14)$$

Example 2. Use Eq. (14) to compute $\ln 2$ to a few decimal places.

Solution. If $y = 2$, then $(y-1)/(y+1) = \frac{1}{3}$ and

$$\ln 2 = 2[\tfrac{1}{3} + \tfrac{1}{3}(\tfrac{1}{3})^3 + \tfrac{1}{5}(\tfrac{1}{3})^5 + \tfrac{1}{7}(\tfrac{1}{3})^7 + \cdots]. \qquad (15)$$

It systematizes the work to compute powers of $\frac{1}{3}$ as follows:

$$\tfrac{1}{3} = 0.3333333\ldots \qquad 1(\tfrac{1}{3}) = 0.3333333\ldots$$
$$(\tfrac{1}{3})^3 = \tfrac{1}{9}(\tfrac{1}{3}) = 0.0370370\ldots \qquad \tfrac{1}{3}(\tfrac{1}{3})^3 = 0.0123457\ldots$$
$$(\tfrac{1}{3})^5 = \tfrac{1}{9}(\tfrac{1}{3})^3 = 0.0041152\ldots \qquad \tfrac{1}{5}(\tfrac{1}{3})^5 = 0.0008230\ldots$$
$$(\tfrac{1}{3})^7 = \tfrac{1}{9}(\tfrac{1}{3})^5 = 0.0004572\ldots \qquad \tfrac{1}{7}(\tfrac{1}{3})^7 = 0.0000653\ldots$$
$$(\tfrac{1}{3})^9 = \tfrac{1}{9}(\tfrac{1}{3})^7 = 0.0000508\ldots \qquad \tfrac{1}{9}(\tfrac{1}{3})^9 = 0.0000056\ldots$$
$$\text{sum} = 0.3465729\ldots$$
$$(\tfrac{1}{3})^{11} = \tfrac{1}{9}(\tfrac{1}{3})^9 = 0.0000056\ldots \qquad 2(\text{sum}) = 0.6931458\ldots$$

Because all terms in the series of Eq. (15) are positive, we know that
$$\ln 2 = 0.6931458 + \epsilon, \tag{16}$$
where ϵ is positive (and presumably small). There are two kinds of error involved in the total error ϵ in Eq. (16):

a) round-off errors, and
b) truncation error.

Round-off errors could be reduced by carrying more decimal places in the computations. Truncation error results from terminating the infinite series (15) after a finite number of terms. In our example, the truncation error is
$$\frac{1}{11}\left(\frac{1}{3}\right)^{11} + \frac{1}{13}\left(\frac{1}{3}\right)^{13} + \frac{1}{15}\left(\frac{1}{3}\right)^{15} + \cdots,$$
and this is less than
$$\frac{1}{11}\left(\frac{1}{3}\right)^{11}\left[1 + \left(\frac{1}{3}\right)^2 + \left(\frac{1}{3}\right)^4 + \cdots\right] = \frac{1}{11}\left(\frac{1}{3}\right)^{11}\frac{1}{1-(1/9)},$$
which is approximately
$$(0.0000005)(\tfrac{9}{8}) \quad \text{or} \quad \approx 6 \times 10^{-7}.$$
Thus,
$$0.6931458 < \ln 2 < 0.6931464,$$
and, to 6 decimal places, we have
$$\ln 2 = 0.693146. \tag{17}$$

Remark 4. Having computed $\ln 2$, we could use Eq. (14) to compute
$$\ln \tfrac{3}{2}, \quad \ln \tfrac{4}{3}, \quad \ln \tfrac{5}{4}, \quad \ln \tfrac{6}{5}, \quad \ldots, \quad \ln \tfrac{10}{9}.$$
Then, by addition, we could build up a table of natural logarithms, $\ln N$, for $N = 2, 3, \ldots, 10$. Thus
$$\ln 2 + \ln \tfrac{3}{2} = \ln (2 \cdot \tfrac{3}{2}) = \ln 3,$$
$$\ln 3 + \ln \tfrac{4}{3} = \ln 4,$$
and so on. These, in turn can be used to compute $\log_{10} N$ from the formula
$$\log_{10} N = \frac{\ln N}{\ln 10}. \tag{18}$$

Example 3. Compute $\ln 1.024$ and use the result to find $\ln 10$.

THE NATURAL LOGARITHMIC FUNCTION

Solution. If $1 + x = 1.024$, then $x = 0.024$, and Eq. (10) gives the result without great effort as follows:

$$\begin{aligned}
x &= 0.024\ 000\ 000, & x &= 0.024\ 000\ 000, \\
x^2 &= 0.000\ 576\ 000, & -\tfrac{1}{2}x^2 &= -0.000\ 288\ 000, \\
x^3 &= 0.000\ 013\ 824, & \tfrac{1}{3}x^3 &= 0.000\ 004\ 608, \\
x^4 &= 0.000\ 000\ 332, & -\tfrac{1}{4}x^4 &= -0.000\ 000\ 083, \\
x^5 &= 0.000\ 000\ 008, & \tfrac{1}{5}x^5 &= 0.000\ 000\ 002,
\end{aligned}$$

$$\ln(1+x) = x - \tfrac{1}{2}x^2 + \tfrac{1}{3}x^3 - \tfrac{1}{4}x^4 + \tfrac{1}{5}x^5 = 0.023\ 716\ 527,$$

to 9 decimals, or, to 6 decimal places, $\ln 1.024 = 0.023\ 717$. We now use the identity $2^{10} = 1024 = 1.024 \times 10^3$ to get

$$10 \ln 2 = \ln 1.024 + 3 \ln 10$$

or

$$\begin{aligned}
\ln 10 &= \tfrac{1}{3}(10 \ln 2 - \ln 1.024) \\
&= \tfrac{1}{3}(6.93146 - 0.02372) \qquad \text{[using Eq. (17)]} \\
&= 2.30258 \qquad \text{(to 5 decimals)}. \tag{19}
\end{aligned}$$

Remark 5. Equations (12) and (13b) show that we could compute $\ln 10$ directly by substituting $x = \tfrac{9}{11}$ in Eq. (12). The arithmetic that this entails is tedious if done without the aid of a computer. We were fortunate in having the assistance of Mr. Robert T. Wiggin of The Brookline High School who composed the following program for a PDP-1 digital computer using the language "Telcomp":

1.0 Set $D(0) = 0$
1.1 Set $L(0) = 0$
1.2 Set $N = 1$
1.3 To part 2

2.1 Set $L(N) = L(N-1) + [2/(2*N-1)]*[9/11]\uparrow(2*N-1)$
2.2 Set $D(N) = LN(10) - L(N)$
2.3 Type $N, L(N), D(N)$
2.4 To part 3 if $D(N) = D(N-1)$
2.5 Set $N = N+1$
2.6 To part 2

3.1 Stop

LOGARITHMIC AND EXPONENTIAL FUNCTIONS

Thus the machine was instructed to compute the sums

$$L(N) = L(N-1) + \frac{2}{2N-1}\left(\frac{9}{11}\right)^{2N-1}$$

for positive integral values of N, starting with $L(0) = 0$. These are just the sums required by Eq. (12) to give ln 10. The rate of convergence can be seen from the following tabulation, taken from the printout from the machine:

Number of terms, N	Sum, $L(N)$
6	2.274798
12	2.301173
18	2.302496
24	2.302579
30	2.302585 (accurate to 6 decimals)

EXERCISES 13–3

1. Prove Eq. (4) for $x = 1$; for $x < 1$.

2. Use Eqs. (8), (9), with $x = 0.5$ and $n = 4$ and the method of Example 1, to determine, to 3 decimal places, values between which ln 1.5 lies. How does their average compare with the actual value of ln 1.5?

3. Compute ln 1.5 to 4 decimal places, using Eqs. (8a), (8b), with $x = 0.5$ and $n = 7$.

4. Compute $\ln \frac{10}{9}$ using Eq. (12). Combine this result with ln 2 from Eq. (17) and $\ln \frac{3}{2}$ from the answer to Exercise 3 to get ln 10 from the equation

$$\ln 10 = \ln 9 + \ln \tfrac{10}{9} = 2 \ln 3 + \ln \tfrac{10}{9}$$
$$= 2(\ln 2 + \ln \tfrac{3}{2}) + \ln \tfrac{10}{9}.$$

5. Use your result for ln 10 and Eqs. (17) and (18) to compute $\log_{10} 2$ to 1 or more decimal places. [*Answer:* $\log_{10} 2 = 0.30103$ to 5 places.]

6. Find a series for $\ln[(N+1)/N]$ by substitution in Eq. (14). Check your result by using Eq. (12) with

$$\frac{1+x}{1-x} = \frac{N+1}{N}$$

and solving for x.

13–4 THE EXPONENTIAL FUNCTION

Does the natural logarithmic function have an inverse, and, if so, what are its main properties? The answers are most easily understood in terms

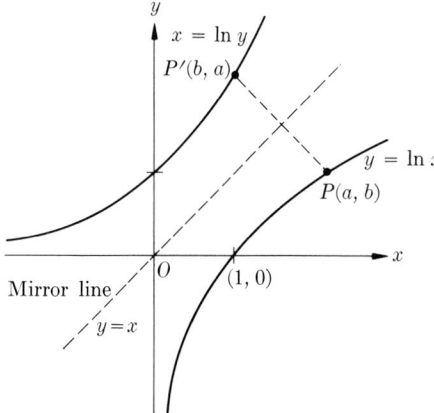

FIG. 13-2. Partial graphs of the related equations $y = \ln x$ and $x = \ln y$. Each curve is the reflection of the other with respect to the line $y = x$. The points $P(a, b)$ with $b = \ln a$ and $P'(b, a)$ are symmetric with respect to the mirror line.

of the graphs of the curves

$$y = \ln x, \qquad x > 0 \tag{1a}$$

and

$$x = \ln y, \qquad y > 0. \tag{1b}$$

These two graphs are shown in Fig. 13-2. We know quite a bit about the graph of Eq. (1a):

i) Its slope at the point $(x, \ln x)$ is $1/x$.
ii) Because the derivative of $\ln x$ is positive, the graph constantly rises from left to right.
iii) $\ln x$ is positive for $x > 1$, zero for $x = 1$, negative for $x < 1$.
iv) Since

$$\ln (2^n) = n \ln 2 \approx (0.693)n,$$

the points having abscissas

$$x_n = 2^n, \qquad n = 1, 2, 3, \ldots,$$

have ordinates y_n that increase without bound as $n \to \infty$. Combining this with the fact that $\ln x$ is continuous and monotone increasing, we see that the portion of the domain where

$$1 \leq x < \infty$$

maps one-to-one onto

$$0 \leq y < \infty$$

under the natural logarithmic mapping $y = \ln x$.

v) Because $\ln(1/x) = -\ln x$, property (iv) above means that the part of the domain where
$$0 < x \le 1$$
maps one-to-one onto
$$-\infty < y \le 0.$$

Thus, as x increases through the domain between 0 and $+\infty$, $\ln x$ increases from $-\infty$ to $+\infty$, and every real number y is the natural logarithm of precisely one positive real number x. The *inverse relation* that we get from ln by interchanging x and y in the set of ordered pairs
$$\ln = \{(x, y) : y = \ln x, \ x > 0\}$$
is, therefore, a function—the *inverse natural logarithmic function* which we shall temporarily denote by $\ln^{-1}$. Thus
$$y = \ln^{-1}(x) \tag{2a}$$
means just the same as
$$x = \ln y, \quad y > 0, \quad -\infty < x < \infty. \tag{2b}$$

The *domain* of $\ln^{-1}$ is the *range* of ln, hence is the set of all real numbers, $-\infty < x < \infty$. Thus, Eqs. (2a) and (2b) both correspond to the upper curve in Fig. 13–2. The number whose natural logarithm is 1 plays a special role, and is the number e referred to in Section 13–2:
$$e = \ln^{-1}(1), \quad 1 = \ln e. \tag{3}$$

Equations (3) *define* e. From Eq. (17) of Section 13–3 we have
$$\ln 2 \approx 0.693, \quad \tfrac{1}{2}\ln 2 \approx 0.346$$
and, adding,
$$0.693 \approx \ln 2 < 1 < \tfrac{3}{2}\ln 2 = \ln(2\sqrt{2}) \approx 1.039.$$

Therefore, the point $(e, 1)$ lies on the lower logarithmic curve of Fig. 13–2 between the points $(2, \ln 2)$ and $(2\sqrt{2}, \tfrac{3}{2}\ln 2)$. This gives us some information about the size of e; it lies between 2 and $2\sqrt{2} = 2.828\ldots$:
$$2 < e < 2\sqrt{2} = 2.828\ldots \tag{4}$$

By studying rational powers of e we get further information. Because
$$\ln(e^x) = x \ln e \quad \text{for } x \text{ rational,}$$

13-4 THE EXPONENTIAL FUNCTION

and
$$\ln e = 1,$$
we have
$$\ln e^x = x \quad \text{for } x \text{ rational}. \tag{5}$$

When we compare Eq. (5) with Eqs. (2b) and (2a) we see that
$$e^x = \ln^{-1}(x) \quad \text{for all rational values of } x. \tag{6}$$

The right-hand side of Eq. (6) has a perfectly well-defined meaning for *any real number* x (not just for the rationals). When x happens to be *rational*, both sides of Eq. (6) are well defined, and are equal. When x is *irrational*, we use Eq. (6) to *define* e^x.

Definition 1. For any real number x, we define e^x to be the same as $\ln^{-1}(x)$:
$$e^x = \ln^{-1}(x). \tag{7}$$

Remark 1. The notation exp (exponential function) is also used to represent the inverse of the natural logarithm:
$$\exp(x) = \ln^{-1}(x), \quad x \text{ real}. \tag{8}$$

Thus Eq. (7) becomes
$$e^x = \exp(x).$$

The use of the exp notation is particularly helpful when working with composite functions, like
$$e^{(\sin^{-1} x)} = \exp(\sin^{-1} x),$$
or other expressions leading to complicated typography.

A summary of what we've done so far:
i) defined ln and studied it,
ii) defined exp $= \ln^{-1}$,
iii) shown that
$$e^x = \exp(x) \quad \text{for } x \text{ rational},$$
iv) *defined* e^x for x irrational to be $\exp(x)$.

Result. The exponential function (base e) is the inverse of the natural logarithmic function. All of its properties are derived from that fact.

LOGARITHMIC AND EXPONENTIAL FUNCTIONS

Theorem 13-1. For every real x,

$$\frac{d}{dx}(e^x) = e^x \tag{9a}$$

or

$$\exp'(x) = \exp(x). \tag{9b}$$

Proof. From the general theory of inverse functions, we can write

$$y = e^x \Leftrightarrow x = \ln y$$

and

$$\frac{dy}{dx} = \frac{1}{dx/dy} = \frac{1}{(1/y)} = y. \qquad \text{Q.E.D.}$$

Theorem 13-2. For all real x_1 and x_2,

$$e^{x_1} e^{x_2} = e^{x_1 + x_2} \tag{10a}$$

or

$$\exp(x_1) \cdot \exp(x_2) = \exp(x_1 + x_2). \tag{10b}$$

Proof. Let

$$y_1 = e^{x_1}, \qquad y_2 = e^{x_2}.$$

Then, by definition,

$$x_1 = \ln y_1, \qquad x_2 = \ln y_2,$$

and

$$x_1 + x_2 = \ln y_1 + \ln y_2 = \ln (y_1 y_2).$$

Therefore,

$$y_1 y_2 = \ln^{-1}(x_1 + x_2) = \exp(x_1 + x_2),$$

so that

$$e^{x_1} e^{x_2} = \exp(x_1 + x_2) = e^{(x_1 + x_2)}. \qquad \text{Q.E.D.}$$

Theorem 13-3. For every real x and rational r,

$$(e^x)^r = e^{rx} \tag{11a}$$

or

$$[\exp(x)]^r = \exp(rx). \tag{11b}$$

Proof. Let
$$y = e^x \quad \text{or} \quad x = \ln y.$$
Then, for any rational r,
$$rx = r \ln y = \ln (y^r),$$
so that
$$y^r = \ln^{-1} (rx) = \exp (rx) = e^{rx}.$$
But $y = e^x$. Therefore,
$$(e^x)^r = e^{rx}. \qquad \text{Q.E.D.}$$

Remark 2. The restriction to *rational* values of r in Eqs. (11a) and (11b) can be removed if we define a^u for any positive base a and real exponent u. Our definition is *motivated* by the thought that, if
$$y = a^u, \tag{12a}$$
then we want it to be true that
$$\ln y = u \ln a. \tag{12b}$$
But (12b) already requires that
$$y = \ln^{-1} (u \ln a) = \exp (u \ln a).$$
Therefore, the only definition of a^u that meets the requirement (12b) is the following:

Definition 2. Let a be a positive number. Then, for any real number u, we define a^u to be $\exp (u \ln a)$:
$$a^u = \exp (u \ln a) \tag{13a}$$
or
$$a^u = e^{u \ln a}. \tag{13b}$$

Since the right sides of Eqs. (13a) and (13b) are well defined for all positive numbers a and all real numbers u, we have certainly defined the general exponential function (base a, $a > 0$) for the domain $-\infty < u < \infty$. It comes as no surprise, but is nevertheless good news, that for any *rational* exponent u, say
$$u = m/n, \quad m, n \text{ integers}, \quad n \neq 0,$$

the earlier definition,
$$a^u = \sqrt[n]{a^m},$$
and the new definition, (13a), agree. In particular, the following results are immediate consequences of Eqs. (13a) and (13b):

$$a^1 = \exp(1 \ln a) = \exp(\ln a) = \ln^{-1}(\ln a) = a,$$
$$a^0 = \exp(0 \ln a) = \exp(0) = \ln^{-1}(0) = 1,$$
$$a^u a^v = \exp(u \ln a) \cdot \exp(v \ln a)$$
$$= \exp[(u + v) \ln a]$$
$$= a^{u+v},$$
$$(a^{m/n})^n = \left[\exp\left(\frac{m}{n} \ln a\right)\right]^n$$
$$= \exp(m \ln a) \quad \text{by Eq. (11b) if } n \text{ is rational}$$
$$= a^m \quad \text{by Eq. (13a).}$$

Remark 3. In Eq. (13b), let a be a fixed positive number different from 1. Then the exponent $u \ln a$ takes all real values between $-\infty$ and $+\infty$ when u does. The mapping

$$u \to a^u, \qquad -\infty < u < \infty,$$

is one-to-one from the domain of all real numbers u onto the range of positive real numbers a^u. That is, for each positive real number x there exists exactly one real number u (given $a > 0$, $a \neq 1$) such that

$$x = a^u. \tag{13c}$$

The exponent u for which Eq. (13c) is true is called the logarithm of x to the base a:

$$\log_a x = u \Leftrightarrow x = a^u. \tag{13d}$$

Comparing Eqs. (13b) and (13d), we see also that

$$\log_a x = u \Leftrightarrow x = e^{u \ln a}$$
$$\Updownarrow \tag{13e}$$
$$\ln x = u \ln a.$$

The statements (13e) lead to the conclusion

$$u = \log_a x = \frac{\ln x}{\ln a}. \tag{13f}$$

Definition 3. Let a be a positive number not equal to one. The logarithmic function, to the base a, is defined on the domain $x > 0$ by Eqs. (13d) or (13f):

$$\log_a x = \frac{\ln x}{\ln a}.$$

Example. $\log_{10} 2 = \dfrac{\ln 2}{\ln 10} \approx \dfrac{0.693146}{2.302585} \approx 0.30103.$

Series for e^x

The generalized mean-value theorem, or Taylor's theorem with remainder, can be applied to the exponential function. The result is this: because the exponential function has continuous derivatives of all orders for all real x, given any positive integer n and any real numbers a and b, there exists c between a and b (c depending on n, a, and b) such that

$$\exp(b) = \exp(a) + \sum_{k=1}^{n} \frac{\exp^{(k)}(a)}{k!} (b-a)^k$$

$$+ \frac{\exp^{(n+1)}(c)}{(n+1)!} (b-a)^{n+1}. \tag{14}$$

Now, the kth derivative of e^x is just e^x, so

$$\exp^{(k)}(x) = \exp(x).$$

If we take $b = x$ and $a = 0$ in Eq. (14),

$$\exp^{(k)}(a) = \exp(0) = e^0 = 1,$$
$$\exp(b) = \exp(x) = e^x,$$

we get

$$e^x = 1 + \sum_{k=1}^{n} \frac{x^k}{k!} + e^c \frac{x^{n+1}}{(n+1)!}, \tag{15}$$

for some c between 0 and x. Now

$$e^c < e^{|x|} \tag{16a}$$

and, for any fixed real number x,

$$\frac{|x|^{n+1}}{(n+1)!} \to 0 \quad \text{as} \quad n \to \infty. \tag{16b}$$

This means that the remainder term in (15) goes to zero as $n \to \infty$, so that we have the following series for e^x:

$$e^x = 1 + \sum_{k=1}^{\infty} \frac{x^k}{k!}$$

$$= 1 + x + \frac{x^2}{2!} + \frac{x^3}{3!} + \frac{x^4}{4!} + \cdots \qquad (17)$$

When $x = 1$, we get an infinite series for e itself:

$$e = 1 + 1 + \frac{1}{2!} + \frac{1}{3!} + \frac{1}{4!} + \cdots \qquad (18)$$

If we want to use (18) to compute e to some specified accuracy, we truncate at the term $1/n!$ with

$$\text{truncation error} = \frac{1}{(n+1)!} + \frac{1}{(n+2)!} + \cdots$$

$$= \frac{1}{(n+1)!}\left[1 + \frac{1}{n+2} + \frac{1}{(n+2)(n+3)} + \cdots\right]$$

$$< \frac{1}{(n+1)!}\left[1 + \frac{1}{2} + \frac{1}{2^2} + \frac{1}{2^3} + \cdots\right],$$

or

$$\text{truncation error} < \frac{2}{(n+1)!}.$$

Therefore, if we want, say, 3 decimal places, we can make

$$\frac{2}{(n+1)!} < \frac{5}{10^4}$$

by making

$$(n+1)! > \frac{2(10^4)}{5} = 4000.$$

Since

$$5! = 120, \qquad 6! = 720, \qquad 7! = 5040,$$

we take $n = 6$,

$$n + 1 = 7.$$

Thus, to 3 decimals, we can use

$$e = 1 + 1 + \frac{1}{2!} + \frac{1}{3!} + \cdots + \frac{1}{6!} + \frac{1}{7!}.$$

13-4 THE EXPONENTIAL FUNCTION

The computation is

$$1 + 1 = 2.00000$$

$$\frac{1}{2!} = 0.50000$$

$$\frac{1}{3!} = \frac{1}{3}\left(\frac{1}{2!}\right) = 0.16667$$

$$\frac{1}{4!} = \frac{1}{4}\left(\frac{1}{3!}\right) = 0.04167$$

$$\frac{1}{5!} = \frac{1}{5}\left(\frac{1}{4!}\right) = 0.00833$$

$$\frac{1}{6!} = \frac{1}{6}\left(\frac{1}{5!}\right) = 0.00139$$

$$\text{sum} = 2.71806.$$

Hence, to 3 decimals,

$$e = 2.718.$$

As our final example, we prove the following theorem:

Theorem 13-4. *e* is irrational.

Proof. Suppose that *e* is rational. Then for some positive integers p and q,

$$e = p/q. \tag{19}$$

We hope to show that Eq. (19) is false. By Eq. (18), we have

$$e = a_n + r_n, \tag{20a}$$

where

$$a_n = 1 + 1 + \frac{1}{2!} + \cdots + \frac{1}{n!} \tag{20b}$$

and

$$r_n = \frac{1}{(n+1)!} + \frac{1}{(n+2)!} + \cdots$$
$$< \frac{2}{(n+1)!}. \tag{20c}$$

Let n be a positive integer that is greater than $q + 1$. Then

$$n + 1 > q + 2 \geq 3. \tag{21}$$

Moreover, if Eq. (19) holds, we have
$$r_n = e - a_n = p/q - a_n$$
and
$$0 < n!r_n = n!e - n!a_n < \frac{2}{n+1} < 1. \qquad (22)$$
But if
$$n > q + 1 \quad \text{and} \quad e = p/q,$$
then
$$n!e = \left(\frac{n!}{q}\right)p$$
is an *integer* because $n!$ is divisible by q. Also, from Eq. (20b), $n!a_n$ is an integer. Therefore,
$$n!e - n!a_n$$
is an integer, which, by (22), is positive but less than 1. There are no such integers. Therefore e is not rational. Q.E.D.

EXERCISES 13–4

In Exercises 1 through 8, find dy/dx.

1. $y = e^{2x}$
2. $y = 3e^{-2x}$
3. $y = e^{\sqrt{x}}$
4. $y = e^{x^3}$
5. $y = \exp(\ln x)$
6. $y = \ln(e^x)$
7. $y = \ln(2 + \cos x)$
8. $y = \exp(\sin x)$

9. Graph the equation $y = \exp(-x^2)$. Where are its inflection points? Comment on extrema, symmetry, asymptotes.

Hyperbolic trigonometry, which can be associated logically with Lobachevskian geometry,* uses the following definitions:

$$\text{hyperbolic sine:} \quad \sinh x = \tfrac{1}{2}(e^x - e^{-x}),$$
$$\text{hyperbolic cosine:} \quad \cosh x = \tfrac{1}{2}(e^x + e^{-x}),$$
$$\text{hyperbolic tangent:} \quad \tanh x = \frac{\sinh x}{\cosh x}.$$

10. a) If $y = \sinh x$, find dy/dx.
    b) If $y = \cosh x$, find dy/dx.

---

* An excellent treatment appears in *Non-Euclidean Geometry*, by Harold E. Wolfe, Henry Holt and Co., 1945.

11. Using the definitions of cosh x and sinh x, deduce the identities:
    a) $\cosh^2 x - \sinh^2 x = 1$;
    b) $\cosh^2 x + \sinh^2 x = \cosh 2x$.

12. If $y = \tanh x$, find dy/dx.

13. The three remaining hyperbolic functions are defined as reciprocals; thus

$$\coth x = \frac{1}{\tanh x}, \quad \tanh x \neq 0,$$

$$\operatorname{sech} x = \frac{1}{\cosh x},$$

$$\operatorname{csch} x = \frac{1}{\sinh x}, \quad \sinh x \neq 0.$$

Setting y equal to each in turn, find dy/dx.

Integrate each of the following:

14. $\int \sinh x \, dx$
15. $\int \cosh x \, dx$
16. $\int \tanh x \, dx$
17. $\int \coth x \, dx$
18. $\int \operatorname{sech} x \, dx$
19. $\int x \sinh x \, dx$

20. Using circular functions, the equations of a unit circle centered at the origin may be parametrized by letting $x = \cos \theta$, $y = \sin \theta$. Show how hyperbolic functions may be used to represent the right-hand branch of a hyperbola centered at the origin. Why not the left-hand branch?

21. Find the area of a hyperbolic sector, that is, the area bounded by a portion of the x-axis, a portion of the hyperbola $x = \cosh \theta$, $y = \sinh \theta$, and the radius OP from the origin to a point on the hyperbola.

22. Sketch the curve

$$y = \frac{a}{2}(e^{x/2} + e^{-x/2}).$$

This is the catenary, or "hanging-chain," curve. It has many interesting applications and properties. See: *Calculus and Analytic Geometry*, 3rd edition, by G. B. Thomas, Jr., pp. 529–31; *A Book of Curves*, by E. H. Lockwood, Cambridge University Press, 1961, pp. 118–124; *A Source Book in Mathematics*, by David Eugene Smith, Dover Publications, 1959, pp. 644–655.

ANSWERS TO EXERCISES

ANSWERS TO EXERCISES

EXERCISES 1-1

1. 3; 2.5; 2.25; 2.1   2. 1.5; 1.7; 1.8; 1.9

3. $m = \dfrac{y_2 - y_1}{x_2 - x_1} = \dfrac{x_2^2 - x_1^2}{x_2 - x_1} = x_2 + x_1$ for $x_2 \neq x_1$

$\lim\limits_{x_2 \to x_1} (x_2 + x_1) = 2x_1$

EXERCISES 1-2

1. (a) 48 (b) 40 (c) 35.2 (d) 33.6 (e) 32.16 (f) 32.016 (g) $16(t_2 + t_1)$
   (h) as $t_2 \to t_1$, $v \to 32t_1$

2. (a) 128 (b) 120 (c) 115.2 (d) 113.6 (e) 112.16 (f) 112.016
   (g) $80 + 16(t_2 + t_1)$ (h) as $t_2 \to t_1$, $v \to 80 + 32t_1$

3. (a) 112 (b) 120 (c) 124.8 (d) 126.4 (e) 127.84 (f) 127.984
   (g) $160 - 16(t_2 + t_1)$ (h) as $t_2 \to t_1$, $v \to 160 - 32t_1$

4. for $t = 2$: $v = 96$ ft/sec, $s = 256$ ft   The body moves upward
   $t = 3$: $v = 64$ ft/sec, $s = 336$ ft   until it reaches 400 ft,
   $t = 4$: $v = 32$ ft/sec, $s = 384$ ft   after 5 seconds, and then
   $t = 5$: $v = 0$ ft/sec, $s = 400$ ft   starts coming down and
   $t = 6$: $v = -32$ ft/sec, $s = 384$ ft   reaches the ground after
   $s = 0$ when $t = 0$ and when $t = 10$ 5 more seconds.

EXERCISES 1-3

1.

2.

3.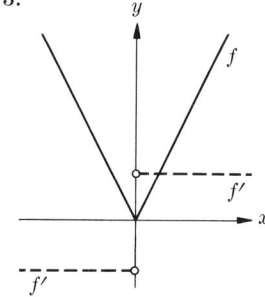

4.

5.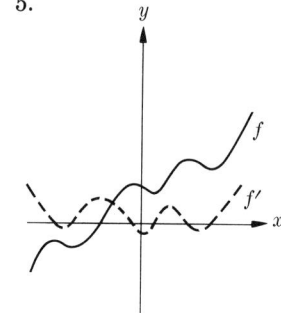

6.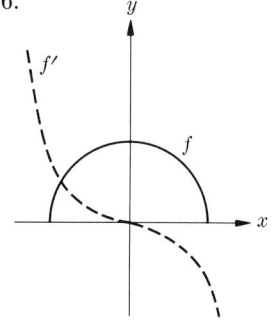

EXERCISES 1-4

1. $f'(x_1) = \lim\limits_{x_2 \to x_1} \dfrac{(2x_2 + 3) - (2x_1 + 3)}{x_2 - x_1} = \lim\limits_{x_2 \to x_1} \dfrac{2(x_2 - x_1)}{x_2 - x_1} = \lim\limits_{x_2 \to x_1} 2 = 2$

2. a 3. $-\dfrac{a}{(ax_1 + b)^2}$ 4. $2x_1$ 5. $2ax_1 + b$ 6. $3x_1^2$

7. $2ax_1 + 3bx_1^2$ 8. $1/2\sqrt{x_1}$ 9. $1/2\sqrt{x_1 + 1}$ 10. $-2/x_1^3$

EXERCISES 1-5

1. $A(x)$ is the area of region under curve from a to x, by definition; $A(a)$ is therefore a line segment, i.e., the area is zero.

2. Bases of trapezoid are $f(b) = mb$, and $f(a) = ma$; the altitude is $b - a$; therefore the area is

$$A(b) = \tfrac{1}{2}(b - a)(mb + ma) = \dfrac{m}{2}(b^2 - a^2).$$

3. $C = -\tfrac{1}{2}ma^2$
$A(x) = \tfrac{1}{2}mx^2 + C = \tfrac{1}{2}mx^2 - \tfrac{1}{2}ma^2$
$A(b) = \tfrac{1}{2}mb^2 - \tfrac{1}{2}ma^2 = \tfrac{1}{2}m(b^2 - a^2)$ as in Exercise 2.

EXERCISES 1-6

1. (a) $x \geq 0$ (b) $x \leq 0$ (c) $x \geq 5$ (d) none
2. (a) $-2 < x < 8$ (b) $x \leq 1$ or $x \geq 3$ (c) $2.49 \leq x \leq 2.51$ (d) none
3. (a) $f(x) = \max(x^2, 4 - x^2)$

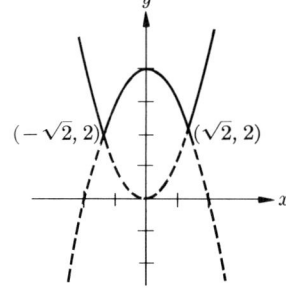

(b) $f(x) = x + |x| = \begin{cases} 0 & \text{if } x < 0 \\ 2x & \text{if } x \geq 0 \end{cases}$

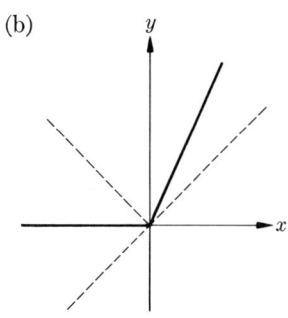

(Cont.)

ANSWERS TO EXERCISES

(c)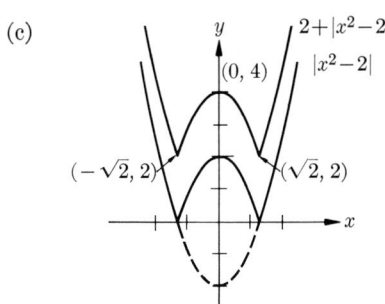

(d)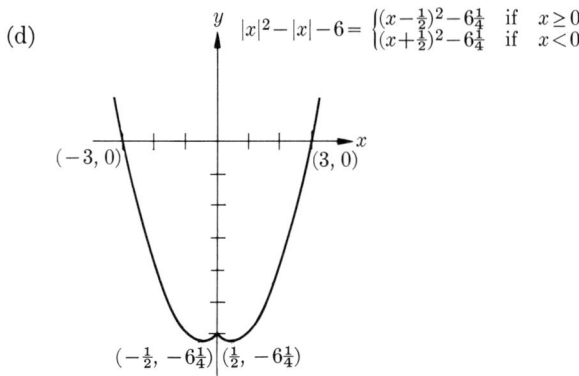

4. (a) c is midpoint of segment from a to b; d is distance from midpoint to either end of segment.
   (b) All values of x from a to b, and including both a and b.
   (c) If we start at c and add d, we reach the right endpoint of the interval, which is also the maximum of a and b. If we start at c and subtract d, we reach the left endpoint, which is the minimum of a and b.
   (d) CASE 1. If $a \geq b$, then $|a - b| = a - b$ and
   $$\frac{a+b}{2} + \tfrac{1}{2}|a-b| = \frac{a+b}{2} + \frac{a-b}{2} = a = \max(a,b),$$
   $$\frac{a+b}{2} - \tfrac{1}{2}|a-b| = \frac{a+b}{2} - \frac{a-b}{2} = b = \min(a,b).$$
   CASE 2. If $b > a$, then $|a - b| = b - a$ and the first formula works out to be equal to $b = \max(a,b)$, while the second is equal to $a = \min(a,b)$.

7. $|a + b + c| \leq |a| + |b| + |c|$

EXERCISES 1-7

1. (a) Yes; $f(1) = 1$
   (b) No; element of X, 1, is assigned to two elements, 1 and 0, of Y.
   (c) 8; $\{(1,1),(2,1)\}$, $\{(1,0),(2,0),(3,0)\}$, $\{(2,1),(3,1)\}$, etc.

ANSWERS TO EXERCISES

2. All define functions. Range:
   (a) $\{0 \leq y \leq 1\}$
   (b) $\{0 \leq y \leq 1\}$
   (c) $\{-1 \leq y \leq 1\}$
   (d) $\{0 \leq y \leq 1\}$
   (e) $\{0 \leq y \leq 1\}$
   (f) $\{0 \leq y < 1\}$
   (g) $\{y \in I\}$
   (h) $\{y \in I \text{ and } y = 2k+1\ k \in I\}$

3.

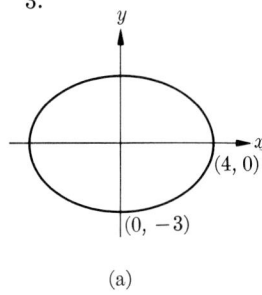
(a)

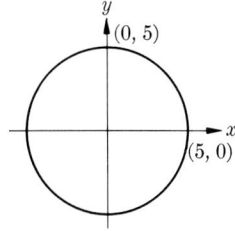
(b)

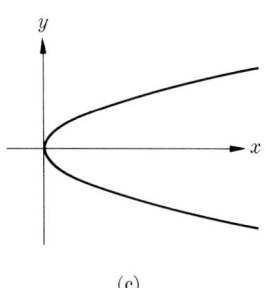
(c)

4. Two choices in each case.
   (a) $D_f: |x| \leq 4$; $R_f: 0 \leq y \leq 3$ or $-3 \leq y \leq 0$
   (b) $D_f: |x| \leq 5$; $R_f: 0 \leq y \leq 5$ or $-5 \leq y \leq 0$
   (c) $D_f: x \geq 0$; $R_f: y \leq 0$ or $y \geq 0$

5. (a) $0 \leq x \leq 4,\ 0 \leq y \leq 3$ (b) $0 \leq x \leq 5,\ 0 \leq y \leq 5$
   (c) same as 4(c)

   Four choices for (a), four choices for (b), two choices for (c).

6. $y = 2^x$ and $-3 \leq x \leq 3$

7. $y = (\tfrac{1}{2})^x$ and $-3 \leq x \leq 3$

8. 2^{-x} is the reflection of 2^x in the y-axis

9. $m = -1$, k any real number

10. $P(-3, 7)$ (a) $(3, 7)$ (b) $(-3, -7)$
    (c) $(3, -7)$ (d) $(7, -3)$

11. (a) y-axis (b) origin (c) x-axis
    (d) $y = x$

12. Graph of
    (a)
    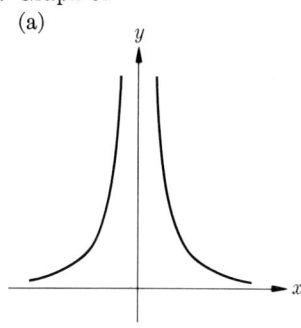

    (b)
    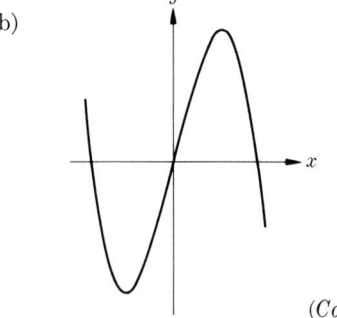

(Cont.)

(c)

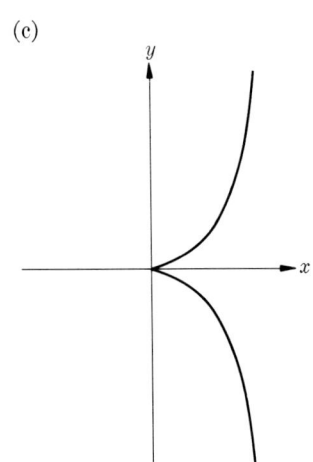

(d) *Note:* $xy(x+y) = 0 \Leftrightarrow x = 0$ or $y = 0$ or $x + y = 0$

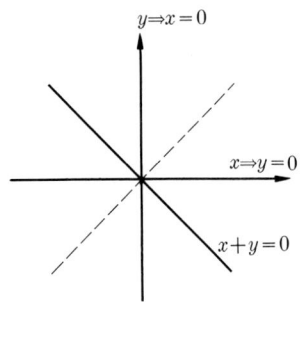

15. $B(2, 9)$ 16. $B(s, r)$ 17. $B(-3, 8)$

EXERCISES 2-1

1. 2   2. 2   3. 0   4. 2

EXERCISES 2-2

1. $L = 2$, N any integer ≥ 101 2. $L = 2$, $N \geq 7$
3. $L = \frac{1}{2}$, $N \geq 8$ 4. $L = \frac{3}{4}$, $N \geq 6$

EXERCISES 2-3

1. N any integer $> 1/\epsilon$ 2. (b) N any integer $> 1/\sqrt{\epsilon}$
3. (a) $\frac{2}{3}$ (b) $\frac{1}{2}$ (c) $\frac{17}{6}$ (d) $\frac{1}{6}$ (e) $\frac{8}{9}$ 4. 0   5. $\frac{1}{2}$ 6. $\sqrt{3}$

EXERCISES 2-4

2. Sequence is monotone increasing and bounded above by 2.
3. $S_n = 1 - 1/(n+1)$, so limit is 1.
4. $T_n < S_n < 1$ and $T_{n+1} > T_n$ 7. Yes; limit is 2.
8. (a) $n^2 + 2n < (n+1)^2$

EXERCISES 3-1

1. $\frac{2}{3}$ 2. 0   3. $\frac{5}{6}$ 4. $\frac{3}{4}$ 5. $\frac{1}{3}$ 6. $\frac{1}{2}$ 7. $\frac{3}{5}$ 8. $\frac{3}{5}$ 9. 0
10. 1   11. $\frac{1}{2}$ 12. 0   13. True for all $\theta \neq 0$.   14. $\frac{2}{5}$ 15. 0
16. 0   17. No limit exists.   18. 0   19. $\sqrt{3/5}$
20. Not defined for $x \geq \frac{1}{5}$; no limit.   21. No limit.   22. 1

ANSWERS TO EXERCISES

EXERCISES 3-2

5. $\dfrac{n}{2}(3n+1)$

6. $\dfrac{n}{3}(n^2+3n+5)$

7. $m_{\text{sec}} = 3$, $m_{\text{tan}} = 3$

8. $m_{\text{sec}} = 3x$, $m_{\text{tan}} = 0$

9. $m_{\text{sec}} = 3(x+x_1)$, $m_{\text{tan}} = 6x_1$

10. $m_{\text{sec}} = -2/x$, $m_{\text{tan}} = -2$

11. $m_{\text{sec}} = -2/xx_1$, $m_{\text{tan}} = -2/x_1^2$

12. $m_{\text{sec}} = 1/(\sqrt{x}+3)$, $m_{\text{tan}} = \tfrac{1}{6}$

13. $m_{\text{sec}} = 1/(\sqrt{x}+\sqrt{x_1})$, $m_{\text{tan}} = 1/2\sqrt{x_1}$

14. $m_{\text{sec}} = 2(x+3)$, $m_{\text{tan}} = 8$

15. $m_{\text{sec}} = a(x+x_1)+b$, $m_{\text{tan}} = 2ax_1+b$

16. $m_{\text{sec}} = x^2+xx_1+x_1^2$, $m_{\text{tan}} = 3x_1^2$

17. $v_{\text{ave}} = t+2$, $v = 4$

18. $v_{\text{ave}} = t+t_1$, $v = 2t_1$

19. $v_{\text{ave}} = a(t+t_1)+b$, $v = 2at_1+b$

20. $v_{\text{ave}} = -1/4(t+1)$, $v = -\tfrac{1}{16}$

21. $v_{\text{ave}} = -1/(t+1)(t_1+1)$, $v = -1/(t_1+1)^2$

22. $v_{\text{ave}} = 2/(\sqrt{2t+1}+\sqrt{2t_1+1})$, $v = 1/\sqrt{2t_1+1}$

EXERCISES 4-1

1. $f(x) = 3x + 2$
$$f'(x_1) = \lim_{x_2 \to x_1} \frac{(3x_2+2)-(3x_1+2)}{x_2-x_1} = \lim_{x_2 \to x_1} \frac{3(x_2-x_1)}{x_2-x_1} = 3$$

2. $6x_1 + 2$ 3. $1/2\sqrt{x_1+1}$ 4. $\tfrac{3}{2}\sqrt{x_1}$ 5. $36x_1^2 - 16x_1 + 21$

6. $4abx_1^3 + 3acx_1^2$ 7. $-8/(x_1-5)^2$ 8. $-(2x_1+3)/(x_1^2+3x_1+5)^2$

9. $-\dfrac{2ax_1+b}{(ax_1^2+bx_1+c)^2}$ 10. $-1/2(x_1+1)^{3/2}$

EXERCISES 4-2

1. 3   2. $6x + 2$

3. Theorem 4–2 does not apply; exponent must be an integer.

4. Same as Exercise 3   5. $36x^2 - 16x + 21$ 6. $4abx^3 + 3acx^2$

7. $-8/(x-5)^2$ 8. $-(2x+3)/(x^2+3x+5)^2$

9. $-(2ax+b)/(ax^2+bx+c)^2$ 10. Same as Exercise 3

12. $18x(3x^2-2)^2$ 13. $6nx(3x^2-2)^{n-1}$

14. $56(2x+5)(x^2+5x+1)^{55}$

15. $2(3x^5-5x^3+2)(15x^4-15x^2)+3(x^3-3x-2)^2(3x^2-3)$

16. $mx^{m-1}(x^m+a)^{n-1}(n+m(x^m+a)^{m-n})$

ANSWERS TO EXERCISES

17. $5x^4 - 12x^2 + 10x$
18. $24x^3 - 3x^2 + 10x + 12$
19. $6x(x^3 - 2)(x^2 + 3)^2(2x^3 + 3x - 2)$
20. $mn(x^m - a)^{n-1}(x^n - a)^{m-1}(2x^{m+n-1} - ax^{m-1} - ax^{n-1})$
21. $-2/(x-2)^2$
22. $18x^2 - 50x - 31$
23. $(x+3)(x-13)/(x-5)^2$
24. $(x+3)^2(x-21)/(x-5)^3$
25. $-3x^3(x+4)/(x^3-2)^2$
26. $-12x(1+x^3)/(x^3-2)^3$
27. $-6x(4x+15)(3x^2+2)^2/(2x^3-5)^3$
28. $-2a/(x-a)^2$

EXERCISES 4–3

1. $f(x) = \cos x$

$$f'(x) = \lim_{\Delta x \to 0} \frac{\cos(x + \Delta x) - \cos x}{\Delta x}$$

$$= \lim_{\Delta x \to 0} \frac{\cos x \cos \Delta x - \sin x \sin \Delta x - \cos x}{\Delta x}$$

$$= \lim_{\Delta x \to 0} \frac{\cos x (\cos \Delta x - 1)}{\Delta x} - \lim_{\Delta x \to 0} \sin x \frac{\sin \Delta x}{\Delta x}$$

$$= -\cos x \lim_{\Delta x \to 0} \frac{(1 - \cos \Delta x)(1 + \cos \Delta x)}{\Delta x (1 + \cos \Delta x)} - \sin x \lim_{\Delta x \to 0} \frac{\sin \Delta x}{\Delta x}$$

$$= -\cos x \lim_{\Delta x \to 0} \frac{\sin \Delta x}{\Delta x} \cdot \lim_{\Delta x \to 0} \frac{\sin \Delta x}{1 + \cos \Delta x} - \sin x \lim_{\Delta x \to 0} \frac{\sin \Delta x}{\Delta x}$$

$$= -\cos x \cdot (1) \cdot \left(\frac{0}{2}\right) - \sin x \cdot (1) = -\sin x$$

$\therefore \cos' x = -\sin x$

2. $f(x) = \sin x$

$$f'(x) = \lim_{\Delta x \to 0} \frac{\sin(x + \Delta x) - \sin x}{\Delta x} = \lim_{\Delta x \to 0} \frac{2 \cos(x + \Delta x/2) \sin \Delta x/2}{\Delta x}$$

$$= \lim_{\Delta x \to 0} \cos(x + \Delta x) \cdot \lim_{\Delta x \to 0} \frac{\sin(\Delta x/2)}{\Delta x/2} = \cos x \cdot 1 = \cos x$$

3. $\tan x = \dfrac{\sin x}{\cos x}$ for $\cos x \neq 0$

$$\tan'(x) = \frac{\cos x (\cos x) - \sin x (-\sin x)}{\cos^2 x} = \frac{\cos^2 x + \sin^2 x}{\cos^2 x} = \sec^2 x$$

4. $\lim\limits_{x \to 0} \sin x = \lim\limits_{\Delta x \to 0} \dfrac{\sin(0 + \Delta x) - \sin 0}{\Delta x} = \lim\limits_{\Delta x \to 0} \dfrac{\sin \Delta x - 0}{\Delta x}$

$$= \lim_{\Delta x \to 0} \frac{\sin \Delta x}{\Delta x} = 1 = \cos 0.$$

5. (a) $f(x) = \sin(x + \pi/2) = \cos x$, $f'(x) = -\sin x$.
   (b) $f'(x) = \cos(x + \pi/2) = -\sin x$
6. $f(x) = \cos(x + \pi/2) = -\sin x$, $f'(x) = -\cos x$

ANSWERS TO EXERCISES

7. $f(x) = \cot x = \dfrac{\cos x}{\sin x}$ for $\sin x \neq 0$

   $f'(x) = \dfrac{-\sin^2 x - \cos^2 x}{\sin^2 x} = -\dfrac{1}{\sin^2 x} = -\csc^2 x$

8. $f(x) = \sec x = 1/\cos x$ for $\cos x \neq 0$

   $f'(x) = \dfrac{0 - 1(-\sin x)}{\cos^2 x} = \dfrac{\sin x}{\cos x} \cdot \dfrac{1}{\cos x} = \tan x \sec x$.

9. $y = |\sin x|$, $-2\pi \leq x \leq 2\pi$; y' fails to exist at all integral multiples of π.
   $y' = \cos x$ $\forall x \neq n\pi$, n any integer. The graph of y' has period π.

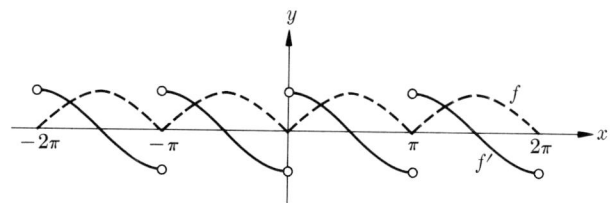

10. $y = \sin|x| = \sin|-x|$
    Curve is symmetric about the y-axis. y' exists $\forall x \neq 0$;
    $y' = -\cos x$ for $x < 0$; $y' = \cos x$ for $x > 0$; y' does not exist for $x = 0$

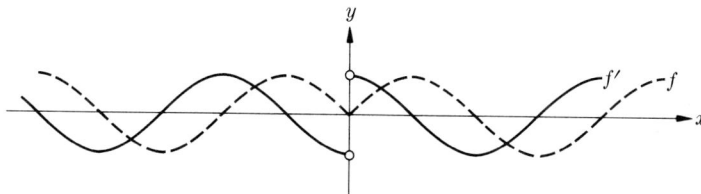

11. Prove: $\dfrac{d(\sin nx)}{dx} = n \cos nx$ and $\dfrac{d(\cos nx)}{dx} = -n \sin nx$

    Proof

    (1) $n = 1$; $\dfrac{d(\sin x)}{dx} = \cos x$ and $\dfrac{d(\cos x)}{dx} = -\sin x$; P_1 holds.

    (2) Assume that P_k is true and show that $P_k \Rightarrow P_{k+1}$.

    $\dfrac{d(\sin(kx))}{dx} = k \cos kx$ and $\dfrac{d(\cos(kx))}{dx} = -k \sin kx$; then

    $\dfrac{d[\sin(k+1)x]}{dx} = \dfrac{d}{dx}[\sin(kx)\cos x + \cos(kx)\sin x]$

    $\qquad = k\cos(kx)\cos x - \sin(kx)\sin x - k\sin(kx)\sin x$
    $\qquad\quad + \cos(kx)\cos x$
    $\qquad = (k+1)\cos(kx)\cos x - (k+1)\sin(kx)\sin x$
    $\qquad = (k+1)[\cos(kx)\cos x - \sin(kx)\sin x]$
    $\qquad = (k+1)\cos(k+1)x$.

    $\therefore P_{k+1}$ holds and the formula is true for all n.

ANSWERS TO EXERCISES

Similarly,

$$\frac{d}{dx}[\cos(k+1)x] = \frac{d}{dx}[\cos(kx)\cos x - \sin(kx)\sin x]$$
$$= -k\sin(kx)\cos x - \cos(kx)\sin x$$
$$\quad - k\cos(kx)\sin x - \sin(kx)\cos x$$
$$= -(k+1)\sin(kx)\cos x - (k+1)\cos(kx)\sin x$$
$$= -(k+1)[\sin(kx)\cos x + \cos(kx)\sin x]$$
$$= -(k+1)\sin(k+1)x. \qquad \text{Q.E.D.}$$

EXERCISES 5-2

1. Absolute max, 0, at $x = 0$; no min
2. Absolute min, -4, at $x = 0$; no max

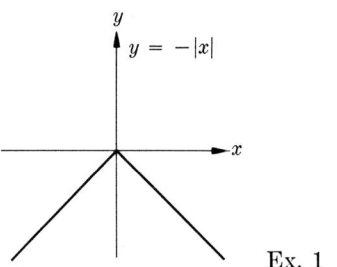

Ex. 1   Ex. 2

3. Local (relative) max, 4, at $x = 0$; no absolute max; absolute min, 0, at $x = -2$ and $x = 2$
4. Absolute min, $-6\frac{1}{4}$, at $x = \frac{1}{2}$; no max

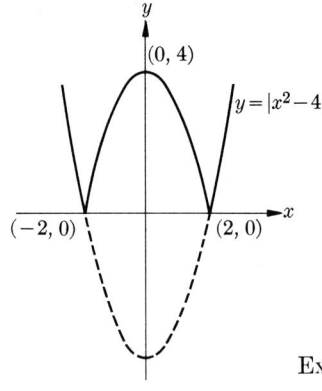
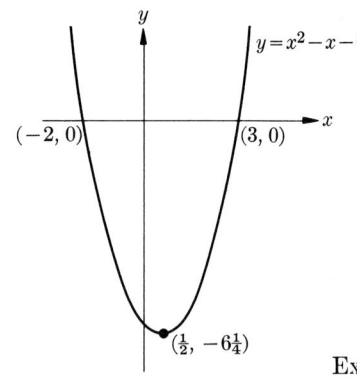

Ex. 3   Ex. 4

5. Absolute min, 0, at $x = -2$ and $x = 3$; relative max, $6\frac{1}{4}$, at $x = \frac{1}{2}$; no absolute max

6. No absolute max; local max, -6, at $x = 0$; absolute min, $-6\frac{1}{4}$, at $x = -\frac{1}{2}$ and $x = \frac{1}{2}$
   (*Note:* $|x|^2 - |x| - 6 = |-x|^2 - |-x| - 6$.)

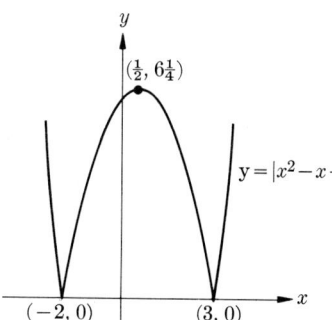

Ex. 5

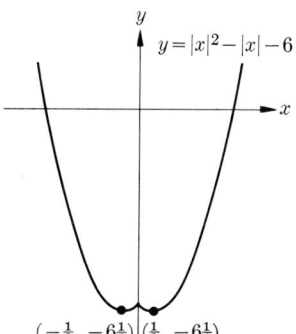

Ex. 6

7. Relative max, 5, at $x = 2$; no absolute max; no absolute min
8. No absolute max; no absolute min
9. Relative min, 3, at $x = 2$; no absolute max; no absolute min

$$y = \frac{x^2 - 4}{x - 2} \quad \text{if} \quad x \neq 2$$

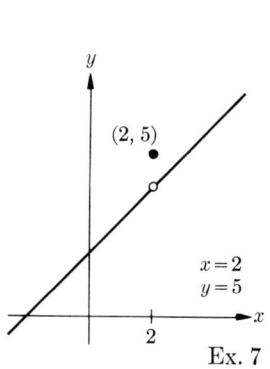

Ex. 7

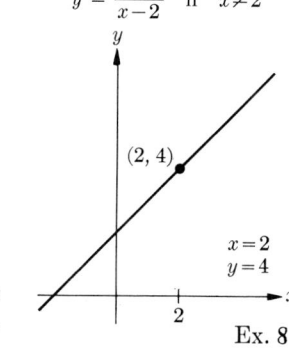

Ex. 8

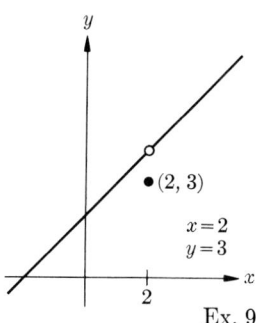

Ex. 9

10. Absolute min, -4, at $x = 0$; no max
11. Absolute max, 4, at $x = 0$; no min

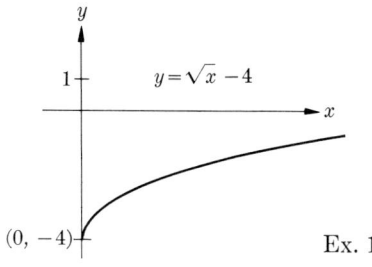

Ex. 10

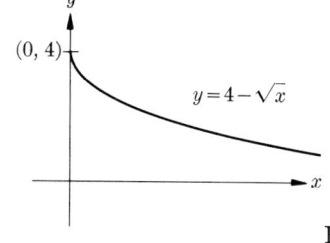

Ex. 11

301

ANSWERS TO EXERCISES

12. Absolute min, 0, at $x = 4$; no max
13. Absolute min, 0, at $x = 4$; no max

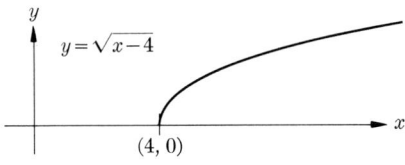
Ex. 12

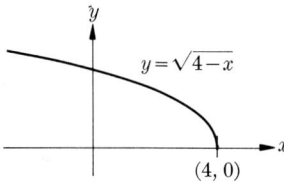
Ex. 13

14. Absolute min, 0, at $x = 0$; no max

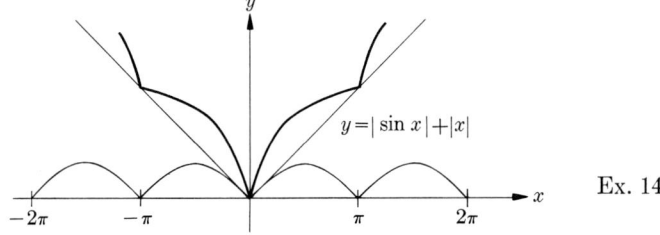
Ex. 14

15. Absolute min, 0, at $x = 0$; no max
16. Absolute max, a, at $x = 0$; absolute min, 0, at $x = a$ and $x = -a$

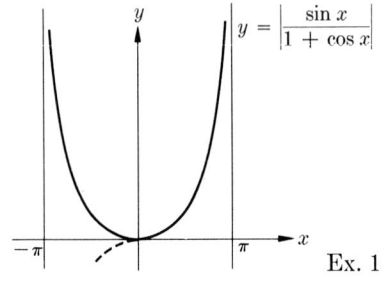
Ex. 15

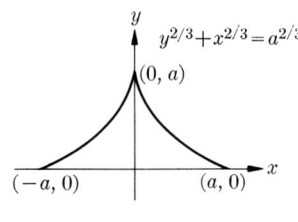
Ex. 16

EXERCISES 5–4

1. (a) Absolute min, -8, at $x = 0$; no max
   (b) Absolute min, -8, at $x = 0$; absolute max, -4, at $x = \pm 2$
   (c) Absolute min, -8, at $x = 0$; no max
   (d) Absolute min, 0, at $x = \pm 2\sqrt{2}$; relative max, 8, at $x = 0$

2. (a) No absolute max; local max, 0, at $x = -1$; local min, $-\frac{32}{27}$, at $x = \frac{1}{3}$; no absolute min
   (b) Absolute max, $6\frac{1}{8}$, at $x = \frac{3}{2}$; local max, 0, at $x = -1$; absolute min, -3, at $x = -2$; local min, $-\frac{32}{27}$, at $x = \frac{1}{3}$
   (c) No absolute max; local max, $\frac{32}{27}$, at $x = \frac{1}{3}$; local min, 1, at $x = 0$; absolute min, 0, at $x = 1$

3. Relative max 0 at $x = 0$, -2 at $x = -1$; relative min 0 at $x = 0$, 2 at $x = 1$; no absolute max or min

4. Absolute min, 1, at $x = 0$; no max

5. Absolute min, 0, at every $x \in I$; $f(x) < 1$; no max

6. Relative max, 4, at $x = 0$; absolute min ≈ 2.53 at $x \approx \pm 1.90$ (use tables)

7. Area is a max where $x = 80$ and is 6400

8. Total length L is a min when $s = (\frac{8}{11}V)^{1/3}$ and $h = (121V)^{1/3}/4$. When $V = 88$, these dimensions are $s = 4$, $h = 5\frac{1}{2}$, and $L = 66$.

9. $s = 2L/33$, $h = L/12$, $V_{max} = L^3/3267$; if $L = 132$, $s = 8$, $h = 11$, $V_{max} = 704$

10. If $s = L/12$, $h = L/12$, and $V = L^3/12^3 = (L/12)^3$; result is a box with equal edges. If $L = 72$, $h = s = 6$, and $V = 216$.

EXERCISES 6–1

1. f is continuous only at $x = 0$. Take $\delta = \epsilon$ when $x = 0$. But for any $x \neq 0$, there are rational and irrational numbers in every neighborhood of x, and for $\epsilon = \frac{1}{2}|x|$ there is no corresponding δ.

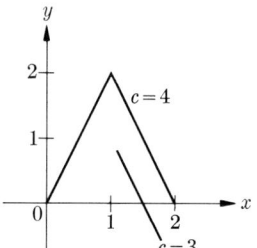

2. Continuous at 1 if $c = 4$.

3. Same graph as for Exercise 2 if $c = 4$.

Ex. 2

4. (a) $\frac{3}{2} < x < \frac{5}{2}$ (b) $2 - c/2 < x < 2 + c/2$, $x \neq 2$
   (c) $-\frac{3}{2}c < x < -\frac{1}{2}c$ (d) $-\frac{3}{2}c \leq x < -c$ or $-c < x \leq -\frac{1}{2}c$

5. (a) $\frac{3}{4} \leq x \leq \frac{5}{4}$

   (b) $f(0.5) = 2$,
   $f(0.6) = 2.2$,
   $f(0.7) = 2.4$,
   $f(0.75) = 2.5$,
   $f(0.8) = 2.6$,
   $f(1.1) = 3.2$,
   $f(1.2) = 3.4$,
   $f(1.25) = 3.5$,
   $f(1.3) = 3.6$,
   $f(1.5) = 4$

   (c) See graph.

   (d) $2 - \epsilon/2 < x < 2 + \epsilon/2$

   (e) $0 < |x - 2| < \epsilon/2$

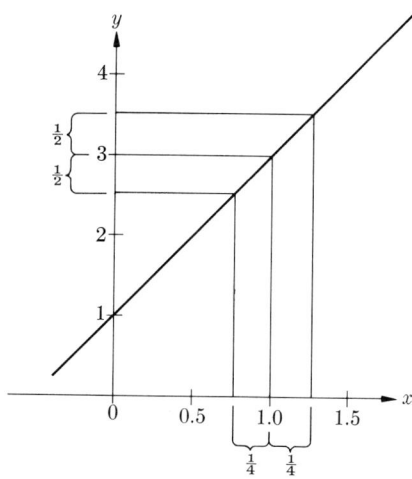

303

ANSWERS TO EXERCISES

6. Formula (8) becomes $q^2 > 1/\epsilon$; the list (9) goes to $[1/\sqrt{\epsilon}]$; no change in (10) if $n = [1/\sqrt{\epsilon}]$. The new function is continuous at every irrational, but not any rational number.

7. (a)

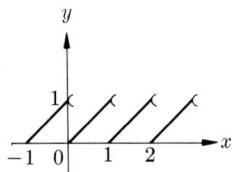

(b)

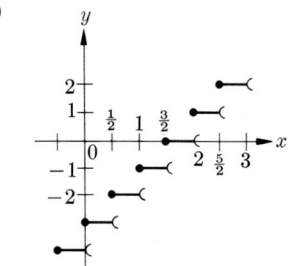

(c)

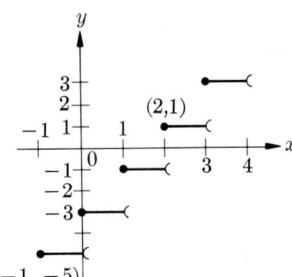

8. Prove by induction on n. Result is true for $n = 1$. If also true for $n = k$, and if S has $k + 1$ elements, let $S = S_1 \cup S_2$, where S_1 has k elements and S_2 has 1 element. By hypothesis S_1 has a maximum, say a, and S_2 has a maximum, say b. Then max $(a, b) = \frac{1}{2}(a + b + |a - b|)$ is the maximum for S.

9. See argument for Ex. 8 above; replace maximum by minimum, and take $a = \min(S_1)$, $b = \min(S_2)$, and $\min(a, b) = \frac{1}{2}(a + b - |a - b|)$ is $\min(S_1 \cup S_2)$.

10. The open interval $0 < x < 1$.

11. Let $h = \min(a, 1 - a)$. Since $0 < a < 1$, both a and $1 - a$ are positive, and $N_h(a)$ is contained in $0 < x < 1$, because $0 \le a - h < a + h \le 1$. In this illustration, a is nearer 1, so $h = 1 - a$ and $a + h = 1$.

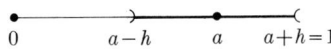

12. Let $a \ne 0$, $\epsilon > 0$. Then

$$\left|\frac{1}{x} - \frac{1}{a}\right| = \frac{|x - a|}{|ax|} < \frac{2|x - a|}{a^2} < \epsilon$$

if

$$|x| > \frac{1}{2}|a| \quad \text{and} \quad |x - a| < \frac{\epsilon a^2}{2}.$$

Take $\delta = \min(\frac{1}{2}|a|, \epsilon a^2/2)$.

ANSWERS TO EXERCISES

13. If f and g are both continuous at a and $\epsilon > 0$, there exist positive numbers δ_1 and δ_2 such that

$$|f(x) - f(a)| < \epsilon/2 \quad \text{if} \quad |x - a| < \delta_1 \quad \text{and} \quad x \text{ is in } D_f,$$

and

$$|g(x) - g(a)| < \epsilon/2 \quad \text{if} \quad |x - a| < \delta_2 \quad \text{and} \quad x \text{ is in } D_g.$$

Take $\delta = \min(\delta_1, \delta_2)$. Note that $D_{f+g} = D_f \cap D_g$.

14. Let f and g both be continuous at a, and let $\epsilon > 0$. There exist positive numbers $\delta_1, \delta_2, \delta_3$ such that

$$|f(x) - f(a)| < 1 \quad \text{if} \quad |x - a| < \delta_1 \quad \text{and} \quad x \text{ is in } D_f,$$

$$|g(x) - g(a)| < \frac{\epsilon}{2(1 + |f(a)|)} \quad \text{if} \quad |x - a| < \delta_2 \quad \text{and} \quad x \text{ is in } D_g,$$

$$|f(x) - f(a)| < \frac{\epsilon}{2(1 + |g(a)|)} \quad \text{if} \quad |x - a| < \delta_3 \quad \text{and} \quad x \text{ is in } D_f.$$

Let $\delta = \min(\delta_1, \delta_2, \delta_3)$ and suppose $|x - a| < \delta$ and x is in $D_f \cap D_g$. Then

$$|f(x)g(x) - f(a)g(a)| = |f(x)g(x) - f(x)g(a) + f(x)g(a) - f(a)g(a)|$$

$$\leq |f(x)| |g(x) - g(a)| + |g(a)| |f(x) - f(a)| < \frac{\epsilon}{2} + \frac{\epsilon}{2} = \epsilon,$$

because $|f(x)| < (1 + |f(a)|)$ and $|g(a)|/(1 + |g(a)|) < 1$.

15. Let f be continuous at a, $f(a) \neq 0$, and $\epsilon > 0$. Then

$$\left|\frac{1}{f(x)} - \frac{1}{f(a)}\right| = \left|\frac{f(x) - f(a)}{f(x)f(a)}\right| \leq \frac{2}{|f(a)|^2} |f(x) - f(a)| < \epsilon$$

provided $|f(x)| \geq \frac{1}{2}|f(a)|$ and $|f(x) - f(a)| < \frac{1}{2}\epsilon|f(a)|^2$. By hypothesis, there exist positive numbers δ_1, δ_2 such that

$$|f(x) - f(a)| < \tfrac{1}{2}|f(a)| \quad \text{if} \quad |x - a| < \delta_1 \quad \text{and} \quad x \text{ is in } D_f,$$

and

$$|f(x) - f(a)| < \tfrac{1}{2}\epsilon|f(a)|^2 \quad \text{if} \quad |x - a| < \delta_2 \quad \text{and} \quad x \text{ is in } D_f.$$

Now take $\delta = \min(\delta_1, \delta_2)$.

EXERCISES 6-2

1. (a) Let

$$f(x) = \begin{cases} 1 - 2x & \text{for } 0 \leq x < \tfrac{1}{2}, \\ \tfrac{1}{2} - 2x & \text{for } \tfrac{1}{2} \leq x \leq 1. \end{cases}$$

(b) Let $f(x) = 1/(1 - 2x)$ for $x \neq \tfrac{1}{2}$ and $f(\tfrac{1}{2}) = 2$.

ANSWERS TO EXERCISES

2. (a) $f(x) = \sin x$, max $= 1$, min $= -1$
   (b) $f(x) = 1/(1 + x^2)$
   (c) $f(x) = 1$ when x is irrational
       $= 0$ when x is rational
       max $= 1$, min $= 0$
   (d) $f(x) = 1 + x^2$ when x is irrational
       $= -x^2$ when x is rational
   Graph consists of points on parabola $y = 1 + x^2$ for x rational and points on parabola $y = -x^2$ for x rational.

EXERCISES 6-3

1. (a) R: $[-\frac{2}{9}\sqrt{3}, \frac{2}{9}\sqrt{3}]$ (b) max $\frac{2}{9}\sqrt{3}$ (c) min $-\frac{2}{9}\sqrt{3}$
2. (a) R: $[0, 2]$ (b) max 2 (c) min 0

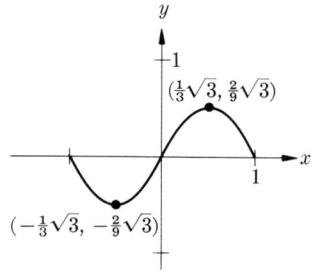
Ex. 1

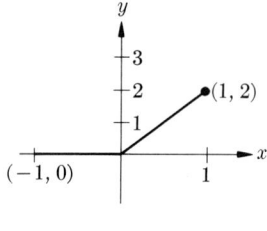
Ex. 2

3. (a) R: $[-4, 5]$ (b) max 5 (c) min -4
4. (a) R: $[0, 4]$ (b) max 4 (c) min 0

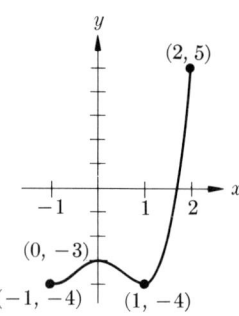
Ex. 3

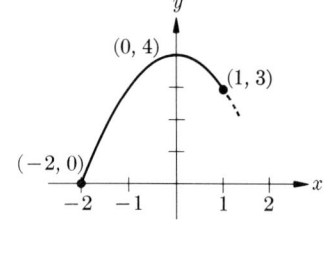
Ex. 4

5. (a) R: $[-\frac{1}{2}, \frac{1}{2}]$ (b) max $\frac{1}{2}$ (c) min $-\frac{1}{2}$

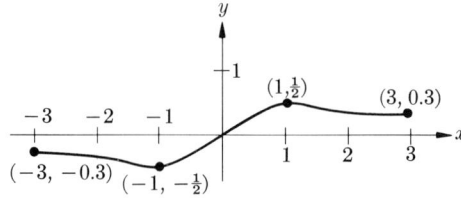
Ex. 5

ANSWERS TO EXERCISES

6. (a) R: $[-1, 5]$ (b) max 5 (c) min -1
7. (a) R: $[-1 - \pi/2, 2\pi]$ (b) max 2π (c) min $-1 - \pi/2$

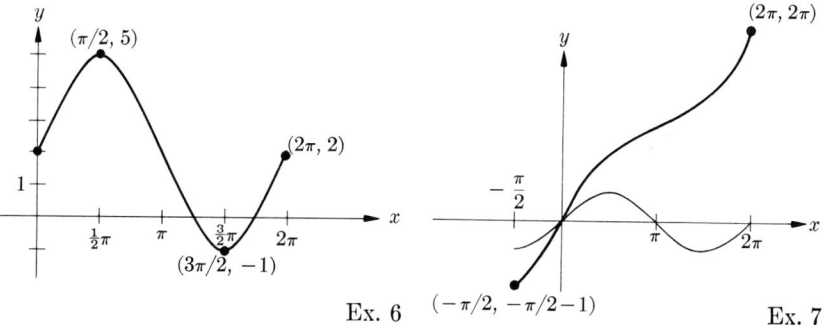

Ex. 6   Ex. 7

8. (a) R: $[-1, \sqrt{2}]$ (b) max $\sqrt{2}$ (c) min -1
9. (a) R: $[-1, 1]$ (b) max 1 (c) min -1
10. (a) R: $[-4\frac{1}{4}, 4\frac{1}{4}]$ (b) max $4\frac{1}{4}$ (c) min $-4\frac{1}{4}$

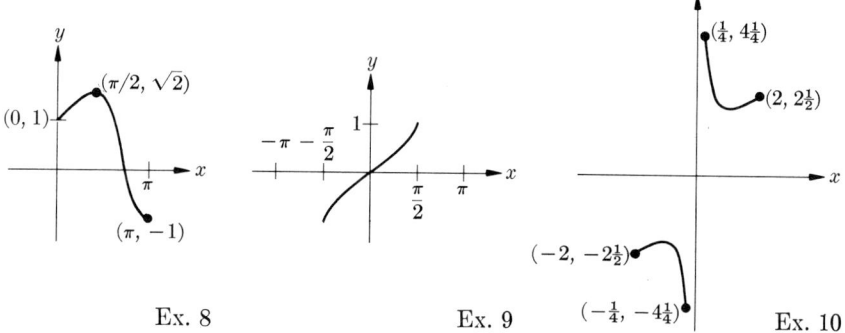

Ex. 8   Ex. 9   Ex. 10

EXERCISES 6-4

1. $n = 888$.
2. $h = 0.02$. Neighborhoods of radius h centered at $0, h, 2h, \ldots, nh = 5$ will be such a finite covering if $n = 250$.

EXERCISES 7-1

1. (a) See answer to Ex. 12 of Section 6-1.
   (b) Indirect proof. Suppose f is uniformly continuous on D. Take $\epsilon = 1$. By hypothesis, there exists $\delta > 0$ such that $|f(x_2) - f(x_1)| < 1$ whenever $|x_2 - x_1| < \delta$ and x_1, x_2 are in D. A finite number, say N, of open intervals of length δ, centered at $\delta/2, \delta, 3\delta/2, \ldots, N\delta/2$ covers D. This implies that the range of f is bounded with

$$|f(x)| < 1 + \max\{f(\delta/2), f(\delta), \ldots, f(N\delta/2)\} = 1 + (2/\delta).$$

ANSWERS TO EXERCISES

But this is false because $f(x) = 1/x$ on the domain $0 < x < 1$, and the range is not bounded. Hence f is not uniformly continuous on D.

2. (a) and (b). Let $\epsilon > 0$, $x_1 \geq 1$, $x_2 \geq 1$. Then $|x_1 x_2| \geq 1$ and
$$\left|\frac{1}{x_1} - \frac{1}{x_2}\right| = \frac{|x_2 - x_1|}{|x_1 x_2|} \leq |x_2 - x_1|.$$
Take $\delta = \epsilon$.

3. (a) First, establish the inequality $|\sin \theta| < |\theta|$ when $\theta \neq 0$. For $0 < \theta < \pi/2$, consider a circular sector AOP of radius 1 and central angle θ radians. The area of the inscribed triangle AOP is $\frac{1}{2} \sin \theta$, while the area of the sector is $\frac{1}{2}\theta$. Hence $\sin \theta < \theta$ for $0 < \theta < \pi/2$. For negative values of θ, between $-\pi/2$ and 0, let $\alpha = -\theta$. Then $\sin \theta = -\sin \alpha$ and
$$\left|\frac{\sin \theta}{\theta}\right| = \frac{\sin \alpha}{\alpha} < 1,$$
so that
$$|\sin \theta| < |\theta|.$$

The inequality is thus established for $0 < |\theta| < \pi/2$. For $|\theta| \geq \pi/2$ there is nothing to prove, because $|\sin \theta| \leq 1 < \pi/2$.

Now take absolute values on both sides of the equation
$$\sin x_1 - \sin x_2 = 2 \cos\left(\frac{x_1 + x_2}{2}\right) \sin\left(\frac{x_1 - x_2}{2}\right),$$
and recall that the absolute value of a product is the product of the absolute values of the factors, and $|\cos \theta| \leq 1$ for any θ. Therefore,
$$|\sin x_1 - \sin x_2| = 2|\cos (x_1 + x_2)/2| \, |\sin (x_1 - x_2)/2|$$
$$\leq 2(1)|(x_1 - x_2)/2| = |x_1 - x_2|.$$

(b) If $\epsilon > 0$, take $\delta = \epsilon$ and apply part (a):
$$|\sin x_1 - \sin x_2| \leq |x_1 - x_2| < \epsilon \quad \text{if} \quad |x_1 - x_2| < \delta = \epsilon.$$

4. Take $\epsilon = 1$. Since f is uniformly continuous on $a < x < b$, there exists $\delta > 0$ such that $|f(x_1) - f(x_2)| < \epsilon$ when $|x_1 - x_2| < \delta$ and $a < x_1$, $x_2 < b$. The open interval $a < x < b$ can be covered by a finite number of intervals $N_\delta(c_i)$, $i = 1, 2, \ldots, N$, $a < c_1 < c_2 < \cdots < c_N < b$. Then for any x, $a < x < b$, there is some c_i such that $|x - c_i| < \delta$ and $|f(x) - f(c_i)| < 1$. Therefore $|f(x)| < 1 + \max\{|f(c_i)| : i = 1, 2, \ldots, N\}$.

5. $\delta = \epsilon/6$ 6. $\delta = \epsilon/12$ 7. $\delta = \epsilon/2$ 8. $\delta = \epsilon/3$ 9. $\delta = \epsilon/3$

10. $\delta = \epsilon^2$. [Hint: First show that if $0 \leq x_1 \leq x_2$, then $0 \leq \sqrt{x_2} - \sqrt{x_1} \leq \sqrt{x_2 - x_1}$.]

EXERCISES 7-2

1. $x_1 = a$, x_2 is abscissa of E, x_3 of A, $x_4 = b$
2. $x_1 = \text{lub }(S)$ 3. $f(x_2) = c$
4. (i) The a's form a monotone increasing sequence that is bounded above; therefore $\lim_{n \to \infty} a_n$ exists.
   (ii) The b's form a monotone decreasing sequence that is bounded below.
   (iii) $\lim (a_n - b_n) = \lim a_n - \lim b_n$
   (iv) Because f is continuous at x and $\lim a_n = \lim b_n = x$, $f(x) = \lim f(a_n) = \lim f(b_n)$. Now, either $f(a_n) \leq 0 \leq f(b_n)$ or $f(b_n) \leq 0 \leq f(a_n)$. Taking limits as $n \to \infty$, we get in either case $f(x) \leq 0 \leq f(x)$. Therefore $f(x) = 0$.

EXERCISES 8-1

1. $f_1(x) = 2\sqrt{x} + 1, x \geq 0$ $f_2(x) = \sqrt{2x + 1}, x \geq -\frac{1}{2}$
2. $f_1(x) = 1/(1+x), x \neq 0, x \neq -1$ $f_2(x) = 1 + (1/x), x \neq 0, x \neq -1$
3. $f_1(x) = \sqrt{(1/x) - 1}, 0 < x \leq 1$ $f_2(x) = 1/\sqrt{x-1}, x > 1$

EXERCISES 8-2

1. $f(x) = 17 - 8x, \delta = \epsilon/8$ 2. $f(x) = 3 \sin x, \delta = \epsilon/3$
3. $f(x) = x, x \neq 0, \delta = \epsilon$ 4. $f(x) = \sin^2 x, \delta = \epsilon/2$
5. $f(x) = \sin(x^2), \delta = \min\{1, \epsilon/(1 + 2|c|)\}$

EXERCISES 8-3

1. $f^{-1}(x) = (x+3)/2, -\infty < x < \infty$
2. $f^{-1}(x) = (x+3)/2, -5 \leq x \leq -1$
3. $f^{-1}(x) = \sqrt{x}, 0 \leq x \leq 4$ 4. $f^{-1}(x) = (1-x)/x, x \neq 0$
5. $f^{-1}(x) = 1/x^2, x > 0$ 6. $f^{-1}(x) = \log 2/\log x = \log_x 2, x > 1$
7. $f^{-1}(x) = -1 - \sqrt{x+4}, x \geq -4$
8. No, because $f(x) = (x+1)^2 - 4$ is not one-to-one from the domain $-3 \leq x \leq 1$ to the range $-4 \leq y \leq 0$. There are two values of x for each $y = x^2 + 2x - 3$ if $-4 < y \leq 0$.
9. $f^{-1}(x) = x/(1-x), x \neq 1$, is a function.
10. f^{-1} is not a function.   11. f^{-1} is not a function.   12. f^{-1} is a function.
13. f^{-1} is a function.   14. f^{-1} is a function.

EXERCISES 8-4

1. (a) If $d = f(a)$, then $g(d) = a$. Let $\epsilon > 0$, $\epsilon_2 = b - a$, and $\epsilon_0 = \min(\epsilon, \epsilon_2)$. There exists a unique x_2 such that $x_2 = f(a + \epsilon_0)$ and $\delta = x_2 - d > 0$. If $d \leq x < d + \delta = x_2$, then $g(d) = a \leq g(x) < g(x_2) = a + \epsilon_0 \leq a + \epsilon$. So $|g(x) - g(d)| < \epsilon$.

ANSWERS TO EXERCISES

(b) If $d = f(b)$, then $g(d) = b$. Let $\epsilon > 0$, $\epsilon_1 = b - a$, and $\epsilon_0 = \min(\epsilon, \epsilon_1)$. There exists a unique x_1 such that $x_1 = f(b - \epsilon_0)$ and $\delta = d - x_1 > 0$. If $x_1 = d - \delta < x \leq d$, then $b - \epsilon_0 = g(x_1) < g(x) \leq g(d) = b$ so that $|g(x) - g(d)| < \epsilon_0 \leq \epsilon$.

2. Reverse the inequality signs in (7) but not in (8). Change (9) to $\delta = \min(x_1 - d, d - x_2)$.

3. Since f is one-to-one and $a < b$, we know that $f(b) \neq f(a)$. Hence there are just two cases to consider: $f(b) > f(a)$ or $f(b) < f(a)$. We consider only the first case in detail. (The second case is handled similarly.) Thus suppose that $f(b) > f(a)$. We show that this implies that f is strictly increasing. In the first place, if $a < x < b$, then $f(a) < f(x) < f(b)$, because otherwise, either:

(a) $f(x) < f(a) < f(b)$ and the intermediate-value theorem assures the existence of x', $x < x' < b$, such that $f(x') = f(a)$, contrary to the fact that f is one-to-one; or else

(b) $f(x) > f(b) > f(a)$ so that there is x'', $a < x'' < x$, such that $f(x'') = f(b)$, and this, too, is false.

Finally, if $a \leq x_1 < x_2 \leq b$, the same type of argument, with x_1 in place of a and x_2 in place of x, leads to the conclusion $f(x_1) < f(x_2) \leq f(b)$. Hence in case $f(b) > f(a)$, f is strictly increasing. In the opposite case, f is strictly decreasing.

4. (a) $c = 2$, $\epsilon_0 = 0.2$, $x_1 = 3.24$, $x_2 = 4.84$, $\delta = 0.76$
   (b) $c = \frac{4}{3}$, $\epsilon_0 = \frac{1}{3}$, $x_1 = 1$, $x_2 = \frac{3}{5}$, $\delta = 0.15$

EXERCISES 9-1

1. (a) discontinuous at $x = 0$ (b) continuous $\forall x \in R$
   (c) discontinuous at every $x \in I$
       continuous on $[n, n+1)$ for $n \in I$ and $-5 \leq n \leq 5$
   (d) continuous   (e) continuous
   (f) continuous on the given interval, except at $\pm\pi/2$, $\pm 3\pi/2$

2. (a) $x = 0$ (b) $x = 1$ (c) $\forall x \in I$ (d) $x = \pm a$
   (e) at $x = -2\pi, -3\pi/2, -\pi/2, \pi/2, 3\pi/2, 2\pi$ (f) same as (e)

3. Yes; $f'(0) = \lim_{\Delta x \to 0} \dfrac{f(0 + \Delta x) - f(0)}{\Delta x} = \lim_{\Delta x \to 0} \dfrac{\Delta x |\Delta x| - 0}{\Delta x} = \lim_{\Delta x \to 0} |\Delta x| = 0$

4. No; $\lim_{\Delta x \to 0^-} \dfrac{|\Delta x|}{\Delta x} = -1$; $\lim_{\Delta x \to 0^+} \dfrac{|\Delta x|}{\Delta x} = 1$

5. Graph of Exercise 3: $f(x) = x^2, x \geq 0$; $f(x) = -x^2, x < 0$
   Graph of Exercise 4: $f(x) = 1, x > 0$; $f(x) = 0, x = 0$; $f(x) = -1, x < 0$

Graph of Ex. 3

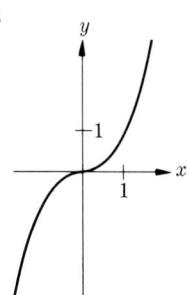

Graph of Ex. 4

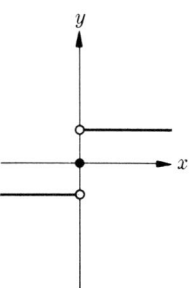

EXERCISES 9-2

1. Rolle's theorem applies: f is continuous and differentiable on $(-\infty, \infty)$; $f(-2) = f(0) = f(3) = 0$. Therefore there exist two real numbers, c_1 and c_2, such that $-2 < c_1 < 0 < c_2 < 3$ and $f'(c_1) = f'(c_2) = 0$; $c_1 \approx -1.12$, $c_2 \approx 1.78$.

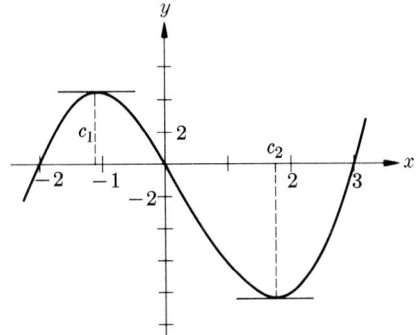

2. (a) $m_L = 1$
   (b) points on graph of f where slope is 1: $\approx (-1.15, 2.92)$ and $(1.15, -.92)$
   (c) Yes
   (d) The range of f for $|x| \leq 2$ is $[-1, 3]$; therefore 4 and 3.2 are excluded.

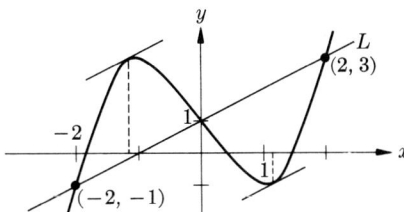

3. No; yes   4. Yes; yes   5. No; no; no   6. No; yes   7. Yes

8. Show: $\sin x < x$ for $x > 0$. The inequality is obviously true if $x > 1$. Thus we need only consider $0 < x \leq 1$. Sin x is continuous and differentiable for all real x. By MVT, taking $a = 0$, $b = x$,

$$\sin x - \sin 0 = (x - 0) \cos c \quad \text{for some } c \text{ in } (0, x).$$

Furthermore, $0 < \cos c < 1$ because $0 < c < 1 < \pi/2$.

$$\therefore \sin x < x.$$

9. Show that $h/(1+h) < \ln(1+h) < h$, $h > -1$ and $h \neq 0$. Let $f(x) = \ln x$, $a = 1$, $b = 1+h$ and apply MVT:

$$\therefore \ln(1+h) - \ln 1 = h\left(\frac{1}{1+\theta h}\right),$$

where $f'(x) = 1/x$ and $f'(c) = 1/(1+\theta h)$, $1 < 1+\theta h < 1+h$, $0 < \theta < 1$.

$$\therefore \ln(1+h) = \frac{h}{1+\theta h}.$$

(i) If $-1 < h < 0$, then $0 < 1+h < 1+\theta h < 1$ and

$$\frac{1}{1+h} > \frac{1}{1+\theta h} > 1, \quad \frac{h}{1+h} < \frac{h}{1+\theta h} < h$$

or

$$\frac{h}{1+h} < \ln(1+h) < h.$$

(ii) If $h > 0$, then $1 < 1+\theta h < 1+h$ and

$$\frac{1}{h} < \frac{1+\theta h}{h} < \frac{1+h}{h}, \quad h > \frac{h}{1+\theta h} > \frac{h}{1+h}$$

or

$$\frac{h}{1+h} < \ln(1+h) < h.$$

10. (i) g linear function, with $g(a) = f(a)$ and $g(b) = f(b)$; $\therefore$ slope of g,

$$g'(x) = \frac{f(b) - f(a)}{b - a}$$

for all real x.

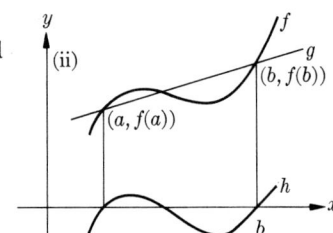

(ii) $h(x) = f(x) - g(x)$

(iii) h is the difference of two functions, each continuous on $[a, b]$ and differentiable on (a, b), and is therefore itself continuous and differentiable on the same domain.

$$h(a) = f(a) - g(a) = 0,$$
$$h(b) = f(b) - g(b) = 0;$$

hence the hypotheses for Rolle's theorem are satisfied.

(iv) There exists a real number c, such that $a < c < b$ for which $h'(c) = 0$:
$$0 = h'(c) = f'(c) - g'(x),$$
since $g'(x) = g'(c)$ by (i), or
$$0 = f'(c) - \frac{f(b) - f(a)}{b - a} \Leftrightarrow f(b) - f(a) = (b - a)f'(c).$$

11. Prove: $e^x > 1 + x$, $x \neq 0$. Let $f(x) = e^x$, $a = 0$, $b = x$. By MVT,
$$\frac{e^x - e^0}{x - 0} = f'(c) \quad \text{for some real number } c,$$
where $0 < c < x$ or $x < c < 0$, or
$$e^x = e^c \cdot x + 1.$$

(i) If $0 < c < x$, $e^c > 1$ and $xe^c > x \Rightarrow xe^c + 1 > 1 + x$. $\therefore e^x > 1 + x$

(ii) If $x < c < 0$, $e^c < 1$, $xe^c > x$, $xe^c + 1 > x + 1$. $\therefore e^x > x + 1$.

12. (a) Sin θ is continuous and differentiable $\forall \theta$. By MVT, with $f(x) = \sin x$,
$$\sin x_1 - \sin x_2 = (x_1 - x_2) \cos x_0,$$
for some x_0, $\exists x_1 < x_0 < x_2$,
$$|\sin x_1 - \sin x_2| = |x_1 - x_2| |\cos x_0|, \quad |\cos x_0| \leq 1,$$
$$|x_1 - x_2| |\cos x_0| \leq |x_1 - x_2|,$$
and
$$|\sin x_1 - \sin x_2| \leq |x_1 - x_2|.$$

(b) Tan θ is continuous $\forall \theta$, $\exists \, 0 \leq \theta < \pi/2$ and differentiable for $0 < \theta < \pi/2$;
$$\frac{d(\tan \theta)}{d\theta} = \sec^2 \theta \geq 1 \quad \text{for} \quad 0 < \theta < \frac{\pi}{2}.$$
By MVT,
$$\tan x - \tan 0 = (x - 0) \sec^2 x_0,$$
for some x_0, $\exists \, 0 < x_0 < x < \pi/2$,
$$\tan x = x \sec^2 x_0;$$
and since $x \sec^2 x_0 > x$,
$$\tan x > x.$$

13. $y = 0$, for $x > 0$, when $\sin(1/x) = 0$; or $x < 0$; or $1/x = n\pi$, $x = 1/n\pi$, n any nonzero integer.
For $|n| \leq 4$, $x = \pm 1/\pi \approx \pm 0.32$; $\pm 1/2\pi \approx \pm 0.16$; $\pm 1/3\pi \approx \pm 0.11$; $\pm 1/4\pi \approx \pm 0.08$.

EXERCISES 9-3

1. (i) Let $f: x \to x^{5/3}$

   (ii) $f(x) = \begin{cases} x^2 \sin(1/x) & \text{if } x \neq 0 \\ 0 & \text{if } x = 0 \end{cases}$

2. (a) $A + B = 3$, A arbitrary (b) $A = 1$, $B = 2$; no

3. (a) $f'(x) = \frac{3}{2} x^{1/2} \sin(1/x) - x^{-1/2} \cos(1/x)$ for $x \neq 0$

   $f'(0) = \lim_{\Delta x \to 0} \frac{(0 + \Delta x)^{3/2} \sin(1/(0 + \Delta x)) - 0}{\Delta x}$

   $= \lim_{\Delta x \to 0} [(\Delta x)^{1/2} \sin(1/\Delta x)] = 0,$

   $f'(0)$ exists and is 0

   (b) not defined for $x = 0$

4. (a) $c = 1$ (b) $c = 1$ (c) $c = \cos^{-1}(3/2\pi)$ (d) $c = (x_1 + x_2)/2$

5. (a) $n = 1$: $p(x) = 1 + \frac{1}{2}(x - 1) = \frac{1}{2}x + \frac{1}{2}$

   (b) $n = 2$: $p(x) = 1 + \frac{1}{2}(x - 1) - \frac{1}{8}(x - 1)^2 = -\frac{1}{8}(x - 3)^2 + \frac{3}{2}$

   (c) $n = 3$: $p(x) = 1 + \frac{1}{2}(x - 1) - \frac{1}{8}(x - 1)^2 + \frac{1}{16}(x - 1)^3$

   $= \frac{1}{16}(x^3 - 5x^2 + 15x + 5)$

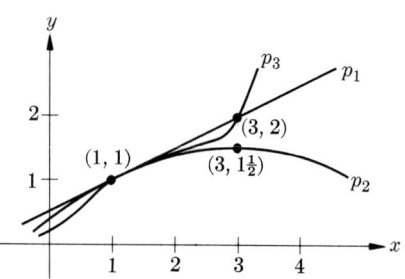

All polynomials agree at $x = 1$ and fit closest at and near $x = 1$.

6. Proof of Theorem 9-6 (hypotheses for MVT hold):
   (a) $\exists$ a real number c, $\exists a < c < b$ and $f(x) - f(a) = f'(c)(x - a)$; but $f'(x) = 0 \; \forall x$ in (a, b), $\therefore f'(c) = 0$ and $f(x) - f(a) = 0$ leads to $f(x) = f(a)$.
   (b) Proof of Theorem 9-7. Let $f(x) = F_1(x) - F_2(x)$; $f(x)$ is continuous on $[a, b]$ and differentiable on (a, b), since $f(x)$ is the difference of two functions, both continuous and differentiable on these intervals. $f'(x) = F_1'(x) - F_2'(x) = 0 \; \forall x$ in (a, b) by hypothesis. Now apply Theorem 9-6: $f(x) = f(a) = C$, some real number. $\therefore F_1(x) - F_2(x) = C$ or $F_1(x) = F_2(x) + C \; \forall x$ in $[a, b]$.
   (c) Proof of Theorem 9-8. By MVT, $f(x_2) - f(x_1) = (x_2 - x_1) f'(c)$ for some $c \ni a \leq x_1 < c < x_2 \leq b$. $x_2 - x_1 > 0$ and $f'(x) > 0 \; \forall x$ in (a, b), given. $\therefore f(x_2) - f(x_1) > 0 \Leftrightarrow f(x_2) > f(x_1)$.

ANSWERS TO EXERCISES

EXERCISES 10-1

1. $\alpha = 2\,\Delta x, \quad \Delta y_{\tan} = 4a\,\Delta x + 2(\Delta x)^2$
2. $\alpha = (3a + \Delta x)\,\Delta x, \quad \Delta y_{\tan} = 3a^2\,\Delta x + 3a\,\Delta x^2 + (\Delta x)^3$
3. $\alpha = -\dfrac{\Delta x}{2\sqrt{a}(\sqrt{a+\Delta x}+\sqrt{a})^2}, \quad \Delta y_{\tan} = \dfrac{\Delta x}{2\sqrt{a}} - \dfrac{(\Delta x)^2}{2\sqrt{a}(\sqrt{a+\Delta x}+\sqrt{a})^2}$
4. $\alpha = \dfrac{(3a+\Delta x)\,\Delta x}{a^3 + (a+\Delta x)^2}, \quad \Delta y_{\tan} = -\dfrac{2}{a^3}\,\Delta x + \dfrac{(3a+\Delta x)\,\Delta x^2}{a^3(a+\Delta x)^2}$
5. $\Delta y = f(x) - f(a) = f'(a)\,\Delta x + \dfrac{f''(c)}{2!}(\Delta x)^2$

   $\Delta y = \cos a \cdot \Delta x - \dfrac{\sin c}{2!}(\Delta x)^2, \quad \text{where } a < c < a + \Delta x.$

EXERCISES 10-2

1. $20(2x+3)^9$ 2. $2 \sin x \cos x = \sin 2x$ 3. $-3x^2 \sin(x^3)$
4. $2(\cos x + \sin x)(\cos x - \sin x) = 2 \cos 2x$
5. $3(5x^2 + 7x - 3)^2 (10x + 7)$ 6. $2(x - \sin x)(1 - \cos x)$
7. $2 \sin 4x$ 8. $2 \sin x \cdot \sin(2 \cos x)$ 9. $\frac{1}{2} \sin x$
10. $\frac{3}{4} \sec^2[(3x+5)/4]$ 11. $\cos x \cdot \cos(\sin x)$ 12. $\frac{2}{5} \sec \frac{2}{5}x \tan \frac{2}{5}x$
13. $-6x \cos^2(x^2+1) \sin^2(x^2+1)$
14. $2x \cos(x+2)(\cos x^2) - (\sin x^2)[\csc^2(x+2)]$
15. $-\dfrac{12}{x^3}\left(\sin \dfrac{2}{x^2}\right)^2 \left(\cos \dfrac{2}{x^2}\right).$

EXERCISES 10-3

1. f increasing on $(-\infty, \infty)$
2. f decreasing on $(-\infty, \infty)$
3. f increasing on $(-\infty, \infty)$
4. f increasing on $[0, \infty)$, decreasing on $(-\infty, 0]$.
5. f increasing on $(-\infty, \infty)$
6. f increasing for $x < 0$, decreasing for $x > 0$
7. on $(-\infty, -1]$ and $[1, \infty)$ f is decreasing; on $[-1, 1]$ f is increasing
8. f is decreasing on $(-\infty, 0]$; f is increasing on $[0, \infty)$
9. f same as Exercise 7
10. f is increasing on $[-2, 3]$; f is decreasing on $[3, 8]$
11. $\frac{1}{3}$ 12. $-\frac{1}{2}$ 13. $1/5x^{4/5}$
14. $1/2\sqrt{x-5}$; restrict D_f to $x \leq 0$ or $x \geq 0$; then f^{-1} is defined for $x \geq 5$.

ANSWERS TO EXERCISES

15. $(f^{-1})' = \frac{1}{6}\left(\frac{x-3}{2}\right)^{-2/3}$

16. $-\frac{1}{2x^2}\sqrt{\frac{x}{1-x}}$;

    restrict D_f to $x < 0$ or $x > 0$; then f^{-1} is defined for $0 < x \le 1$.

17. $\frac{dx}{dy} = \frac{1+x^2}{1-2xy}$;

    restrict D_f to one of three intervals: $-\infty < x \le 1$, $-1 \le x \le 1$, $1 \le x < \infty$.

18. $(f^{-1})' = \frac{3}{2}x^{1/2}$; restrict D_f to $x \ge 0$.

19. At $(2, \frac{2}{5})$, $dy/dx = -\frac{3}{25}$; if D_f is $|x| \le 1$, $-1 \le x \le 1$, or $|x| \ge 1$, then f^{-1} exists and

    $f^{-1}: x \to \dfrac{1+\sqrt{1-4x^2}}{2x}$ or $\dfrac{1-\sqrt{1-4x^2}}{2x}$, $f^{-1}(\tfrac{2}{5}) = -\dfrac{25}{3}$.

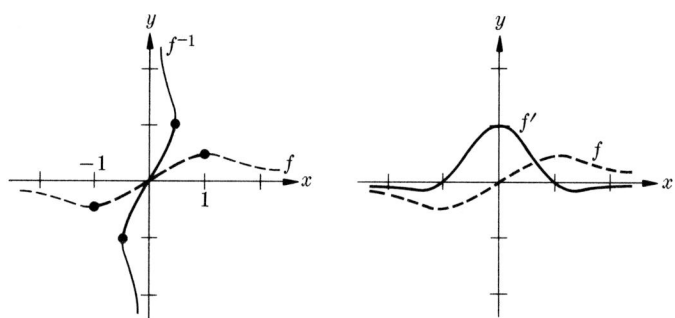

Graph of f and f^{-1} for $(-\infty, -1]$, $[-1, 1]$, $[1, \infty)$

Graph of f and f'

20. $dy/dx = (3-x)/(y-1)$; at $(6, 5)$, $y' = -\frac{3}{4}$; at $(-1, 4)$, $y' = \frac{4}{3}$.

EXERCISES 10-4

1. True for $(1, 1)$, $(-1, 2)$, $(-1, -1)$, $(0, \pm\sqrt{3})$, $(\pm\sqrt{3}, 0)$, etc., and x and y not both zero;

    $\dfrac{dy}{dx} = -\dfrac{2x+y}{x+2y}$; $\left.\dfrac{dy}{dx}\right|_{(1,1)} = -1$

2. x and y cannot both be zero; true for $(2, -4+\sqrt{22})$, $(\pm\sqrt{10}, 0)$, $(0, \pm\sqrt{10})$, etc.;

    $\dfrac{dy}{dx} = -\dfrac{x(2+3xy)}{x^3+2y}$; $\left.\dfrac{dy}{dx}\right|_{(2,-4+\sqrt{22})} = \sqrt{22}-6$

3. $x \neq 0$ and $y \neq 0$; true for $(\pi/2, \pi/2)$, $(3\pi/2, -3\pi/2)$;

$$\frac{dy}{dx} = -\frac{y \cos x + \sin x}{\sin x + x \cos y}; \quad \frac{dy}{dx}\bigg|_{(\pi/2, \pi/2)} = -1$$

4. x and y not both zero; $(1, 1)$, $(1, -3)$, $(0, 2)$, $(-2, 0)$, etc.;

$$\frac{dy}{dx} = \frac{1 - 2xy - y^2}{x^2 + 2xy + 1}; \quad \frac{dy}{dx}\bigg|_{(1,1)} = -\frac{1}{2}$$

5. True for $(0, 0)$, $(\pi/3, \sqrt{\pi/15})$, etc.

$$\frac{dy}{dx} = \frac{1 + 3y^3 \sin 6x}{10y - 3y^2 \sin 3x}; \quad \frac{dy}{dx}\bigg|_{(\pi/3, \sqrt{\pi/15})} = \frac{\sqrt{15\pi}}{10\pi}$$

$\left(\text{Note: } \frac{dy}{dx} \text{ not defined at } (0,0).\right)$

6. $\left(\frac{\pi}{4}, \frac{\pi}{4}\right)$ on curve; $\frac{dy}{dx} = \frac{\sin 2x + 2x \cos 2x}{\sin 2y + 2y \cos 2y}$; at $\left(\frac{\pi}{4}, \frac{\pi}{4}\right)$, $\frac{dy}{dx} = 1$

7. $xy > 0$; true for $(1, 3 - 2\sqrt{2})$, $(1, 3 + 2\sqrt{2})$, etc.
$f_y(x, y) = \frac{1}{2}x^{-1/2}y^{-3/2}(y - x)$; $f_x(x, y) = \frac{1}{2}x^{-3/2}y^{-1/2}(-y + x)$
$\phi'(x) = -x/y$; $\phi'(x)|_{(1, 3+2\sqrt{2})} = 2\sqrt{2} - 3$

8. $|x| \leq a, |y| \leq a$, x and y not both zero; $\frac{dy}{dx} = -\left(\frac{y}{x}\right)^{1/3}$;

for $a = 8$ at $(1, 3^{3/2})$, $\frac{dy}{dx} = -\sqrt{3}$

9. x and y not both zero; $dy/dx = -1$

10. $(2, -2)$ on graph; $\frac{dy}{dx} = \frac{3x^2 + 4y}{3y^2 - 4x}$; $\frac{dy}{dx}\bigg|_{(2,-2)} = 1$

11. $x > 0$ and $y > 0$; $(2, 3)$ on graph; $\frac{dy}{dx} = -\frac{2\sqrt{3y}}{3\sqrt{2x}}$; $\frac{dy}{dx}\bigg|_{(2,3)} = -1$

12. $(0, 0)$, $(1, 1)$ on graph; $\frac{dy}{dx} = \frac{3x^2 - 2y}{2x - 3y^2}$; $\frac{dy}{dx}\bigg|_{(1,1)} = -1$

13. $x \neq 0, y \neq 0, x \neq -y, y \neq \pm 1$; $(-\frac{10}{3}, 2)$, $(\frac{10}{3}, -2)$, etc.

$$\frac{dy}{dx} = \frac{1 - y^2}{3y^2 + 2xy + 1}; \quad \frac{dy}{dx}\bigg|_{(-10/3, 2)} = 9$$

14. $-\frac{3}{2}$ 15. $\frac{11}{2}\sqrt{2}$

EXERCISES 11-1

1. $4 \cos 4x$ 2. $-6 \cos 3x \sin 3x = -3 \sin 6x$ 3. $3 \tan^2 x \sec^2 x$
4. $-3 \csc^2 (3x)$ 5. $-6 \cot (3x) \csc^2 (3x)$
6. $3 \sec (3x) \tan^2 (3x) + 3 \sec^3 (3x) = 3 \sec (3x)[\tan^2(3x) + \sec^2 (3x)]$

ANSWERS TO EXERCISES

7. $10 \sec^2 (2x) - 3 \sin 3x$ 8. $\dfrac{4 \cos 2x}{(1 - \sin 2x)^2}$ 9. $2 \cos 4x - \cos 2x$

10. $\sec 3x (2 \sec^2 2x + 3 \tan 2x \tan 3x)$

11. $-3 \csc^2 (3x) \sin 2x + 2 \cot (3x) \cos 2x$ 12. $-10 \csc^2 (5x) \cot (5x)$

13. The graph has no maximum or minimum since the slope is everywhere greater than or equal to zero; and $y' = 0$ only at isolated points.
    (a) Slope is zero at $x = \pi/2 + 2n\pi$, $n = 0, \pm 1, \pm 2, \ldots$
    (b) Slope is a max at $x = -\pi/2 + 2n\pi$, n any integer.
    (c) min slope at $x = \pi/2 + 2n\pi$, n any integer.

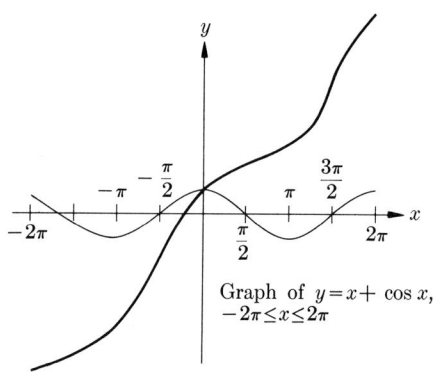

Graph of $y = x + \cos x$, $-2\pi \leq x \leq 2\pi$

14. Yes   15. See Exercise 12(b), Section 9-2.

16. Prove that $(\sin x)/x$ is a strictly decreasing function for $0 < x < \pi/2$.

$$f'(x) = \frac{x \cos x - \sin x}{x^2} = \frac{\cos x}{x^2} (x - \tan x) < 0$$

because $\cos x > 0$ and $\tan x > x$ by the result of Exercise 15.

17. Prove that $x > \sin x > 2x/\pi$ for $0 < x < \pi/2$. Compare the functions:
    $f_1: y = x$ has slope 1 $\forall x$
    $f_2: y = 2x/\pi$ has slope $2/\pi < 1$, and contains the points $(0, 0)$ and $(\pi/2, 1)$;
    $f_3: y = \sin x$ has slope $\cos x$ and contains also points $(0, 0)$, $(\pi/2, 1)$.
    $d^2y/dx^2 = -\sin x < 0$ for $0 < x < \pi/2$; $\sin x$ is concave downward on given interval; $f_2(0) = f_3(0)$ and $f_2(\pi/2) = f_3(\pi/2)$ but $f_2(x) < f_3(x)$ for $0 < x < \pi/2$; hence $x > \sin x > 2x/\pi$.

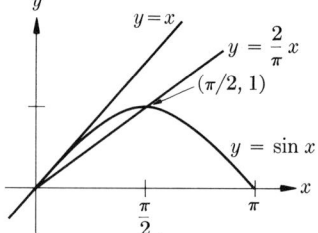

18. $\dfrac{d(\cot u)}{dx} = \dfrac{d(\tan (\pi/2 - u))}{dx} = \sec^2\left(\dfrac{\pi}{2} - u\right)\left(-\dfrac{du}{dx}\right)$

$= \csc^2 u \left(-\dfrac{du}{dx}\right) = -\csc^2 u \dfrac{du}{dx}$

19. $\dfrac{d(\cot u)}{dx} = \dfrac{d(\cos u/\sin u)}{dx} = \dfrac{\sin u(-\sin u) - (\cos u \cdot \cos u)}{\sin^2 u}\dfrac{du}{dx}$

$= -\dfrac{\sin^2 u + \cos^2 u}{\sin^2 u}\dfrac{du}{dx} = -\csc^2 u \dfrac{du}{dx}$

20. $\dfrac{d(\csc u)}{dx} = \dfrac{d(\sec (\pi/2 - u))}{dx} = \sec\left(\dfrac{\pi}{2} - u\right)\tan\left(\dfrac{\pi}{2} - u\right)\left(-\dfrac{du}{dx}\right)$

$= \csc u \cot u \left(-\dfrac{du}{dx}\right) = -\csc u \cot u \dfrac{du}{dx}$

21. $\dfrac{d(\csc u)}{dx} = \dfrac{d(1/\sin u)}{dx} = \dfrac{-\cos u}{\sin^2 u}\dfrac{du}{dx} = -\dfrac{\cos u}{\sin u}\cdot\dfrac{1}{\sin u}\dfrac{du}{dx} = -\cot u \csc u \dfrac{du}{dx}$

EXERCISES 11-2

1. $\dfrac{2}{\sqrt{1 - 4x^2}}$ 2. $\dfrac{2x}{\sqrt{1 - 4x^2}}$ 3. $-\dfrac{3}{\sqrt{1 - 9x^2}}$ 4. $\dfrac{6}{1 + 9x^2}$

5. $-\dfrac{6}{1 + 4x^2}$ 6. $\dfrac{1}{a^2 + x^2}$ 7. $\dfrac{1}{\sqrt{a^2 - x^2}}$ 8. $\dfrac{\cos x}{|\cos x|}$

9. $\dfrac{-\sin x}{\sqrt{1 - \cos^2 x}} = \dfrac{-\sin x}{|\sin x|}$ 10. $\dfrac{-\csc^2 x}{1 + \cot^2 x} = -1$ 11. $\dfrac{-\sec^2 x}{1 + \tan^2 x} = -1$

12. $\dfrac{\sec^2 x}{|\tan x|\sqrt{\tan^2 x - 1}}$ 13. $\dfrac{-\sec x \tan x}{|\sec x|\sqrt{\sec^2 x - 1}} = \dfrac{-\sec x}{|\sec x|}$

14. $\dfrac{dy}{dx} = \dfrac{1}{x\sqrt{x^2 - 1}}$

Ex. 15

15. Let $y = \tan^{-1} x$. Then $\sin y = x/\sqrt{1 + x^2}$ or $y = \sin^{-1} x/\sqrt{1 + x^2}$).

$\dfrac{d}{dx}(\tan^{-1} x) = \dfrac{d}{dx}\left(\sin^{-1} \dfrac{x}{\sqrt{1 + x^2}}\right) = \dfrac{\sqrt{1 + x^2} - x^2/\sqrt{1 + x^2}}{\sqrt{1 - x^2/(1 + x^2)}}$

$= \dfrac{(1 + x^2 - x^2)/(1 + x^2)\sqrt{1 + x^2}}{(1 + x^2 - x^2)/\sqrt{1 + x^2}} = \dfrac{1}{1 + x^2}$

319

16. (a)

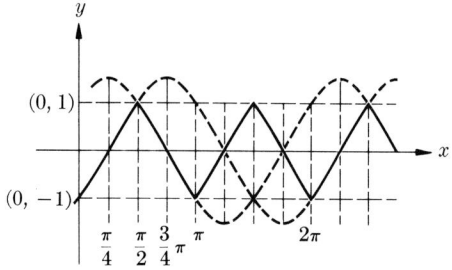

Note: The portion of the curve from $(\pi/2, 1)$ to $(\pi, -1)$ is *not* a straight line, but coincides with the dashed sinusoidal curve, etc.

(b) $D_f = \{x : x \in R\}$; $R_f = \{y : y \leq 1\}$; period $= \pi$
(c) f' is not a continuous function.

17.

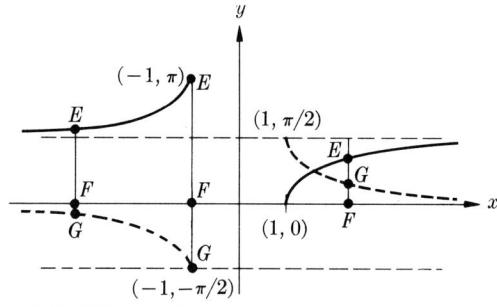

$EF + FG = \pi/2$, $|x| \geq 1$

EXERCISES 12-1

1. See Fig. 12-3. $M_1 \leq M$, $M_2 \leq M$;
$$M_1(c - a) + M_2(b - c) \leq M(c - a) + M(b - c);$$
$$M_1(c - a) + M_2(b - c) \leq M(b - a)$$
2. $L_f(P_5) = U_f(P_5) = 3$ 3. $L_f(P_5) = \frac{7}{5}$ $U_f(P_5) = \frac{8}{5}$
4. $L_f(P_5) = \frac{7}{5}$; $U_f(P_5) = \frac{8}{5}$
5. $L_f(P_5) \approx 0.646$; $U_f(P_5) \approx 0.746$
6. $L_f(P_5) = 19.5$; $U_f(P_5) = 21.0$
7. $L_f(P_5) = 0.16$; $U_f(P_5) = 0.34$
8. (a) b^2/n (b) $b^2/2 = \lim_{n \to \infty} U_f(P_n)$
   (c) $L_f(P_m) < L_f(P_n) < U_f(P_n) < U_f(P_m)$

EXERCISES 12-2

1. (a) $\sum_{k=1}^{5} (7a_k) = 7a_1 + 7a_2 + 7a_3 + 7a_4 + 7a_5$
$$= 7(a_1 + a_2 + a_3 + a_4 + a_5) = 7 \sum_{k=1}^{5} a_k$$

(b) $\sum_{k=1}^{4}(a_k + b_k) = (a_1 + b_1) + (a_2 + b_2) + (a_3 + b_3) + (a_4 + b_4)$
$= (a_1 + a_2 + a_3 + a_4) + (b_1 + b_2 + b_3 + b_4)$
$= \sum_{k=1}^{4} a_k + \sum_{k=1}^{4} b_k$

(c) $\sum_{k=1}^{n}(a_{k+1} - a_k) = (a_2 - a_1) + (a_3 - a_2) + (a_4 - a_3) + \cdots$
$+ (a_n - a_{n-1}) + (a_{n+1} - a_n) = a_{n+1} - a_1$

(d) $\sum_{k=1}^{n}[(k+1)^2 - k^2] = [(2^2 - 1^2) + (3^2 - 2^2) + \cdots$
$+ ((n-1)^2 - (n-2)^2) + (n^2 - (n-1)^2)$
$+ ((n+1)^2 - n^2)]$
$= (n+1)^2 - 1^2 = n^2 + 2n$

(e) $\sum_{k=1}^{n}[(k+1)^2 - k^2] = \sum_{k=1}^{n}(k^2 + 2k + 1 - k^2) = \sum_{k=1}^{n}(2k+1)$

(f) $\sum_{k=1}^{n}(2k+1) = \sum_{k=1}^{n} 2k + \sum_{k=1}^{n} 1 = 2\sum_{k=1}^{n} k + n$ [see (a) and (b)]

2. $\sum_{k=1}^{n}[(k+1)^2 - k^2] = n^2 + 2n$ [from (d)]
$= \sum_{k=1}^{n}(2k+1)$ [from (e)]
$= 2\sum_{k=1}^{n} k + n$ [from (f)]

Hence
$$n^2 + 2n = 2\sum_{k=1}^{n} k + n \Leftrightarrow n^2 + n = 2\sum_{k=1}^{n} k;$$

finally, if $2\sum_{k=1}^{n} k = n^2 + n$,

$$\sum_{k=1}^{n} k = \frac{n^2 + n}{2} \quad \text{or} \quad 1 + 2 + 3 + 4 + \cdots + n = \frac{n}{2}(n+1).$$

3. Show that
$$\sum_{k=1}^{n}[(k+1)^3 - k^3] = (n+1)^3 - 1, \quad \text{etc.}$$

$\sum_{k=1}^{n}[(k+1)^3 - k^3] = (2^3 - 1) + (3^3 - 2^3) + (4^3 - 3)^3 + \cdots$
$+ [n^3 - (n-1)^3] + [(n+1)^3 - n^3] = (n+1)^3 - 1.$

But
$$(n+1)^3 - 1 = n^3 + 3n^2 + 3n + 1 - 1$$
$$= n^3 + 3n^2 + 3n, \quad \text{(i)}$$

$$\sum_{k=1}^{n}[(k+1)^3 - k^3] = \sum_{k=1}^{n}(k^3 + 3k^2 + 3k + 1 - k^3)$$
$$= \sum_{k=1}^{n}(3k^2 + 3k + 1). \quad \text{(ii)}$$

Hence
$$n^3 + 3n^2 + 3n = \sum_{k=1}^{n}(3k^2 + 3k + 1) \quad \text{[from (i) and (ii)]}.$$

Rewriting the result,
$$3\sum_{k=1}^{n}k^2 + 3\sum_{k=1}^{n}k + n = n^3 + 3n^2 + 3n$$

$$\sum_{k=1}^{n}k^2 = \frac{n^3 + 3n^2 + 3n}{3} - \sum_{k=1}^{n}k - \frac{n}{3}$$

$$\sum_{k=1}^{n}k^2 = \frac{n^3 + 3n^2 + 3n}{3} - \frac{n^2 + n}{2} - \frac{n}{3}$$

$$= \frac{2n^3 + 6n^2 + 6n - 3n^2 - 5n}{6}$$

$$= \frac{n}{6}(2n^2 + 3n + 1)$$

and
$$1^2 + 2^2 + 3^2 + \cdots + n^2 = \frac{n}{6}(2n+1)(n+1).$$

4. $L_f(P_n) = (b^3/3)(1 - 1/n)(1 - 1/2n)$;
   $U_f(P_n) = (b^3/3)(1 + 1/n)(1 + 1/2n)$; $E_n = b^3/n$.

5. $L_f(P) = (\pi/4n)[\cot(\pi/4n) - 1]$; $U_f(P) = (\pi/4n)[\cot(\pi/4n) + 1]$;
   $U_f(P) - L_f(P) = \pi/2n$.

6. $L_f(P_n) \leq L_f(P_{2n})$; if $f(x)$ is constant, $L_f(P_n) = L_f(P_{2n})$.

7. If f is constant $U_f(P_n) = U_f(P_{2n})$, otherwise $U_f(P_n) > U_f(P_{2n})$.

8. Graphically, the column of rectangles, base Δx, height $f(b) - f(a)$, represents the difference $U_f(P_n) - L_f(P_n)$.

9. Total difference,
$$U_f(P_n) - L_f(P_n) = [f(a) - f(b)]\Delta x = -[f(b) - f(a)]\Delta x = |f(b) - f(a)|\Delta x.$$

10. Rectangle $ABCD = |f(a) - f(b)| \max \Delta x$; shaded portion of $ABCD$ shows $U_f(P_n) - L_f(P_n) = E_n$. $E_n < |f(a) - f(b)| \max \Delta x$.

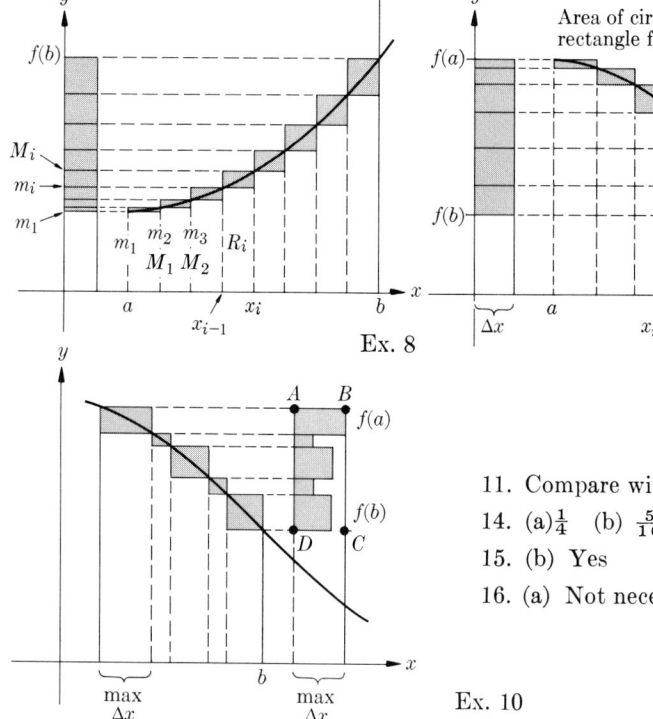

Ex. 8   Ex. 9   Ex. 10

11. Compare with Exercises 6-9.
14. (a) $\frac{1}{4}$ (b) $\frac{5}{16}$ (c) $\frac{1}{2}(1 - 1/n)$
15. (b) Yes (c) Yes
16. (a) Not necessarily

EXERCISES 12-3

1. (a) $c_k = (x_k + x_{k-1})/2$ (b) $|c_k| = \sqrt{(x_k^2 + x_k x_{k-1} + x_{k-1}^2)/3}$
   (c) $c_k = \sqrt[3]{(x_k^2 + x_{k-1}^2)(x_k + x_{k-1})/4}$
   (d) $c_k = (x_k + x_k^{1/2} + x_{k-1}^{1/2} + x_{k-1})/4$

2. The answer is a special case of Eq. (19).
3. The answer is a special case of Eq. (19).
4. $b^4/4$ 5. $2(\sqrt{b} - 1)$ 6. 8   7. $\frac{7}{3}$ 8. $\frac{8}{3}$ 9. $\frac{13}{3}$
10. 2   11. 0   12. $\sqrt{2} - 1$ 13. $\frac{1}{2}$ 14. $\frac{1}{2}\pi$ 15. π/ω
16. $\frac{2}{9}$ 17. 72   18. 0.692

19. $\lim_{n \to \infty} \frac{1}{n}\left[f\left(\frac{1}{n}\right) + f\left(\frac{2}{n}\right) + \cdots + f\left(\frac{n}{n}\right)\right]$

$= \lim_{n \to \infty}\left[f\left(\frac{1}{n}\right) \cdot \frac{1}{n} + f\left(\frac{2}{n}\right) \cdot \frac{1}{n} + \cdots + f\left(\frac{n}{n}\right) \cdot \frac{1}{n}\right]$

$= \lim_{n \to \infty} \sum_{k=1}^{n} f\left(\frac{k}{n}\right)\frac{1}{n} = \lim_{n \to \infty} \sum_{k=1}^{n} f(c_k)\,\Delta x = \int_0^1 f(x)\,dx$

ANSWERS TO EXERCISES

20. (a) $\int_0^1 x^{15}\, dx = x^{16}/16 = \frac{1}{16}$ (b) $\int_0^1 x^{1/2}\, dx = \frac{2}{3}$

(c) $\int_0^1 \sin(\pi x)\, dx = 2/\pi$

21. $U_f(P) = \sum_{i=1}^n (a_i^2 - a_i a_{i-1})$ and $L_f(P) = \sum_{i=1}^n (a_{i-1} a_i - a_{i-1}^2)$ (7', 8')

$$U_f(P) - L_f(P) = \sum_{i=1}^n (a_i^2 - 2a_{i-1} a_i + a_{i-1}^2)$$

$$= \sum_{i=1}^n (a_i - a_{i-1})^2 = \sum_{i=1}^n \delta_i^2 = \delta_1^2 + \delta_2^2 + \cdots + \delta_n^2$$

Let $\max(a_i - a_{i-1}) = \|P\|$; then

$$\delta_1^2 + \delta_2^2 + \delta_3^2 + \cdots + \delta_n^2 \leq \delta_1\|P\| + \delta_2\|P\| + \cdots + \delta_n\|P\|,$$

$$\delta_1^2 + \delta_2^2 + \cdots + \delta_n^2 \leq (\delta_1 + \delta_2 + \delta_3 + \cdots + \delta_n)\|P\|.$$

$\therefore U_f(P) - L_f(P) \leq b\|P\|$, since $\delta_1 + \delta_2 + \cdots + \delta_n = b$.

22. If $\frac{1}{2}b^2(1 - 1/n) \leq I \leq \frac{1}{2}b^2(1 + 1/n)$, for every positive integer n, then we can let $n \to \infty$, and get $b^2/2 \leq I \leq b^2/2 \Rightarrow I = b^2/2$. (See Theorem 3–4.)

23. $U_f(P) \approx 1.1414$; $L_f(P) \approx 1.0580$; average is 1.0997 (table value: $\ln 3 = 1.0986$; difference 0.0011)

24. $U_f(P) \approx 1.7144$; $L_f(P) \approx 1.5144$; average is 1.6144; (table value: 1.6094; difference 0.0050)

25. The sum lies between 656.7 and 666.7. The average of these is 661.7, so the sum differs from 662 by less than 6 units, therefore by less than 1%.

EXERCISES 12–4

1. $V = \frac{73}{18}\pi$ 2. $\frac{289}{72}\pi$; difference $\frac{1}{72}\pi$ 3. 4π; difference $\frac{1}{18}\pi$
4. $\frac{1681}{256}\pi$; $\approx 6.57\pi$ 5. $\frac{26385}{4096}\pi \approx 6.44\pi$
6. $\frac{32}{5}\pi = 6.40\pi$; with $\Delta x = \frac{1}{4}$, $V = 6.57\pi$; with $\Delta x = \frac{1}{8}$, $V = 6.44\pi$; by integration, $V = 6.40\pi$; difference is 0.17π and 0.04π
7. $\sqrt{13}$ 8. $\sqrt{13}$ 9. $L \approx 8.272$ 10. ≈ 3.815 11. 15,990 ft³
12. 62 ft.   13. $61\frac{1}{2}$ ft   14. $61\frac{1}{3}$ ft
15. $L \approx 3.8149$; $L \approx 3.815$ by Simpson's rule and 3.815 by trapezoidal rule
16. $61\frac{1}{3}$ ft   17. $6.401\pi \approx 20.105$
18. (a) $T \approx 0.6970$ (b) 0.6932   (c) to 5 terms: 0.7833; to 8 terms: 0.6345; to 10 terms: 0.6456
19. See Example 5, Section 6–1.
20. (a) 0.90794; $0.90794 < \sin 2 < 0.90935$ (b) 1.42

EXERCISES 12-5

2. $\sqrt{1+x^2}$ 3. $1/x$ 4. $-\sqrt{1-x^2}$ 5. $1/(1+x^2)$
6. $2\cos 4x^2$ 7. $2x/(1+\sqrt{1-x^2})$ 8. $-\cos x/(2+\sin x)$
9. $\sqrt{x^2+1} - x$ 10. $x_1 f(x_1)$ 11. $d^2y/dx^2 = 4y$; $k = 4$
12. $f(x) = 1/2\sqrt{x}$
13. (a) $L_f(P) = 6.750$ (b) $S_f(P,Q) = 8.625$ (c) $U_f(P) = 10.750$;
    $\int_1^3 x^2\,dx \approx 8.67$
14. $L_f(P) = 6.5$; $S_f(P,Q) = 8.625$; $U_f(P) = 11.00$
15. $\int_0^{\pi/2} \sqrt{\sin x}\,dx = -\int_{\pi/2}^0 \sqrt{\sin(\pi/2 - t)}\,dt = -\int_{\pi/2}^0 \sqrt{\cos t}\,dt = \int_0^{\pi/2} \sqrt{\cos t}\,dt$

EXERCISES 13-1

1. 5, $\tfrac{5}{3}$, -5 2. 4, $\tfrac{3}{2}$, $\tfrac{1}{2}$

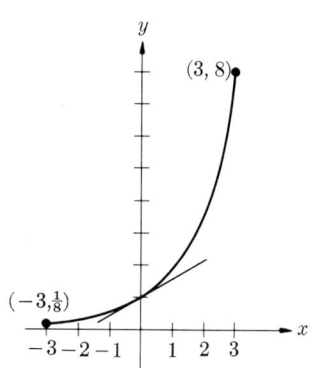

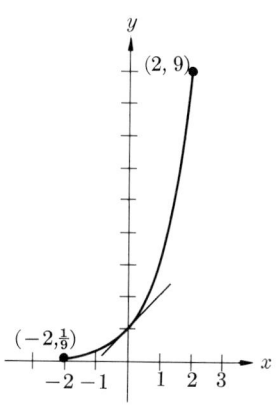

3. Slope at 0 is about 0.6.   4. Slope at $x = 0$ is about 1.

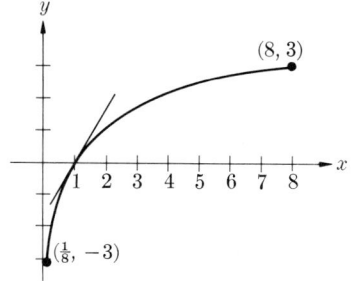

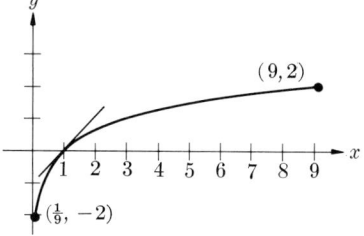

5. Slope at $x = 1$ is about $\tfrac{5}{3}$.   6. Slope at $x = 1$ is about 1.4.
7. $y_1 \approx 0.6$, $y_2 = 1$, $y_3 \approx 1.6$ 8. $y_1 \approx 0.6$, $y_2 = 1$, $y_3 \approx 1.6$

ANSWERS TO EXERCISES

9. Slope at $x = 0$ is about -0.6.

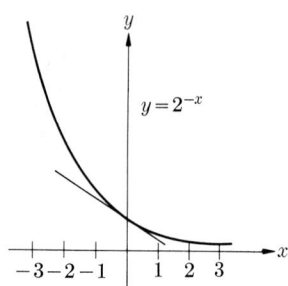

10. $a > 0$; $a > 0$ and $a \neq 1$.
11. Domain: all reals; range: all real $y > 0$
12. m_a is a function with domain of all reals and range all nonzero reals. $m_a(0) = 1$ if $a = e$. The graphs of Exercises 3 and 4 suggest the answer to this exercise.

EXERCISES 13-2

1. 0.6970 2. 0.4061 3. 1.1167 4. 1.6833
5. 2.6291 6. -0.6970
7. (a) $G(1, 2) \approx \frac{1}{2}(1 + \frac{1}{2})$; $G(5, 10) \approx \frac{1}{2}(1 + \frac{1}{2})$
   (b) $G(\frac{1}{2}, 1) \approx \frac{1}{2} \cdot \frac{1}{2}(2 + 1)$; $G(1, 2) \approx \frac{1}{2}(1 + \frac{1}{2})$
   (c) $G(1, 3) = 1 + \frac{1}{3}$; $G(2, 6) = 1 + \frac{1}{3}$

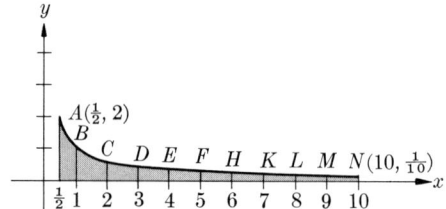

8. $R_1 = 0.2(1/1\frac{1}{5}) = 0.2(\frac{5}{6}) = \frac{1}{6}$, $R'_1 = 1 \cdot \frac{1}{6} = \frac{1}{6}$;
   $R_2 = 0.2(1/1\frac{2}{5}) = 0.2(\frac{5}{7}) = \frac{1}{7}$, $R'_2 = 1 \cdot \frac{1}{7} = \frac{1}{7}$;
   $R_3 = 0.2(1/1\frac{3}{5}) = 0.2(\frac{5}{8}) = \frac{1}{8}$, $R'_3 = 1 \cdot \frac{1}{8} = \frac{1}{8}$;
   $R_4 = 0.2(1/1\frac{4}{5}) = 0.2(\frac{5}{9}) = \frac{1}{9}$, $R'_4 = 1 \cdot \frac{1}{9} = \frac{1}{9}$;
   $R_5 = 0.2(\frac{1}{2}) = \frac{1}{10}$, $R'_5 = 1 \cdot \frac{1}{10} = \frac{1}{10}$.
   $\therefore$ There exists a one-to-one correspondence between R_i and R'_i, $i = 1, \ldots, 5$, and $R_i = R'_i$

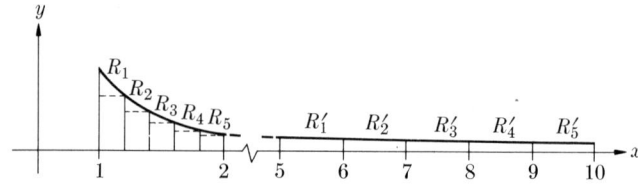

326

9. $L_e(2) \approx 0.6970$, $L_e(1.5) \approx 0.4061$, sum ≈ 1.1031; $L_e(3) \approx 1.1167$
10. $L_e(2) \approx 0.6970$, $L_e(5) \approx 1.6833$, sum ≈ 2.3803; $L_e(10) \approx 2.30259$
11. $L_e(2) \approx 0.6970$, $L_e(0.5) \approx -0.6970$, sum $= 0$; $L_e(1) = 0$
12. $L_e(3) \approx 1.1167$, $L_e(0.5) \approx -0.6970$, sum $= 0.4197$; $L_e(1.5) \approx 0.4061$
13. (a) 0.09523
    (b) (i) $L_e(1.1)^2 \approx 0.19046$ (ii) $0.38092 \approx L_e(1.1)^4$
        (iii) $0.47615 \approx L_e(1.1)^5$ (iv) $0.76184 \approx L_e(1.1)^8$
    (d) $L_e(2) \approx 0.7$; $4(0.17314) = 0.69256 \approx 0.7$
    (e) $L_e(1.19) \approx 0.17314 \approx L_e(\sqrt[4]{2})$

EXERCISES 13-3

2. Let $x = 0.5$; $\ln(1 + 0.5) = 0.401042$. $\ln 1.5$ lies between 0.401042 and 0.407292.
3. Let $x = 0.5$; $\ln 1.5 \approx 0.4055$. $0.4054 < \ln 1.5 < 0.4055$
4. $\ln 10 = 2(\ln 2 + \ln \frac{3}{2}) + \ln \frac{10}{9}$, $\ln \frac{3}{2} \approx 0.4054650$, $\ln \frac{10}{9} \approx 0.1053600$,
$\ln 10 \approx 2(0.6931458 + 0.4054648) + 0.1053600 = 2.3025812$
5. $\log_{10} 2 \approx \dfrac{0.693146}{2.302585}$; $\log_{10} 2 \approx 0.30103$
6. $\ln\left(\dfrac{N+1}{N}\right) = 2\left(\dfrac{1}{2N+1} + \dfrac{1}{3}\left(\dfrac{1}{2N+1}\right)^3 + \dfrac{1}{5}\left(\dfrac{1}{2N+1}\right)^5 + \cdots\right)$
for $0 < \dfrac{N+1}{N} < \infty$ or $0 < N < \infty$.

EXERCISES 13-4

1. $2e^{2x}$ 2. $-6e^{-2x}$ 3. $\dfrac{1}{2\sqrt{x}}e^{\sqrt{x}}$ 4. $3x^2 e^{x^3}$ 5. $\dfrac{1}{x}\ln x$
6. 1   7. $-\dfrac{\sin x}{2 + \cos x}$ 8. $(\cos x)e^{\sin x}$
9. Points of inflection: $(-\sqrt{2}/2, e^{-1/2})$, $(\sqrt{2}/2, e^{-1/2}) \approx (\pm 0.71, 0.61)$; absolute max at $(0, 1)$; curve is symmetric about y-axis; x-axis is an asymptote.

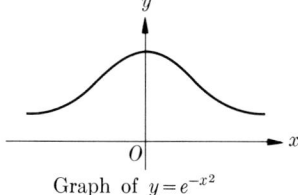

Graph of $y = e^{-x^2}$

10. (a) $\cosh x$ (b) $\sinh x$
11. (a) $\cosh^2 x = \frac{1}{4}(e^{2x} + 2 + e^{-2x})$ $\sinh^2 x = \frac{1}{4}(e^{2x} - 2 + e^{-2x})$
      $\cosh^2 x - \sinh^2 x = 1$

(b) $\cosh^2 x + \sinh^2 x = \frac{1}{4}(2e^{2x} + 2e^{-2x}) = \frac{1}{2}(e^{2x} + e^{-2x}) = \cosh 2x$

12. $\operatorname{sech}^2 x$ 13. (i) $-\operatorname{csch}^2 x$ (ii) $-\tanh x \operatorname{sech} x$ (iii) $-\coth x \operatorname{csch} x$
14. $\cosh x + C$ 15. $\sinh x + C$ 16. $\ln \cosh x + C$
17. $\ln |\sinh x| + C$ 18. $2 \tan^{-1}(e^x) + C$ 19. $x \cosh x - \sinh x + C$
20. In Exercise 11 we derived the identity $\cosh^2 \theta - \sinh^2 \theta = 1$. Let $x = \cosh \theta$ and $y = \sinh \theta$; then, as θ varies from $-\infty$ to ∞, $P(x, y)$ describes the right-hand branch of the hyperbola $x^2 - y^2 = 1$. Since e^θ and $e^{-\theta}$ are always positive, $x = \frac{1}{2}(e^\theta + e^{-\theta}) > 0$. To get the left-hand branch let $x = -\cosh \theta$.

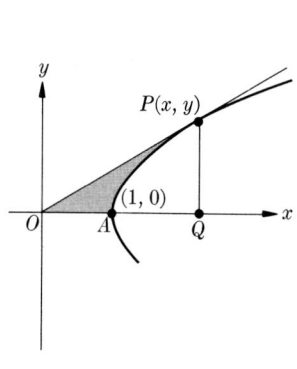

Ex. 21

Ex. 22

21. Area OAP = Area $\triangle OPQ$ − Area AQP. On AP, let $x = \cosh \theta$, $y = \sinh \theta$, $dx = \sinh \theta \, d\theta$, and $x^2 - y^2 = 1$; at A, $\theta = 0$; at P, $\theta = u$.

$$AQP = \int_A^Q y \, dx = \int_0^u \sinh \theta (\sinh \theta \, d\theta)$$
$$= \int_0^u \sinh^2 \theta \, d\theta = \frac{1}{2} \int_0^u (\cosh 2\theta - 1) \, d\theta = \frac{1}{4} \sinh 2u - \frac{1}{2}u.$$
$\therefore$ Area $OAP = \frac{1}{2} \sinh u \cdot \cosh u - (\frac{1}{4} \sinh 2u - \frac{1}{2}u) = \frac{1}{2}u.$
$A = \frac{1}{2}u.$

22. Graph of $y = \dfrac{a}{2}(e^{x/a} + e^{-x/a}) = a \cosh \dfrac{x}{a}$

TABLE 1
Natural trigonometric functions

Angle		Sine	Co-sine	Tan-gent	Angle		Sine	Co-sine	Tan-gent
Degree	Radian				Degree	Radian			
0°	0.000	0.000	1.000	0.000					
1°	0.017	0.017	1.000	0.017	46°	0.803	0.719	0.695	1.036
2°	0.035	0.035	0.999	0.035	47°	0.820	0.731	0.682	1.072
3°	0.052	0.052	0.999	0.052	48°	0.838	0.743	0.669	1.111
4°	0.070	0.070	0.998	0.070	49°	0.855	0.755	0.656	1.150
5°	0.087	0.087	0.996	0.087	50°	0.873	0.766	0.643	1.192
6°	0.105	0.105	0.995	0.105	51°	0.890	0.777	0.629	1.235
7°	0.122	0.122	0.993	0.123	52°	0.908	0.788	0.616	1.280
8°	0.140	0.139	0.990	0.141	53°	0.925	0.799	0.602	1.327
9°	0.157	0.156	0.988	0.158	54°	0.942	0.809	0.588	1.376
10°	0.175	0.174	0.985	0.176	55°	0.960	0.819	0.574	1.428
11°	0.192	0.191	0.982	0.194	56°	0.977	0.829	0.559	1.483
12°	0.209	0.208	0.978	0.213	57°	0.995	0.839	0.545	1.540
13°	0.227	0.225	0.974	0.231	58°	1.012	0.848	0.530	1.600
14°	0.244	0.242	0.970	0.249	59°	1.030	0.857	0.515	1.664
15°	0.262	0.259	0.966	0.268	60°	1.047	0.866	0.500	1.732
16°	0.279	0.276	0.961	0.287	61°	1.065	0.875	0.485	1.804
17°	0.297	0.292	0.956	0.306	62°	1.082	0.883	0.469	1.881
18°	0.314	0.309	0.951	0.325	63°	1.100	0.891	0.454	1.963
19°	0.332	0.326	0.946	0.344	64°	1.117	0.899	0.438	2.050
20°	0.349	0.342	0.940	0.364	65°	1.134	0.906	0.423	2.145
21°	0.367	0.358	0.934	0.384	66°	1.152	0.914	0.407	2.246
22°	0.384	0.375	0.927	0.404	67°	1.169	0.921	0.391	2.356
23°	0.401	0.391	0.921	0.424	68°	1.187	0.927	0.375	2.475
24°	0.419	0.407	0.914	0.445	69°	1.204	0.934	0.358	2.605
25°	0.436	0.423	0.906	0.466	70°	1.222	0.940	0.342	2.748
26°	0.454	0.438	0.899	0.488	71°	1.239	0.946	0.326	2.904
27°	0.471	0.454	0.891	0.510	72°	1.257	0.951	0.309	3.078
28°	0.489	0.469	0.883	0.532	73°	1.274	0.956	0.292	3.271
29°	0.506	0.485	0.875	0.554	74°	1.292	0.961	0.276	3.487
30°	0.524	0.500	0.866	0.577	75°	1.309	0.966	0.259	3.732
31°	0.541	0.515	0.857	0.601	76°	1.326	0.970	0.242	4.011
32°	0.559	0.530	0.848	0.625	77°	1.344	0.974	0.225	4.332
33°	0.576	0.545	0.839	0.649	78°	1.361	0.978	0.208	4.705
34°	0.593	0.559	0.829	0.675	79°	1.379	0.982	0.191	5.145
35°	0.611	0.574	0.819	0.700	80°	1.396	0.985	0.174	5.671
36°	0.628	0.588	0.809	0.727	81°	1.414	0.988	0.156	6.314
37°	0.646	0.602	0.799	0.754	82°	1.431	0.990	0.139	7.115
38°	0.663	0.616	0.788	0.781	83°	1.449	0.993	0.122	8.144
39°	0.681	0.629	0.777	0.810	84°	1.466	0.995	0.105	9.514
40°	0.698	0.643	0.766	0.839	85°	1.484	0.996	0.087	11.43
41°	0.716	0.656	0.755	0.869	86°	1.501	0.998	0.070	14.30
42°	0.733	0.669	0.743	0.900	87°	1.518	0.999	0.052	19.08
43°	0.750	0.682	0.731	0.933	88°	1.536	0.999	0.035	28.64
44°	0.768	0.695	0.719	0.966	89°	1.553	1.000	0.017	57.29
45°	0.785	0.707	0.707	1.000	90°	1.571	1.000	0.000	

TABLE 2
Exponential functions

x	e^x	e^{-x}	x	e^x	e^{-x}
0.00	1.0000	1.0000	2.5	12.182	0.0821
0.05	1.0513	0.9512	2.6	13.464	0.0743
0.10	1.1052	0.9048	2.7	14.880	0.0672
0.15	1.1618	0.8607	2.8	16.445	0.0608
0.20	1.2214	0.8187	2.9	18.174	0.0550
0.25	1.2840	0.7788	3.0	20.086	0.0498
0.30	1.3499	0.7408	3.1	22.198	0.0450
0.35	1.4191	0.7047	3.2	24.533	0.0408
0.40	1.4918	0.6703	3.3	27.113	0.0369
0.45	1.5683	0.6376	3.4	29.964	0.0334
0.50	1.6487	0.6065	3.5	33.115	0.0302
0.55	1.7333	0.5769	3.6	36.598	0.0273
0.60	1.8221	0.5488	3.7	40.447	0.0247
0.65	1.9155	0.5220	3.8	44.701	0.0224
0.70	2.0138	0.4966	3.9	49.402	0.0202
0.75	2.1170	0.4724	4.0	54.598	0.0183
0.80	2.2255	0.4493	4.1	60.340	0.0166
0.85	2.3396	0.4274	4.2	66.686	0.0150
0.90	2.4596	0.4066	4.3	73.700	0.0136
0.95	2.5857	0.3867	4.4	81.451	0.0123
1.0	2.7183	0.3679	4.5	90.017	0.0111
1.1	3.0042	0.3329	4.6	99.484	0.0101
1.2	3.3201	0.3012	4.7	109.95	0.0091
1.3	3.6693	0.2725	4.8	121.51	0.0082
1.4	4.0552	0.2466	4.9	134.29	0.0074
1.5	4.4817	0.2231	5	148.41	0.0067
1.6	4.9530	0.2019	6	403.43	0.0025
1.7	5.4739	0.1827	7	1096.6	0.0009
1.8	6.0496	0.1653	8	2981.0	0.0003
1.9	6.6859	0.1496	9	8103.1	0.0001
2.0	7.3891	0.1353	10	22026	0.00005
2.1	8.1662	0.1225			
2.2	9.0250	0.1108			
2.3	9.9742	0.1003			
2.4	11.023	0.0907			

TABLE 3
Natural logarithms of numbers

n	$\log_e n$	n	$\log_e n$	n	$\log_e n$
0.0	*	4.5	1.5041	9.0	2.1972
0.1	7.6974	4.6	1.5261	9.1	2.2083
0.2	8.3906	4.7	1.5476	9.2	2.2192
0.3	8.7960	4.8	1.5686	9.3	2.2300
0.4	9.0837	4.9	1.5892	9.4	2.2407
0.5	9.3069	5.0	1.6094	9.5	2.2513
0.6	9.4892	5.1	1.6292	9.6	2.2618
0.7	9.6433	5.2	1.6487	9.7	2.2721
0.8	9.7769	5.3	1.6677	9.8	2.2824
0.9	9.8946	5.4	1.6864	9.9	2.2925
1.0	0.0000	5.5	1.7047	10	2.3026
1.1	0.0953	5.6	1.7228	11	2.3979
1.2	0.1823	5.7	1.7405	12	2.4849
1.3	0.2624	5.8	1.7579	13	2.5649
1.4	0.3365	5.9	1.7750	14	2.6391
1.5	0.4055	6.0	1.7918	15	2.7081
1.6	0.4700	6.1	1.8083	16	2.7726
1.7	0.5306	6.2	1.8245	17	2.8332
1.8	0.5878	6.3	1.8405	18	2.8904
1.9	0.6419	6.4	1.8563	19	2.9444
2.0	0.6931	6.5	1.8718	20	2.9957
2.1	0.7419	6.6	1.8871	25	3.2189
2.2	0.7885	6.7	1.9021	30	3.4012
2.3	0.8329	6.8	1.9169	35	3.5553
2.4	0.8755	6.9	1.9315	40	3.6889
2.5	0.9163	7.0	1.9459	45	3.8067
2.6	0.9555	7.1	1.9601	50	3.9120
2.7	0.9933	7.2	1.9741	55	4.0073
2.8	1.0296	7.3	1.9879	60	4.0943
2.9	1.0647	7.4	2.0015	65	4.1744
3.0	1.0986	7.5	2.0149	70	4.2485
3.1	1.1314	7.6	2.0281	75	4.3175
3.2	1.1632	7.7	2.0412	80	4.3820
3.3	1.1939	7.8	2.0541	85	4.4427
3.4	1.2238	7.9	2.0669	90	4.4998
3.5	1.2528	8.0	2.0794	95	4.5539
3.6	1.2809	8.1	2.0919	100	4.6052
3.7	1.3083	8.2	2.1041		
3.8	1.3350	8.3	2.1163		
3.9	1.3610	8.4	2.1282		
4.0	1.3863	8.5	2.1401		
4.1	1.4110	8.6	2.1518		
4.2	1.4351	8.7	2.1633		
4.3	1.4586	8.8	2.1748		
4.4	1.4816	8.9	2.1861		

* Deduct 10 from each of the first nine values in column two.

INDEX

INDEX

(Numbers in parentheses refer to exercises on the indicated pages.)

A
Absolute maximum (minimum), 91
Absolute value(s),
   definition, 13
   geometric interpretation of, 15
   and maximum/minimum of (a, b), 16 (4)
   of product, 14
   properties of, 14
   of quotient, 14
   of sum, 14
Absolute value function,
   derivative of, 9
   domain of derivative, 9
Accessible point, 120
Algebraic functions,
   derivative of, 193
American Mathematical Monthly, 169
Approximation,
   of sum of square roots, 244
Area,
   additivity of, 217
   definition of, 218
   existence of, 227, 228
   and integration, 214
   of parabolic region, 72
Area function,
   derivative of, 12
Area problem, 11

B
Boundedness,
   and maximum/minimum, 114
Bounded sequence, 33
Bounded set, 113

C
Catenary, 289 (22)
Chain rule, 180
Circular functions,
   derivative of, 198
Closed set, 120
Completeness postulate, 47, 113
Composite functions, 138
   continuity of, 142
   derivative of, 180

Computer program
   for ln 10, 277
Continuity and discontinuity,
   of ruler function, 107, 108
Continuity,
   epsilon-delta definition, 104
   of linear function, 104
   at a point, 104
   of $\sqrt{x}$, 106
   uniform, 124
Continuous function(s), 103
   bounded range, 116
   integration of, 220, 225
   maximum/minimum, 116
Convergent sequence, 33
   and boundedness, 34
Courant, R., and Robbins, H.,
   "What Is Mathematics?", 71
Covering,
   open, 119
Critical value(s), 95
   for maximum/minimum, 95
Cunningham, F., Jr., 258

D
Deleted neighborhood, 57
Derivative(s), 6
   of algebraic function, 193
   of circular functions, 198
   of composite functions, 180
   of constant function, 78
   of cosine, 87
   of $\cos^{-1} u$, 206
   of $\cot^{-1} u$, 208
   of $\csc^{-1} u$, 211
   of cu, 78
   definition of, 7, 76
   of inverse functions, 184, 185
   left-hand, 8
   of polynomials, 80
   of power function, x^n, 78
   and primitive, 240
   of products, 78
   of quotients, 80
   of rational functions, 80, 81
   right-hand, 8

335

INDEX

of $\sec^{-1} u$, 210
of sine, 84
of $\sin^{-1} u$, 204
of sum, 78
of $\tan^{-1} u$, 207
of trigonometric functions, 84, 198
of x^n, n rational, 189
Derived function, 8, 77
Differentiability, 10
Differential calculus,
   fundamental problems of, 10
Differentiable function(s), 10, 158
Discontinuity
   of greatest integer function, 105
   removable, 216
Discontinuous (see also discontinuity), 105, 106
Divergent sequence, 33

E

e, 49
   series for, 173
Exponential function(s), 263, 278
   base a, 283
   derivative of, 282
   series for, 286
Extrema,
   tests for, 97

F

Fermat, Pierre De (1601–1665),
   theorem of, 74 (27)
Fine, Henry B., 191
Finite sets,
   maximum/minimum of, 109 (8) (9)
Function(s), 17
   composite, 138
   composition of, 139
   continuous, 103
   continuous, integration of, 220, 225
   continuous on a closed bounded interval, 103
   decreasing, 23
   definition of, 17
   derived, 77
   differentiability and continuity, 159
   domain of, 18
   exponential, 278
   greatest integer, $[x]$, 21
   greatest integer, continuity and discontinuity of, 105
   increasing, 23, 154
   increment of, 177
   and inverse, 23
   inverse of, 145
   left inverse of, 146
   linear, continuity of, 104
   logarithmic and exponential, 263
   one-to-one, 23
   piecewise continuous, 215
   range of, 18
   real-valued, 19
   Riemann integrable, 223
   ruler, 107, 108
   square-root, continuity of, 106
   strictly increasing, strictly decreasing, 157 (3)
   uniformly continuous, 104
   ways of representing, 19, 20
Fundamental theorem(s) of calculus, 240, 258

G

Geometric series, 27
Greatest integer function, $[x]$, 21
Greatest lower bound, 46
   of a set, 113
Gregory, James (1638–1675), 169

H

Heine-Borel theorem, 118, 120
Hyperbola,
   rectangular, 2
Hyperbolic functions, 288, 289 (10–22)

I

Implicit differentiation, 190
Increasing function,
   inverse of, 154
Increment of a function, 177
Infimum, 46
Inner point, 119
Integral,
   derivative of, 258

INDEX

Integration,
  of continuous functions, 220, 225
Interior point, 7
Intermediate value(s), 104
Intermediate-value theorem, 129, 130
  for integrals, 256
Inverse function(s), 145
  continuity of, 153
  derivatives of, 184, 185
  graphical test, 151
Inverse relation, 150
Inverse trigonometric functions,
  derivatives of, 201
Irrationality of e, 287
Isolated point,
  maximum of function at, 91

L

Lang, S., 264
Least upper bound, 46
  of a set, 112
Limit(s), 25
  of $(1 - \cos \theta)/\theta^2$, 67 (11)
  of $(1 - \cos \theta)/\theta$, 67 (12)
  of Δu, 82
  of functions, 57
  of a polynomial, 61
  of a product, 35, 44 (7), 60, 66
  of a quotient, 42, 60, 66
  of a reciprocal, 39
  of a sequence, 26, 28
  of a sequence, definition of, 30
  of $\sin \theta/\theta$, 62, 65
  of $\sqrt[n]{n}$, 51
  of a sum, 32, 60, 66
  one-sided, 7
  uniqueness of, 43
Linear function,
  continuity of, 104
Local maximum, 90
Local minimum, 91
Lockwood, E. H., 289 (22)
Logarithmic function(s), 263
  base a, 285

M

Mapping, 19
Mathematical induction, 70, 71

Maximum,
  absolute, 91
  local (relative), 90
Maxima and minima, 88, 110, 113
  and boundedness, 114
  and critical values, 94
  at endpoints, 94
  existence of, 104
  of finite sets, 109 (8) (9)
  first-derivative test, 99
  at isolated point, 91
  nonexistence of, 92
  where derivative fails to exist, 94, 95
  and zero derivative, 94
Mean-value theorem, 164, 166 (10)
  extensions of, 167
Menger, Karl, 260
Method of false position, 136 (4)
Minimum,
  absolute, 91
  local (relative), 91
Monotone sequence,
  convergence of, 48
Monotonic functions,
  integrability of, 230

N

Natural logarithm(s), 270
  computation of, 272
Natural logarithmic function, 270
Neighborhood(s),
  collection of, 118
  deleted, 57
  in domain and range of function, 104
  $N_h(a)$, 15, 16

O

Olmsted, John M. H., 191, 229, 248
Open covering, 119
  finite, 120
Open set, 119

P

Partition,
  norm of, 221
  refinement of, 227
  regular, 220

337

INDEX

π, calculation of, to 100,000 decimals, 269
Piecewise continuous function, 215
Primitive,
  and derivative, 240

R

Rational number system,
  not complete, 113
Relation, 17
Relative maximum/minimum, 90, 91
Removable discontinuity, 216
Riemann integral(s),
  definition of, 223
  properties of, 251
  upper and lower, 229
Riemann sums,
  lower and upper, 221
Robbins, H., and Courant, R.,
  "What Is Mathematics?", 71
Rolle, Michel (1652–1716), 166
Rolle's theorem, 163
Ruler function, 230
  continuity and discontinuity
    of, 107, 108

S

Sequence,
  bounded, 33
  bounded above (below), 46
  convergent, 33
  definition of, 30
  divergent, 33
  graphical representation of, 45
  greatest lower bound of, 46
  least upper bound of, 46
  monotone, 46
  supremum of, 46
Series,
  for e^x, 286
  for $\ln(1+x)/(1-x)$, 275
  for $\ln(1+x)$, 273
  geometric, 27
Set(s),
  bounded, 113
  closed, 120
  finite, maximum/minimum of,
    109 (8), (9)

  greatest lower bound of, 113
  least upper bound, 112
  lower bound, 113
  open, 119
  upper bound, 112
Shanks, Daniel, 206, 268
Smith, David Eugene, 289 (22)
Square-root function,
  continuity of, 106
Squares,
  sum of first n, 70
Sum(s),
  first n cubes, 73 (2)
  first n integers, 73 (1)
  first n squares, 70
Supremum, 46
Symmetry,
  of a curve, tests for, 23
  properties of, 23

T

Tangent, 2
Taylor, Brook (1685–1731), 169
Taylor's theorem, 169
Trapezoidal approximation, 247
  error in, 248
Triangle inequality, 14, 16 (5)
Trigonometric functions,
  derivatives of, 198
Trigonometric identities,
  $\sin(A+B)$, $\sin(A-B)$,
  $\sin x_1 - \sin x_2$, 128 (3)

U

Uniform continuity, 124

V

Vance, E. P., 71
Velocity, 4
  at time t_1, 5
  average, 4

W

"What Is Mathematics?", 71
Wiggin, Robert T., 277
Wolfe, Harold E., 288
Wolfe, James, 169
Wrench, John W., Jr., 206, 269